MECHANICS OF MACHINES

Mechanics of Machines

Advanced Theory and Examples

JOHN HANNAH
B.Sc.(Eng.), C.Eng., F.I.Mech.E., F.I.Prod.E.

FORMERLY HEAD OF DEPARTMENT OF ENGINEERING AND SCIENCE,
SOUTHGATE TECHNICAL COLLEGE

AND

R. C. STEPHENS
M.Sc.(Eng.), C.Eng., M.I.Mech.E.

FORMERLY PRINCIPAL LECTURER IN MECHANICAL ENGINEERING,
WEST HAM COLLEGE OF TECHNOLOGY

EDWARD ARNOLD

© John Hannah and R. C. Stephens 1972

First published 1963
by Edward Arnold (Publishers) Ltd
41 Bedford Square
London WC1B 3DQ
Reprinted 1964, 1966
Second edition 1972
Reprinted 1972, 1974, 1976, 1978, 1979

Boards ISBN : 0 7131 3253 1
Paper ISBN : 0 7131 3254 X

Printed in Great Britain by Butler & Tanner Ltd, Frome and London

PREFACE

This book is intended for students taking the final years of an engineering degree or diploma course and follows on from the authors' earlier work, *Mechanics of Machines—Elementary Theory and Examples*. Students taking Higher National Certificate courses should also find this book useful within the restricted syllabuses of such courses.

Each chapter consists of a concise but thorough statement of the theory, followed by a number of worked examples in which the theory is amplified and extended. A large number of unworked examples are also included.

For the sake of completeness, much of the theory in the elementary book has been repeated here, being abbreviated or expanded as appropriate, and the whole of the authors' *Examples in Mechanical Vibrations* has been included in revised form.

The authors acknowledge with thanks the permission granted by the Senates of the Universities of London and Glasgow and the Council of The Institution of Mechanical Engineers to use questions set at their examinations. These have been designated U. Lond., U. Glas., and I. Mech. E. respectively.

The authors are indebted to their colleagues and students for constructive criticism and help in checking the solutions to the 600 problems included.

PREFACE TO THE SECOND EDITION

Following the adoption of the *Système International d'Unités*, the text and numerical questions of the First Edition have been converted to SI units and the opportunity has been taken to add some additional questions, both worked and unworked.

The units and abbreviations are as specified in *The Use of SI Units*, British Standards Institution, PD 5686, January 1969.

JOHN HANNAH
R. C. STEPHENS

NOTE ON SI UNITS

The fundamental units in the Système International d'Unités are the metre, kilogramme and second, with the newton as the derived unit. Where mixed quantities are involved in a problem, the solution has generally been worked throughout in the basic units ; for example, for a given power of 2 MW the figure 2×10^6 W has been substituted, and for a density of 7·8 Mg/m³ the figure $7·8 \times 10^3$ kg/m³ has been substituted. In some cases, however, it has been found convenient to work throughout in multiples of the basic unit or to make preliminary calculations in such units.

v

CONTENTS

CHAPTER 1

DYNAMICS

1.1 Mass, force, weight and momentum. The *mass* of a body is determined by comparison with a standard mass, using a beam-type balance. Thus mass is independent of gravitational acceleration since any variation in g will have an equal effect on the standard mass.

Force is that which tends to change the state of rest or uniform motion of a body. Unit force is that required to give unit acceleration to unit mass.

The *weight* of a body is the force of attraction which the earth exerts upon it and is determined by a suitably-calibrated spring-type balance. Thus the weight varies from place to place as g varies but is standardized at a point where g has the value 9·806 65 m/s². For normal engineering purposes, however, g is taken as 9·81 m/s².

The *momentum* of a body is the product of its mass and velocity.

$$\text{Force} \propto \text{rate of change of momentum}$$

$$\propto \text{mass} \times \text{rate of change of velocity}$$

i.e. $\quad\quad\quad\quad P = kmf \quad$ where k is a constant.

The units of the quantities are chosen so as to make the value of k unity,

i.e. $\quad\quad\quad\quad\quad\quad\quad P = mf \quad . \quad\quad . \quad\quad . \quad\quad . \quad$ (1.1)

In the *Système International d'Unités*, the fundamental quantities are mass, length and time, force being a derived quantity. The unit of mass is the kilogramme, the unit of length is the metre and the unit of force is the newton, which is the force required to give a mass of 1 kg an acceleration of 1 m/s².

Then $\quad\quad\quad\quad P \text{ (N)} = m \text{ (kg)} \times f \text{ (m/s}^2)$

Since a body falling freely under the earth's gravitational force has an acceleration g, the weight

$$W = mg$$

1.2 Impulse. The *impulse* of a constant force P acting for a time t is the product Pt. If, during this time, the velocity changes from u to v,

then $\quad\quad\quad\quad\quad\quad P = mf = \dfrac{m(v-u)}{t}$

1

or $\qquad\qquad\qquad Pt = m(v - u)$ . . . . (1.2)

i.e. $\qquad\qquad$ impulse of force = change of momentum

A force which acts for a very short time is referred to as an *impulsive force*.

1.3 Conservation of linear momentum. The total momentum of a system of masses in any one direction remains constant unless acted upon by an external force in that direction.

1.4 Circular motion. Consider a body of mass m moving in a circular path of radius r with constant speed v, Fig. 1.1. If it moves from

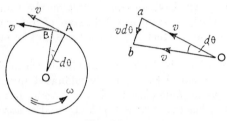

FIG. 1.1

A to B in time dt and the angle AOB is $d\theta$, then, from the relative velocity diagram, the change of velocity is represented by ab.

Thus $\qquad\qquad$ change of velocity $= v\, d\theta$

$$\therefore \text{ acceleration} = v\frac{d\theta}{dt}$$

i.e. $\qquad\qquad\qquad\qquad f = v\omega$

where ω is the angular velocity of AO, and since $v = \omega r$,

$$f = \omega^2 r \quad \text{or} \quad \frac{v^2}{r} \qquad . \quad . \quad (1.3)$$

This acceleration is directed towards the centre of rotation, O, and is called the *centripetal acceleration*. The radially inward, or centripetal, force required to produce this acceleration is given by

$$P = mf = m\omega^2 r \quad \text{or} \quad m\frac{v^2}{r} \qquad . \quad . \quad (1.4)$$

If a body rotates at the end of an arm, this force is provided by the tension in the arm. The reaction to this force acts at the centre of rotation and is called the *centrifugal force*. It represents the inertia force of the body, resisting the change in the direction of its motion.

A common concept of centrifugal force in engineering problems is to regard it as the radially outward force which must be applied to a body to convert the dynamical condition to the equivalent static condition ; this is known as *d'Alembert's Principle*. This concept is particularly useful in problems on engine governors and balancing of rotating masses.

1.5 Work and energy. *Work* is the product of the average force and the distance moved in the direction of the force by its point of application. The unit of work is the joule (J), which is the work done by a force of 1 N moving through a distance of 1 m.

If a constant force P moves through a distance x, work done $= Px$.

If the force varies linearly from zero to a maximum value P, work done $= \frac{1}{2}Px$.

In the general case where $P = f(x)$, work done $= \int_0^x f(x)\, dx.$

Power is the rate of doing work. The unit of power is the watt (W), which is 1 J/s or 1 N m/s. Thus the power developed by a force P N moving at v m/s is Pv W.

Energy is the capacity to do work, mechanical energy being equal to the work done on a body in altering either its position or its velocity.

The *potential energy* (P.E.) of a body is the energy it possesses due to its position and is equal to the work done in raising it from some datum level. Thus the P.E. of a body of mass m at a height h above datum level is mgh.

The *kinetic energy* (K.E.) of a body is the energy it possesses due to its velocity. If a body of mass m attains a velocity v from rest under the influence of a force P and moves a distance s, then

$$\text{work done by } P = P \times s$$

i.e. $$\text{K.E. of body} = mf \times \frac{v^2}{2f} = \tfrac{1}{2}mv^2 \qquad . \qquad . \qquad (1.5)$$

The *strain energy* of a body is the energy stored when the body is deformed. If an elastic body of stiffness S is extended a distance x by a force P,

$$\text{work done = strain energy} = \tfrac{1}{2}Px = \tfrac{1}{2}Sx^2 \ . \qquad . \qquad (1.6)$$

1.6 Conservation of energy. Energy can neither be created nor destroyed. It may exist in a variety of forms, such as mechanical, electrical or heat energy, but a loss of energy in any one form is always accompanied by an equivalent increase in another form. When work is done on a rigid body, the work is converted into kinetic or potential energy or is used in overcoming friction. If the body is elastic, some of the work will also be stored as strain energy.

1.7 Moment of inertia. The moment of inertia of a particle of mass dm, Fig. 1.2, at a distance r from an axis through the centre of

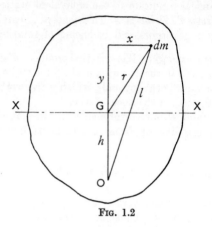

FIG. 1.2

gravity G, perpendicular to the plane of the paper is $dm.r^2$. Hence the moment of inertia of the whole body about G is

$$\int dm.r^2$$

If the total mass of the body is m, this may be written $I = mk^2$; k is termed the *radius of gyration* and is the radius at which the mass would have to be concentrated to give the same value of I.

The moment of inertia of the body about a parallel axis through O, distant h from G is

$$\int dm.l^2$$

$$= \int dm\{x^2 + (h + y)^2\}$$

$$= \int dm\{r^2 + h^2 + 2hy\}$$

$$= I_{\mathrm{G}} + mh^2 + 2h \times \int dm.y$$

$\int dm.y$ is the total moment of the mass about XX and since XX passes through the centre of gravity, G, this term is zero.

Hence $I_{\mathrm{O}} = I_{\mathrm{G}} + mh^2$.   .   .   .   (1.7)

The units of I are kg m².

1.8   Torque and angular acceleration. Let the body shown in Fig. 1.3 rotate about an axis through O and let the angular acceleration produced by a torque T be α.

Then acceleration of particle of mass dm at radius $l = \alpha l$

$\therefore$ force required to accelerate particle $= dm \cdot \alpha l$

$\therefore$ torque required to accelerate particle $= dm \cdot \alpha l^2$

$\therefore$ total torque required to accelerate the body $= \int dm \cdot \alpha l^2$

i.e. $$T = I_0 \alpha = m k_0^2 \alpha \quad . \qquad . \qquad . \qquad . \qquad (1.8)$$

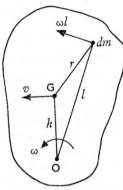

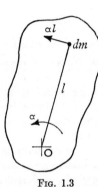

FIG. 1.3   FIG. 1.4

1.9   Angular momentum or moment of momentum. The *angular momentum* of a body about its axis of rotation is the moment of its linear momentum about that axis.

If the body shown in Fig. 1.4 is rotating about an axis through O with angular velocity ω, then

velocity of particle of mass dm at radius $l = \omega l$

$\therefore$ momentum of particle $= dm \cdot \omega l$

$\therefore$ moment of momentum about O $= dm \cdot \omega l^2$

$\therefore$ total moment of momentum of body about O $= \int dm \cdot \omega l^2$

$$= I_0 \omega \qquad . \qquad (1.9)$$

If G is the centre of gravity, then $I_0 = I_G + mh^2$

so that moment of momentum about O $= I_G \omega + mh^2 \omega$

$$= I_G \omega + mvh \qquad . \qquad (1.10)$$

i.e. the angular momentum about any axis O is equal to the angular momentum about a parallel axis through G, together with the moment of the linear momentum about O.

The units of linear and angular momentum are different, being kg m/s and kg m²/s respectively, so that, for a body possessing both linear and angular momentum, these quantities are not additive.

1.10 Conservation of angular momentum. The total angular momentum of a system of masses about any one axis remains constant unless acted upon by an external torque about that axis.

1.11 Angular impulse. The *angular impulse* of a constant torque T acting for a time t is the product Tt. If, during this time, the angular velocity changes from ω_1 to ω_2, then

$$T = I\alpha = I\frac{(\omega_2 - \omega_1)}{t}$$

or
$$Tt = I(\omega_2 - \omega_1) \qquad . \qquad . \qquad . \qquad . \quad (1.11)$$

i.e.   impulse of torque = change of angular momentum

A torque which acts for a very short time is referred to as an *impulsive torque*.

If an impulsive force acts on a body, causing rotation, the angular momentum of the body about the line of action of the force, immediately before and after the impact, remains unchanged, since the force has no moment about that line.

If two rotating gear wheels or friction discs are suddenly meshed together, the same impulsive force acts at the circumference of both wheels, but if the radii are different, the wheels are subjected to different impulsive torques and thus the changes in angular momentum of the wheels are different. The external impulsive torque causing the change in total momentum is supplied by the reactions at the bearings. When the wheels have equal radii, however, the impulsive torques are equal and there is no change in the total angular momentum.

1.12 Work done by a torque. If a constant torque T moves through an angle θ,

$$\text{work done} = T\theta$$

If the torque varies linearly from zero to a maximum value T,

$$\text{work done} = \tfrac{1}{2}T\theta$$

In the general case where $T = f(\theta)$,

$$\text{work done} = \int_0^\theta f(\theta)\, d\theta$$

The power developed by a torque T N m moving at ω rad/s is

$$T\omega \quad \text{or} \quad \frac{2\pi NT}{60}\text{W} \quad \text{where } N \text{ is the speed in rev/min.}$$

1.13 Angular kinetic energy. For a body rotating about axis O, Fig. 1.4,

$$\text{K.E. of particle of mass } dm = \frac{dm}{2}(\omega l)^2$$

$$\therefore \text{ total K.E. of body} = \frac{\omega^2}{2}\int dm \cdot l^2$$

$$= \tfrac{1}{2}I_0\omega^2 \quad . \qquad . \qquad . \quad (1.12)$$

$$= \tfrac{1}{2}(I_G + mh^2)\omega^2$$

$$= \tfrac{1}{2}I_G\omega^2 + \tfrac{1}{2}mv^2 \qquad . \quad (1.13)$$

1.14 Total K.E. and rate of change of K.E. If the c.g. of a body is moving with linear velocity v and the body is also rotating about the c.g. with angular velocity ω,

then total K.E. = K.E. of translation + K.E. of rotation

$$= \tfrac{1}{2}mv^2 + \tfrac{1}{2}I\omega^2 \quad . \qquad . \qquad . \qquad . \qquad . \quad (1.14)$$

where I is the moment of inertia of the body about the c.g.

$$\text{Rate of change of K.E.} = \frac{d}{dt}(\tfrac{1}{2}mv^2 + \tfrac{1}{2}I\omega^2)$$

$$= mvf + I\omega\alpha$$

$$= Pv + T\omega \qquad . \qquad . \qquad . \quad (1.15)$$

$$= \text{work done in unit time}$$

1.15 Equivalent mass of a rotating body. Consider a body of mass m rotating about an axis through O, Fig. 1.5 ; let the radius of gyration about this axis be k. If a tangential force P, acting at radius r

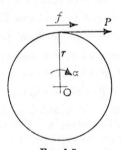

Fɪɢ. 1.5

produces an angular acceleration α, then the equation of angular motion of the body is

$$P \times r = I_0\alpha = mk^2\frac{f}{r}$$

where f is the linear acceleration of the force P.

$$\therefore P = m\left(\frac{k}{r}\right)^2 f \qquad . \qquad . \qquad . \qquad . \quad (1.16)$$

which is the equation of linear motion of the body, assumed concentrated at a radius r.

The quantity $m\left(\dfrac{k}{r}\right)^2$ is the equivalent mass of the body, referred to the line of action of P. In problems concerning both linear and angular accelerations such as vehicle dynamics, this equivalent mass may be added to the actual mass in order to obtain the total equivalent mass having linear acceleration only.

1.16. Acceleration of a geared system. Let two gear wheels A and B, having moments of inertia I_a and I_b respectively, mesh with a speed

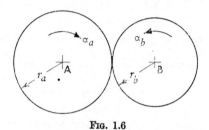

FIG. 1.6

ratio $\omega_b/\omega_a = n$, Fig. 1.6. If a torque T is applied to wheel A to accelerate the system and α_a and α_b are the accelerations of A and B respectively, then

torque required on B to accelerate B $= I_b\alpha_b = I_b n\alpha_a$

$\therefore$ torque required on A to accelerate B $= n^2 I_b\alpha_a$

since the torques are inversely proportional to the speeds.

Torque required on A to accelerate A $= I_a\alpha_a$

$\therefore$ total torque required on A to accelerate A and B $= I_a\alpha_a + n^2 I_b\alpha_a$

i.e. $$T = (I_a + n^2 I_b)\alpha_a \qquad . \qquad . \qquad . \qquad . \quad (1.17)$$

The quantity $I_a + n^2 I_b$ may be regarded as the equivalent moment of inertia of the gears referred to wheel A. This principle may be extended to any number of wheels geared together, the moment of inertia of each wheel in the train being multiplied by the square of its gear ratio relative to the reference wheel. Thus, in problems on hoists, the moments of inertia of the various gears may be reduced to an equivalent moment of inertia on the motor shaft or drum shaft.

The above result can also be obtained from the general principle that the net energy supplied to a system in unit time is equal to the rate of change of its kinetic energy (see Art. 1.14).

Thus
$$T\omega_a = \frac{d}{dt}(\tfrac{1}{2}I_a\omega_a{}^2 + \tfrac{1}{2}I_b\omega_b{}^2)$$

$$= \tfrac{1}{2}(I_a + n^2I_b) \times \frac{d}{dt}(\omega_a)^2$$

$$= \tfrac{1}{2}(I_a + n^2I_b) \times 2\omega_a\alpha_a$$

i.e.
$$T = (I_a + n^2I_b)\alpha_a$$

This torque is in addition to any torque required on A to overcome external resisting torques applied to A and/or B.

Care must be taken in using the equivalent inertia in problems in which efficiencies are given. If the efficiency of the gearing is η, the torque required on A to accelerate B is $\dfrac{n^2I_b\alpha_a}{\eta}$ so that the total torque required on A to accelerate A and B $= \left(I_a + \dfrac{n^2I_b}{\eta}\right)\alpha_a$.

1.17 Maximum acceleration and retardation of vehicles. The maximum possible acceleration or retardation of a vehicle is limited by the friction force between the wheels and the road.

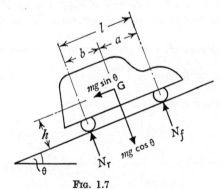

FIG. 1.7

Consider a vehicle accelerating up a gradient inclined at an angle θ to the horizontal, Fig. 1.7. Let the tractive force be F and the reactions at the front and rear wheels be N_f and N_r respectively.

Then, for maximum acceleration,

$$F = \mu N_f, \text{ for front wheel drive,}$$

$$F = \mu N_r, \text{ for rear wheel drive,}$$

and
$$F = \mu(N_f + N_r), \text{ for four-wheel drive.}$$

For this latter ideal condition to be realized, however, the ratio of the tractive forces applied to the front and rear wheels must be in the ratio of the reactions, N_f and N_r, respectively, otherwise slipping will not occur at all wheels simultaneously.

Resolving forces perpendicular and parallel to the incline,

$$N_f + N_r = mg \cos \theta \quad . \quad . \quad . \quad . \quad (1.18)$$

$$F = mg \sin \theta + mf \quad . \quad . \quad . \quad (1.19)$$

Taking moments about G,

$$Fh = N_r b - N_f a \quad . \quad . \quad . \quad . \quad (1.20)$$

Hence f may be determined from equations (1.18), (1.19) and (1.20).

For maximum retardation, the braking force F is obtained in the same way as for the tractive force but acts in the opposite direction.

(a) General Dynamics

1. *A uniform plank rests upon a horizontal bench with one end of the plank projecting over the sharp edge of the bench, the plank being at right angles to this edge. The plank is pulled out horizontally until the centre of gravity overhangs the edge by a distance a and is then released. The plank rotates about the edge and then slides down.*

(a) Determine the angular velocity and angular acceleration of the plank after it has turned through an angle θ, assuming that no sliding has taken place.

(b) If sliding begins when the plank has turned through an angle α, show that the coefficient of friction is $\dfrac{k^2 + 3a^2}{k^2} \tan \alpha$, *where k is the radius of gyration of the plank about its centre of gravity.*   (U. Lond.)

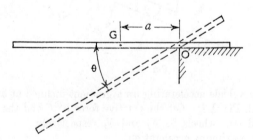

FIG. 1.8

(a) When the plank has rotated through an angle θ, Fig. 1.8,

$$\text{loss in P.E.} = \text{gain in K.E.}$$

i.e. $$mga \sin \theta = \tfrac{1}{2} I_0 \omega^2 = \frac{m(k^2 + a^2)\omega^2}{2}$$

i.e. angular velocity, $\omega = \sqrt{\dfrac{2ga \sin \theta}{k^2 + a^2}}$

Angular acceleration $= \dfrac{d\omega}{dt} = \dfrac{d\omega}{d\theta} \times \dfrac{d\theta}{dt} = \dfrac{d\omega}{d\theta} \times \omega$

$$= \tfrac{1}{2} \cdot \times \left(\dfrac{2ga \sin \theta}{k^2 + a^2}\right)^{-\frac{1}{2}} \times \dfrac{2ga}{k^2 + a^2} \cos \theta \times \left(\dfrac{2ga \sin \theta}{k^2 + a^2}\right)^{\frac{1}{2}}$$

$$= \dfrac{ga \cos \theta}{k^2 + a^2}$$

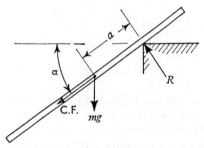

FIG. 1.9

(b) When the plank has turned through an angle α, Fig. 1.9, the force acting along the plank is the sum of the centrifugal force and the component of the weight parallel to the plank.

i.e. tangential force, $F = m\omega^2 a + mg \sin \alpha$

$$= m \cdot \dfrac{2ga \sin \alpha}{k^2 + a^2} a + mg \sin \alpha$$

$$= mg \sin \alpha \cdot \dfrac{k^2 + 3a^2}{k^2 + a^2}$$

The normal reaction, R, is given by

$$Ra = I_G \dfrac{d\omega}{dt} = mk^2 \cdot \dfrac{ga \cos \alpha}{k^2 + a^2}$$

i.e. $R = mg \dfrac{k^2 \cos \alpha}{k^2 + a^2}$

$$\therefore \mu = \dfrac{F}{R} = \dfrac{mg \sin \alpha \dfrac{k^2 + 3a^2}{k^2 + a^2}}{mg \dfrac{k^2 \cos \alpha}{k^2 + a^2}} = \dfrac{k^2 + 3a^2}{k^2} \tan \alpha$$

2. *In Fig. 1.10 C is a square-threaded screw rigidly attached to a base-plate B which rests on a plane horizontal surface D ; the disc A is bored and screwed to run freely on the screw C. The disc A has a mass of 15 kg and a*

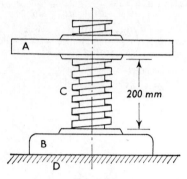

FIG. 1.10

radius of gyration of 140 mm, while the base-plate and screw have a mass of 20 kg and a radius of gyration of 80 mm. The coefficient of friction at the screw-thread is 0·10, and that between the base-plate and surface D is 0·30. The screw-thread has a mean diameter of 50 mm and a pitch of 25 mm. The under-surface of the base B is 120 mm diameter.

Determine the angle through which the base B will turn if A is allowed to fall through 200 mm ; assume there is no plastic distortion on impact of A on B.   (U. Lond.)

$$I_a = 15 \times 0{\cdot}14^2 = 0{\cdot}294 \text{ kg m}^2$$

and $$I_b = 20 \times 0{\cdot}08^2 = 0{\cdot}128 \text{ kg m}^2$$

Accelerating torque on A,

$$T = \frac{mgd}{2}\left\{\frac{\tan\alpha - \mu}{1 + \mu\tan\alpha}\right\}^*$$

$$\tan\alpha = \frac{p}{\pi d} \text{ where } \alpha \text{ is the helix angle}$$

$$= \frac{25}{50\pi} = 0{\cdot}159$$

$$\therefore T = \frac{15 \times 9{\cdot}81 \times 0{\cdot}05}{2} \cdot \frac{0{\cdot}159 - 0{\cdot}10}{1 + 0{\cdot}10 \times 0{\cdot}159}$$

$$= 0{\cdot}2135 \text{ N m}$$

* See the authors' *Mechanics of Machines—Elementary Theory and Examples,* Art. 8.3.

$$\therefore 0 \cdot 2135 = 0 \cdot 294 \alpha_a$$

$$\therefore \alpha_a = 0 \cdot 726 \text{ rad/s}^2$$

In falling through 200 mm, angle turned through,

$$\theta_a = \frac{200}{25} . \times 2\pi$$

$$= 50 \cdot 25 \text{ rad}$$

$\therefore$ final angular velocity, $\omega_a = \sqrt{2\alpha_a \theta_a}$

$$= \sqrt{2 \times 0 \cdot 726 \times 50 \cdot 25}$$

$$= 8 \cdot 54 \text{ rad/s}$$

If ω_b is the initial angular velocity of the base when A and B make contact, then by conservation of angular momentum,

$$0 \cdot 294 \times 8 \cdot 54 = (0 \cdot 294 + 0 \cdot 128)\omega_b$$

$$\therefore \omega_b = 5 \cdot 94 \text{ rad/s}$$

Friction torque on base $= \frac{2}{3}\mu mgr$,* assuming uniform pressure

$$= \frac{2}{3} \times 0 \cdot 30 \times (15 + 20) \times 9 \cdot 81 \times 0 \cdot 06$$

$$= 4 \cdot 12 \text{ N m}$$

$$\therefore 4 \cdot 12 = (0 \cdot 294 + 0 \cdot 128)\alpha_b$$

$$\therefore \alpha_b = 9 \cdot 75 \text{ rad/s}^2$$

$$\therefore \theta_b = \frac{\omega_b{}^2}{2\alpha_b}$$

$$= \frac{5 \cdot 94^2}{2 \times 9 \cdot 75}$$

$$= \underline{1 \cdot 813 \text{ rad}}$$

* See the authors' *Mechanics of Machines — Elementary Theory and Examples* Art. 9.1.

3. *Fig. 1.11 shows an inertia starter for an internal combustion engine.
E is the engine flywheel, having moment of inertia 125 kg m². It is geared to*

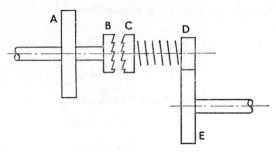

Fig. 1.11

*pinion D so that the speed of E is $\frac{1}{10}$ of the speed of D. Pinion D is con-
nected to the plate C of a dog clutch by a torsion spring which requires 200 N m
torque to twist C through 1 radian relative to D. The flywheel A, having
moment of inertia 0·6 kg m², is made to rotate at 1500 rev/min and then the
dog clutch is engaged. The clutch can transmit torque from B to C only in the
direction in which A is initially rotating, or from C to B in the opposite
direction.*

*Neglecting all losses and the inertia of parts B, C and D, and neglecting
also any additional energy supplied by the motor used to give A its initial
velocity, find the final angular velocities of A and of E, and the maximum
torque transmitted by the clutch.*   (U. Lond.)

Equivalent moment of inertia of E referred to D

$$= 125/10^2 = 1·25 \text{ kg m}^2$$

When the clutch is engaged, the angular momentum of the system
remains constant since no external torque is applied. If N is the common
speed, then

$$0·6 \times 1500 \times \frac{2\pi}{60} + 1·25 \times 0 = (0·6 + 1·25) \times N \times \frac{2\pi}{60}$$

$$\therefore N = 486 \text{ rev/min}$$

After reaching this speed, the spring unwinds, transmitting torque from
C to B in the opposite direction. During unwinding, the same impulse
acts on A and D as during winding up, so that the changes in angular
momentum are the same during these periods. Thus, if N_a and N_d are
the final speeds of A and D,

$$0.6 \times (486 - 1500) \times \frac{2\pi}{60} = 0.6 \times (N_a - 486) \times \frac{2\pi}{60}$$

i.e. $$N_a = -528 \text{ rev/min}$$

$$1.25 \times (486 - 0) \times \frac{2\pi}{60} = 1.25 \times (N_d - 486) \times \frac{2\pi}{60}$$

i.e. $$N_d = 972 \text{ rev/min}$$

Thus the final speed of E is $972/10 = 97.2$ rev/min, in the opposite direction to the initial motion of A.

Alternatively, the total angular momentum of the system before and after clutching is the same,

i.e. $0.6 \times 1500 \times \dfrac{2\pi}{60} + 1.25 \times 0 = 0.6 \times N_a \times \dfrac{2\pi}{60} + 1.25 \times N_d \times \dfrac{2\pi}{60}$

i.e. $$N_a + 2.083N_d = 1500 \ . \qquad . \qquad . \qquad . \qquad . \qquad (1)$$

Also the total K.E. of the system before and after clutching is the same, i.e.

$$\frac{0.6}{2} \times \left(1500 \times \frac{2\pi}{60}\right)^2 + \frac{1.25}{2} \times 0$$

$$= \frac{0.6}{2} \times \left(N_a \times \frac{2\pi}{60}\right) + \frac{1.25}{2} \times \left(N_d \times \frac{2\pi}{60}\right)^2$$

i.e. $\quad N_a{}^2 + 2.083N_d{}^2 = 2\,250\,000 \quad . \qquad . \qquad . \qquad . \qquad (2)$

Therefore, from equations (1) and (2),

$$N_a = -528 \text{ rev/min} \quad \text{and} \quad N_d = 972 \text{ rev/min}$$

Strain energy stored at maximum twist

= loss of K.E. of system

$$= \left\{\frac{0.6}{2} \times \left(\frac{2\pi}{60} \times 1500\right)^2 + \frac{1.25}{2} \times 0\right\} - \frac{(0.6 + 1.25)}{2} \times \left(\frac{2\pi}{60} \times 486\right)^2$$

$$= 5020 \text{ J}$$

If T is the maximum torque transmitted by the clutch and θ the maximum angle of twist of the spring,

$$\text{strain energy in spring} = \tfrac{1}{2}T\theta = \tfrac{1}{2}T \times \frac{T}{200}$$

i.e. $$\frac{T^2}{2 \times 200} = 5020$$

$$\therefore T = 1415 \text{ N m}$$

4. *Two flywheels A and B each 340 mm diameter, with radius of gyration 140 mm, and each of mass 55 kg are mounted on frictionless bearings at the same level, 450 mm apart, as shown in Fig. 1.12. Initially B is stationary*

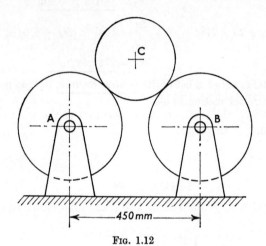

FIG. 1.12

and A is rotating clockwise at 100 rev/min. A third wheel C, 280 mm diameter, with radius of gyration 115 mm and mass 36 kg, initially not rotating, is now placed on A and B as shown.

Find the final angular velocities of A, B and C after slipping between the wheels has ceased.

Assuming that there is no slip between B and C, and that slipping between A and C ceases after 1 s, find

(a) the tangential friction forces acting between A and C and between B and C;

(b) the coefficient of friction between A and C.   (U. Lond.)

$$I_a = I_b = 55 \times 0.14^2 = 1.078 \text{ kg m}^2$$

$$I_c = 36 \times 0.115^2 = 0.476 \text{ kg m}^2$$

$$r_a = r_b = 0.17 \text{ m}, \quad r_c = 0.14 \text{ m}$$

$$\theta = \sin^{-1}\frac{0.225}{0.31} = 46° 30'$$

Let the impulses of the forces acting on A and B during slipping be X and Y respectively, Fig. 1.13, let ω be the final angular velocities of A and B and let Ω be the final angular velocity of C.

Then $$\Omega = \frac{0.17}{0.14}\omega$$

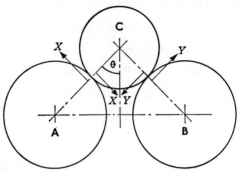

Fig. 1.13

$$X \times 0 \cdot 17 = I_a\left(100 \times \frac{2\pi}{60} - \omega\right)$$

i.e. $\qquad X = 6 \cdot 34(10 \cdot 46 - \omega) \qquad . \qquad . \qquad$ (1)

$$Y \times 0 \cdot 17 = I_b\omega$$

i.e. $\qquad Y = 6 \cdot 34\omega \qquad . \qquad . \qquad . \qquad$ (2)

$$(X - Y) \times 0 \cdot 14 = I_c\Omega = I_c \times \frac{0 \cdot 17}{0 \cdot 14}\omega$$

i.e. $\qquad X - Y = 4 \cdot 12\omega \qquad . \qquad . \qquad . \qquad$ (3)

Hence, from equations (1), (2) and (3),

$$\omega = \underline{3 \cdot 95 \text{ rad/s}}$$

and $\qquad \Omega = \underline{4 \cdot 79 \text{ rad/s}}$

From equation (1),

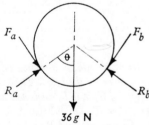

impulse, $X = 6 \cdot 34(10 \cdot 46 - 3 \cdot 95) = 41 \cdot 3 \text{ N s}$

∴ tangential *force* between A and C,

$\qquad F_a = \underline{41 \cdot 3 \text{ N}}$ since this force acts

for 1 s

From equation (2),

Fig. 1.14

impulse, $Y = 6 \cdot 34 \times 3 \cdot 95 = 25 \cdot 1 \text{ N s}$

∴ tangential *force* between B and C,

$\qquad F_b = \underline{25 \cdot 1 \text{ N}}$ since this force acts for 1 s

The forces acting on C are shown in Fig. 1.14. For equilibrium of vertical forces,

$$(R_a + R_b) \cos \theta - (F_a + F_b) \sin \theta = 36 \times 9 \cdot 81 = 353 \text{ N}$$

i.e. $(R_a + R_b) - (41 \cdot 3 + 25 \cdot 1) \tan 46° 30' = 353 \sec 46° 30'$

i.e. $$R_a + R_b = 582 \text{ N}$$

For equilibrium of horizontal forces,

$$(R_a - R_b) \sin \theta + (F_a - F_b) \cos \theta = 0$$

i.e. $(R_a - R_b) + (41 \cdot 3 - 25 \cdot 1) \cot 46° 30' = 0$

i.e. $$R_a - R_b = -15 \cdot 4 \text{ N}$$

$$\therefore R_a = 298 \cdot 7 \text{ N}$$

$$\therefore \mu = \frac{F_a}{R_a} = \frac{41 \cdot 3}{298 \cdot 7} = \underline{0 \cdot 138}$$

5. *A belt, of total mass 12 kg, connects two pulleys A and B as shown in Fig. 1.15. These pulleys are identical, each of mass 18 kg, and each having a radius of gyration of 260 mm, and an external radius of 300 mm. A third pulley C, of mass 14 kg, and having radius of gyration 220 mm, and external*

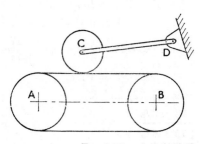

Fig. 1.15

radius 240 mm, is mounted on a light arm CD so that it can be lifted from or lowered on to the belt. The thickness of the belt, also its deflection under the weight of C, may be neglected, and it may be assumed that the belt does not slip on either of the pulleys A or B.

Initially A and B and belt are stationary, whilst C is just clear of the belt and spinning at 200 rev/min. C is now lowered on to the belt.

(a) Determine the final speeds of each pulley after C has ceased to slip relatively to the belt.

(b) If the coefficient of friction between C and the belt is 0·25, find the time during which C slips relatively to the belt.   (U. Lond.)

Retarding torque on C $= \mu m_c g r_c = \mu \times 14 \times 9 \cdot 81 \times 0 \cdot 24$

$$= 33\mu \text{ N m}$$

$\therefore$ for pulley C, $33\mu = I_c\alpha_c = 14 \times 0\cdot22^2\alpha_c$

$$\therefore \mu = 0\cdot0205\alpha_c \qquad . \qquad . \qquad . \qquad . \qquad (1)$$

Equivalent mass of pulleys referred to line of action of accelerating force

$$= 2 \times m_a \times \left(\frac{k_a}{r_a}\right)^2$$

$$= 2 \times 18 \times \left(\frac{0\cdot26}{0\cdot30}\right)^2$$

$$= 27 \text{ kg}$$

$\therefore$ total equivalent mass to be accelerated $= 27 + 12 = 39$ kg

$\therefore$ for belt and pulleys A and B, $\mu m_c g = mf$

i.e. $\qquad\qquad\qquad\qquad 14 \times 9\cdot81\mu = 39 \times 0\cdot3\alpha_a$

$$\therefore \mu = 0\cdot0852\alpha_a \qquad . \qquad (2)$$

$\therefore$ from equations (1) and (2), $0\cdot0205\alpha_c = 0\cdot0852\alpha_a$. $\qquad$ (3)

If t is the time of slipping and ω_a and ω_c are the final angular velocities of A and C, then

$$\alpha_a = \frac{\omega_a}{t}, \quad \alpha_c = \frac{\left(200 \times \dfrac{2\pi}{60} - \omega_c\right)}{t} \quad \text{and} \quad \omega_a = \frac{0\cdot24}{0\cdot30}\omega_c$$

Substituting in equation (3),

$$0\cdot0205\frac{\left(200 \times \dfrac{2\pi}{60} - \omega_c\right)}{t} = 0\cdot0852 \times \frac{\omega_a}{t}$$

$$= 0\cdot0852 \times \frac{0\cdot24}{0\cdot30} \times \frac{\omega_c}{t}$$

from which $\omega_c = 4\cdot82$ rad/s $\quad$ or $\quad$ $\underline{46 \text{ rev/min}}$

$$\therefore \omega_a = 4\cdot82 \times \frac{0\cdot24}{0\cdot30}$$

$$= 3\cdot85 \text{ rad/s} \quad \text{or} \quad \underline{36\cdot8 \text{ rev/min}}$$

$$\alpha_a = \frac{\omega_a}{t} = \frac{3\cdot85}{t}$$

$\therefore$ from equation (2), $\quad 0\cdot25 = 0\cdot0852 \times \dfrac{3\cdot85}{t}$

$$\therefore t = \underline{1\cdot313 \text{ s}}$$

6. *A heavy cylinder rolls across a horizontal surface with a linear velocity of 5 m/s, mounts a step 90 mm high without rebound or slip and continues to roll up an inclined surface of slope $\sin^{-1} 0.2$. If the outside diameter of the cylinder is 600 mm and the radial thickness is 45 mm, find the linear velocity immediately after the impact and the distance travelled after mounting the step before coming to rest.*   (U. Lond.)

$$I = m\left(\frac{0.60^2 + 0.51^2}{8}\right) = 0.0775m \text{ kg m}^2$$

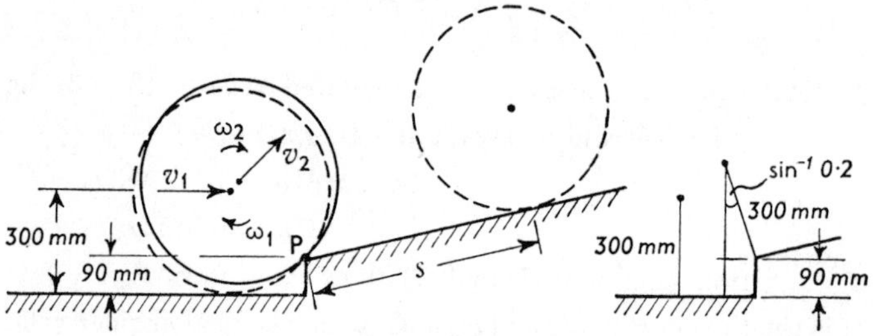

Fig. 1.16

In Fig. 1.16, let v_1 and v_2 be the linear velocities of the cylinder immediately before and after impact at P and let ω_1 and ω_2 be the angular velocities of the cylinder immediately before and after impact at P.

During impact, the moment of momentum of the cylinder about P remains constant since the impulsive force at P has no moment about that point.

Hence $$I\omega_1 + mv_1 \times 0.21 = I\omega_2 + mv_2 \times 0.3$$

But $$\omega_1 = \left(\frac{5}{0.30}\right) = 16.67 \text{ rad/s}$$

and $$\omega_2 = \frac{v_2}{0.30} = 3.33v_2 \text{ rad/s}$$

$\therefore \ 0.0775m \times 16.67 + m \times 5 \times 0.21 = 0.0775m \times 3.33v_2 + mv_2 \times 0.3$

from which $$v_2 = \underline{4.2 \text{ m/s}}$$

When the cylinder comes to rest, gain of P.E. = loss of K.E. If the distance travelled along the slope is s m, then height risen $= 0.2s$ so that

$$mg\{0.09 + 0.30 \cos(\sin^{-1} 0.2) - 0.3 + 0.2s\} = \frac{m}{2}v_2^2 + \frac{I}{2}\omega_2^2$$

i.e. $$mg(0.084 + 0.2s) = \frac{m}{2} \times 4.2^2 + 0.0775\frac{m}{2} \times (3.33 \times 4.2)^2$$

$$\therefore \ s = \underline{7.95 \text{ m}}$$

7. *A uniform beam ABCD, 6 m long, rests horizontally on supports at B and C such that AB = CD. End A is raised by turning the beam about C and is then released.*

(a) *Assuming that there is no slipping at the supports, show that end D will rise above the original horizontal position of the beam if BC is less than 3·464 m.*

(b) *If BC = 3 m and A is raised a vertical distance of 0·6 by turning the beam about C, determine the vertical distance D will rise when the beam is released.* (U. Lond.)

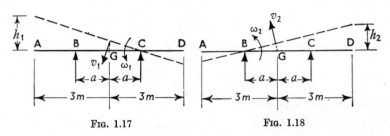

FIG. 1.17   FIG. 1.18

(a) In Figs. 1.17 and 1.18, let ω_1 and ω_2 be the angular velocities of the beam immediately before and after impact at B, let v_1 and v_2 be the linear velocities of G immediately before and after impact at B and let m be the mass of the beam.

During impact, the moment of momentum of the beam about B remains constant since the impulsive force at B has no moment about that point.

i.e. $$I_G\omega_1 - mv_1a = I_G\omega_2 + mv_2a$$

taking anticlockwise momentum as positive.

i.e. $$m \times \frac{6^2}{12}\omega_1 - m\omega_1a^2 = m \times \frac{6^2}{12}\omega_2 + m\omega_2a^2$$

$$\therefore \omega_2 = \omega_1\left(\frac{3 - a^2}{3 + a^2}\right) \qquad . \qquad . \qquad . \qquad (1)$$

Therefore, if the end D is to rise,

$$3 > a^2$$

or $$a < 1·732 \text{ m}$$

i.e. $$\underline{BC < 3·464 \text{ m}}$$

(b) If h_1 is the initial height of A above the supports and h_2 is the final height of D above the supports, then

$$\text{initial height of G above supports} = \frac{h_1a}{3 + a}$$

and   final height of G above supports $= \dfrac{h_2 a}{3 + a}$

In falling from the height h_1, loss of P.E. = gain of K.E.

i.e. $mg \times \dfrac{h_1 a}{3 + a} = \tfrac{1}{2} I_G \omega_1{}^2 + \tfrac{1}{2} m v_1{}^2$

$$= \frac{m}{2}\frac{6^2}{12}\omega_1{}^2 + \frac{m}{2}\omega_1{}^2 a^2$$

$$\therefore \; \omega_1 = \sqrt{\frac{2 g h_1 a}{(3 + a)(3 + a^2)}}$$

Similarly, in rising to height h_2 above D,

$$\omega_2 = \sqrt{\frac{2 g h_2 a}{(3 + a)(3 + a^2)}}$$

Substituting in equation (1),

$$\frac{2 g h_2 a}{(3 + a)(3 + a^2)} = \frac{2 g h_1 a}{(3 + a)(3 + a^2)}\left(\frac{3 - a^2}{3 + a^2}\right)^2$$

or $h_2 = h_1 \left(\dfrac{3 - a^2}{3 + a^2}\right)^2$

But $h_1 = 0.6$ m and $a = 1.5$ m,

$$\therefore \; h_2 = 0.6\left(\frac{3 - 1.5^2}{3 + 1.5^2}\right)^2 = \underline{0.012\ 26\ \text{m}}$$

8. *A 0·6-m square downward opening trapdoor has a horizontal frictionless pin joint along one edge. The door has a mass of 6 kg, its centre of gravity is at its centre of symmetry, its mass is uniformly distributed, and, when closed, lies in a horizontal plane. When the catch is released, the door accelerates downwards about the horizontal pin joint until it finally hangs vertically. At this instant the centre of the lower edge of the door comes into contact with a horizontal coil spring, of stiffness 20 kN/m, which acts as a buffer, absorbing the kinetic energy of the door by compression. The angle through which the door turns beyond the vertical position in compressing the spring is small.*

If the door catch is suddenly released when the door is in a closed position, determine—

(i) the instantaneous acceleration of the centre of gravity of the door at the moment of release;

(ii) the instantaneous force on the pin joint at the moment of release;

(iii) the maximum compression of the horizontal spring;

(iv) the horizontal force on the pin joint at the moment of maximum compression of the spring.   (U. Lond.)

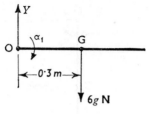

FIG. 1.19

(*i*) Referring to Fig. 1.19, couple about O,

$$T_O = I_O \alpha_1$$

where α_1 is the instantaneous angular acceleration of the door.

i.e.
$$6g \times 0.3 = 6 \times \frac{0.6^2}{3} \times \alpha_1$$

$$\therefore \alpha_1 = 24.5 \text{ rad/s}^2$$

$\therefore$ instantaneous linear acceleration of centre of gravity of door

$$= 24.5 \times 0.3$$

$$= \underline{7.35 \text{ m/s}^2}$$

(*ii*) Couple about G,

$$T_G = I_G \alpha_1$$

i.e.
$$Y \times 0.3 = 6 \times \frac{0.6^2}{12} \times 24.5$$

where Y is the vertical reaction at the hinge.

$$\therefore Y = \underline{14.7 \text{ N}}$$

(*iii*) At impact,

$$\text{loss in P.E.} = \text{strain energy given to spring}$$

i.e.
$$6g \times 0.3 = \tfrac{1}{2}Sx^2 = \tfrac{1}{2} \times 20 \times 10^3 \times x^2$$

where x m is the compression of the spring.

$$\therefore x = 0.042 \text{ m} \quad \text{or} \quad \underline{42 \text{ mm}}$$

(*iv*) In Fig. 1.20, let X be the horizontal reaction at the hinge, P be the force in the spring and f the linear acceleration of G.

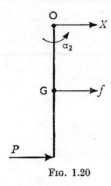

FIG. 1.20

Then at the point of maximum compression,

$$P = 20 \times 10^3 \times 0\cdot042 = 840 \text{ N}$$
$$P + X = 6 \times f = 6 \times (0\cdot3 \times \alpha_2) \quad . \qquad . \qquad . \qquad (1)$$

Also
$$P \times 0\cdot6 = I_O\alpha_2 = 6 \times \frac{0\cdot6^2}{3} \times \alpha_2$$

i.e.
$$\alpha_2 = \frac{P}{1\cdot2}$$

Substituting in equation (1),

$$P + X = \tfrac{3}{2}P$$
$$\therefore X = \tfrac{1}{2}P = \underline{420 \text{ N}}$$

9. *The mass of the pendulum of an impact testing machine of the Charpy type is 22 kg, its centre of gravity is 0·7 m from the axis of suspension and the point at which the test piece is struck is 0·8 m from the axis. When allowed to swing freely through a small angle the pendulum makes 34 complete swings per minute.*

In a test, the angle of descent of the pendulum is 150°, and the angle of ascent after the test piece is broken is 120°. If the fracture of the test piece occurs in 0·01 s, find the mean reaction on the bearings due to the impact.

(U. Lond.)

Referring to Fig. 1.21, periodic time,

$$t_p = 2\pi\sqrt{\frac{k^2 + h^2}{gh}} = 2\pi\sqrt{\frac{I_O}{mgh}}$$

i.e.
$$\frac{60}{34} = 2\pi\sqrt{\frac{I_O}{22 \times 9\cdot81 \times 0\cdot7}}$$
$$\therefore I_O = 11\cdot92 \text{ kg m}^2$$

Let ω_1 and ω_2 be the angular velocities of the pendulum just before and after impact.

Before impact, loss of P.E. = gain of K.E.

i.e. $22 \times 9\!\cdot\!81 \times 0\!\cdot\!7 \times (1 + \cos 30°) = \frac{1}{2} \times 11\!\cdot\!92 \times \omega_1{}^2$

$$\therefore \omega_1 = 6\!\cdot\!91 \text{ rad/s}$$

After impact, gain of P.E. = loss of K.E.

i.e. $22 \times 9\!\cdot\!81 \times 0\!\cdot\!7 \times (1 + \cos 60°) = \frac{1}{2} \times 11\!\cdot\!92 \times \omega_2{}^2$

$$\therefore \omega_2 = 6\!\cdot\!17 \text{ rad/s}$$

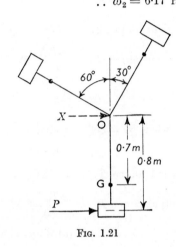

Fig. 1.21

If P is the mean impulsive force at the test piece, then

$$P \times 0\!\cdot\!8 = I_{\mathrm{O}}\alpha$$

where α is the angular deceleration of the pendulum during impact.

$$\alpha = \frac{6\!\cdot\!91 - 6\!\cdot\!17}{0\!\cdot\!01} = 74 \text{ rad/s}^2$$

$$\therefore P = \frac{11\!\cdot\!92 \times 74}{0\!\cdot\!8} = 1103 \text{ N}$$

If X is the mean horizontal reaction on the bearings at O due to impact, then

$$P + X = mf$$

where f is the linear deceleration of the c.g. during impact.

$$\therefore 1103 + X = 22 \times 74 \times 0\!\cdot\!7$$

$$\therefore X = \underline{37 \text{ N}}$$

B

10. *The torque required to drive a ventilating fan is proportional to the square of its shaft speed, 76 kW being absorbed when the speed is 383 rev/min. It is driven through a 2 : 1 speed reduction gear from a motor for which the output torque is zero at 750 rev/min increasing uniformly as the speed decreases to 675 rev/min when it generates 90 kW.*

Sketch the 'torque-speed' curves for the gear output and fan and find the speed at which the fan will run at the corresponding power.   Comment briefly on the running stability.   (I. Mech. E.)

Let suffix f denote the fan, suffix m the motor and suffix g the gear output.

$$\text{Fan input torque, } T_f = aN_f^2$$

and $$\text{fan input power, } P_f = \frac{2\pi T_f N_f}{60} \text{ W}$$

$$= \frac{\pi}{30}aN_f^3$$

$P_f = 76 \times 10^3$ W when $N_f = 383$ rev/min, so that $a = 0{\cdot}012\,92$

$$\therefore \; T_f = 0{\cdot}012\,92 N_f^2 \text{ N m} \qquad (1)$$

$$\text{Motor output torque, } T_m = b - cN_m$$

and $$\text{motor output power, } P_m = \frac{2\pi T_m N_m}{60} \text{ W}$$

$$= 0{\cdot}1048(b - cN_m)N_m$$

$P_m = 0$ when $N_m = 750$ rev/min and $P_m = 90 \times 10^3$ when $N_m = 675$ rev/min

$$\therefore \; b = 750c \quad \text{and} \quad c = 17$$

$$\therefore \; T_m = 17(750 - N_m) \text{ N m}$$

But $T_m = \frac{1}{2}T_g$ and $N_m = 2N_g = 2N_f$

$$\therefore \; T_g = 68(375 - N_f) \text{ N m} \qquad . \qquad (2)$$

From equations (1) and (2), the torque-speed curves for the gear output and fan are plotted, Fig. 1.22.   From the intersection of these curves, the running speed of the fan is found to be 300 rev/min.

For all fan speeds below 300 rev/min the driving torque is greater than the resisting torque and for all fan speeds above 300 rev/min the resisting torque is greater than the driving torque. Therefore, if displaced from a speed of 300 rev/min, the fan will always return to this speed and the running is stable.

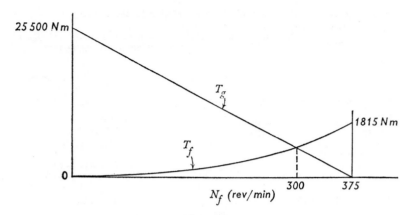

FIG. **1.22**

11. A piston moves horizontally in a cylinder with the rod passing through each end as shown in Fig. 1.23. The cylinder is filled with oil which can pass through small holes from one side of the piston to the other at the rate of $q = c \times p$ m³/s, where c is a constant and p is the pressure difference between the two sides in N/m². If the parts moving with the piston have a mass of m kg and the effective area of one side in s m², obtain an expression for the displacement t s after starting from rest under the action of a constant force P N. If $m = 50$, $c = 25 \times 10^{-9}$, $s = 6\cdot5 \times 10^{-3}$ and $P = 5 \times 10^3$, evaluate this for $t = 0\cdot05$ s.

The solution of the equation

$$\frac{d^2x}{dt^2} + k\frac{dx}{dt} = b \text{ is } x = A + Be^{-kt} + \left(\frac{b}{k}\right)t \quad (I. \text{ Mech. E.})$$

$$\left(Ans.: \frac{Pc^2m}{s^4}\left(-1 + e^{-\frac{s^2}{cm}t}\right) + \frac{Pc}{s^2}t \, ; \, 75\cdot7 \text{ mm}\right)$$

12. Two wagons of mass 3 tonne and 9 tonne rest on a straight horizontal track. The wagons are connected by a light inextensible chain which is rigidly attached to one wagon and passes through a friction grip on the other so that if the pull exceeds 10 kN the chain will slip. The rolling resistance of each wagon is 180 N per tonne mass and initially both wagons are at rest with the chain slack. If a force of $(2\cdot7 + 0\cdot18t)$ kN, where t is the time in seconds, is applied to the 3-tonne wagon, and the chain becomes taut 2 s after this is applied, find :

(a) the speed of both wagons after the chain has stopped slipping ;
(b) the time of slip. (U. Lond.)

(Ans.: 0·428 m/s ; 0·46 ₁

13. A packing case of mass 36 kg rests on the floor of a lorry. Any forward movement of the case is resisted by a compression spring of stiffness 900 N/m and friction, the coefficient of friction being 0·4. When travelling at 72 km/h, the lorry is braked suddenly with a retardation 7·5 m/s².

Find (a) the displacement of the case relative to the lorry,
   (b) the time taken for this movement,
   (c) the absolute displacement of the case during this time.   (U. Lond.)

(Ans. : 0·285 m ; 0·628 s ; 11·37 m)

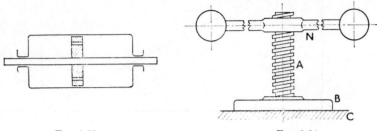

FIG. 1.23   FIG. 1.24

14. To determine the moment of inertia of a flywheel it is supported with the inner face of the rim resting on a horizontal knife-edge parallel to the axis of the wheel. The radius of the inner face of the rim is 0·75 m and the time for a complete small oscillation is 2·61 s. Find the radius of gyration of the wheel about its axis, which contains the mass centre.

A resisting torque, varying directly as the speed, is applied to the wheel of mass 1·75 Mg. This torque has a magnitude of 36 N m when the speed is 150 rev/min. Find the time taken to increase the speed from 150 to 300 rev/min when a constant driving torque of 100 N m is applied to the wheel.   (U. Lond.)

(Ans. : 0·842 m ; 448 s)

15. A flywheel, $m = 80$ kg, $k = 400$ mm, is accelerated from rest to a steady speed of 75 rad/s by the application of a constant torque, the friction torque opposing the motion being $13·5\omega$ N m, where ω is the speed in rad/s at any instant.

Find the applied torque. Obtain the expression for the speed at any time t s from the start, and calculate the speed after 1 s.

Also find the final steady speed and obtain the expression for the speed at time t if the wheel had been acted upon by a torque of magnitude $120 (75 - \omega)$ N m. Calculate for this case the speed after 0·1 s.

Roughly sketch on the same time base curves showing how ω varies with t for each of the two cases.

Assume that the solution of $\dfrac{d\omega}{dt} + p\omega = q$ is $\omega = Ae^{-pt} + q/p$.   (I. Mech. E.)

(Ans. : 1014 N m ; $75(1 - e^{-1·055t})$; 48·9 rad/s ; 67·3 rad/s ; 67·3
$(1 - e^{-10·44t})$; 43·6 rad/s)

16. In Fig. 1.24 the square-threaded screw A is attached rigidly to the circular plate B which rests on a plane horizontal surface C; two weighted arms are attached to the nut N which can run freely on the screw.

The screw and base have a mass of 23 kg and radius of gyration of 115 mm ; the under-surface of B has a mean contact radius of 150 mm. The nut N with attached arms has a mass of 16 kg and a radius of gyration of 200 mm. The screw has a mean diameter of 60 mm and a pitch of 30 mm. The coefficient of friction between base-plate B and surface C is 0·35, and that between nut and screw may be neglected.

If the nut and arms are given a rotational speed of 60 rev/min when 225 mm from the base, find :

(a) the angular velocity of the nut and arms just before reaching the base ;

(b) the angular velocity of the nut and arms just after impact with the base, assuming that the nut does not rebound ;

(c) the angle through which the base will turn before coming to rest.

(*U. Lond.*)

(*Ans. :* 12·25 rad/s ; 8·32 rad/s ; 1·625 rad)

17. Two co-axial shafts, A and B, may be coupled together by means of a dog clutch, one element of which is rigidly connected to shaft B whilst the other element is connected to shaft A through a torsionally flexible spring of stiffness 560 N m/rad. The rotating masses on shaft A have a moment of inertia of 34 kg m² and those on shaft B 93·5 kg m².

With shaft A rotating at 300 rev/min and shaft B at rest the dog clutch is suddenly engaged.

Determine (a) the final angular velocities of each shaft,

(b) the maximum energy stored in the spring,

(c) the maximum couple exerted by the spring,

(d) the maximum angle of twist of the spring. (*U. Lond.*)

(*Ans. :* −140 and +160 rev/min ; 12·3 kJ ; 3·71 kN m ; 379°)

18. Two flywheels A and B are free to rotate on parallel shafts as shown in Fig. 1.25. Initially A rotates at 240 rev/min clockwise, and B is stationary.

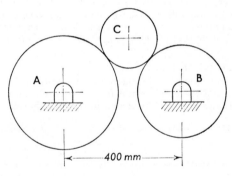

Fɪɢ. 1.25

A third wheel C, in contact with B and initially not rotating, is lowered until it rests also on A. It may be assumed that no slip occurs between B and C.

The particulars of the wheels A, B and C respectively are: outer diameter 375 mm, 300 mm, 200 mm, radius of gyration 180 mm, 120 mm, 90 mm, mass 75 kg, 60 kg, 25 kg.

(a) Determine the final velocities of each wheel, after slip between A and C has ceased.

(b) If slipping between A and C lasts for 2 s after initial contact, find the coefficient of friction between A and C.

(A partly graphical solution is suggested for this part.) (*U. Lond.*)

(*Ans. :* 129·8 rev/min ; 162·4 rev/min ; 243·5 rev/min ; 0·328)

19. A solid uniform cylinder of radius 150 mm rolls without slipping down a plane AO inclined at 45° to the horizontal. It then rolls up a plane OB inclined at $\theta°$ to the horizontal. The centre of gravity of the cylinder descends a vertical distance of 0·6 m while the cylinder is rolling down AO.

Calculate :

(a) the minimum value of θ which will cause the cylinder to be brought to rest when it strikes OB ;

(b) the vertical distance through which the cylinder will rise as it rolls up OB if $\theta = 45°$. (*U. Lond.*)

(*Ans.* : 75° ; 66·7 mm)

20. A cylindrical vessel 1·2 m diameter, of mass 100 kg and radius of gyration 450 mm, is made to roll on a horizontal floor, without slipping, by means of a horizontal force of 70 N acting 1·2 m from the floor in a plane which is at right angles to and passes through the mid-point of the axis of the cylinder.

Calculate the linear acceleration of the centre of the cylinder and the magnitude and direction of the frictional force between the cylinder and the floor.

If the applied force remains horizontal and in the same plane as before, calculate by how much its line of action must be lowered so that the cylinder may be given a truly rolling acceleration when the coefficient of friction between the cylinder and floor is zero.

Any formula must be proved. (*U. Lond.*)

(*Ans.* : 0·896 m/s² ; 19·6 N in direction of applied force ; 0·2625 m)

21. Two similar rollers A and B have their axes parallel and distance $2a$ apart in a horizontal plane. Each roller has a radius r and moment of inertia I ; initially B is at rest and A has a speed of ω rad/s. A uniform beam of mass m is then lowered on to the rollers, making contact with them simultaneously and with its centre of gravity midway between A and B. Assuming no slip between B and the beam, show that after slipping at A has ceased the velocity of the beam is $\dfrac{Ir\omega}{mr^2 + 2I}$.

During the period of slip at A, if the centre of gravity of the beam is between the roller centres and the coefficient of friction between the beam and A is μ, show that the distance S moved by the beam from rest in time t is

$$S = a(1 - \cos \alpha t), \text{ where } \alpha^2 = \frac{\mu mg}{2a\left(m + \dfrac{I}{r^2}\right)} \qquad (U.\ Lond.)$$

22. Two parallel shafts are connected by an open belt. The driving shaft A runs initially at 500 rev/min, it has a mass of 180 kg with radius of gyration 125 mm and the belt pulley is 375 mm diameter. The driven shaft B has a mass of 67·5 kg with radius of gyration 75 mm and its fast and loose pulleys are 150 mm diameter. Its initial speed is 150 rev/min with the same direction as A, the belt being on the loose pulley. The belt is then moved to the fast pulley. Make a sketch indicating the tension in each side of the belt during slip, and by considering the change in moment of momentum of each shaft about its own axis, find the shaft speeds when slip ceases. (*I. Mech. E.*)

(*Ans.* : 299 rev/min ; 748 rev/min)

23. The torque-speed characteristic of an electric motor is given below:

Speed, rev/min .	.	0	250	500	750	1000
Torque, N m .	.	34	77·5	95	105	120

A flywheel is connected to the motor and it is found that the time taken for the speed to increase from 250 to 1000 rev/min is 34 s. If the moment of inertia of the motor armature and shaft is 5·35 kg m² what is the moment of inertia of the flywheel ? Neglect friction. (*U. Lond.*) (*Ans.*: 75·1 kg m²)

24. A rigid beam AB of uniform cross-section, and of mass 36 kg, is hinged at A to a fixed support and is maintained in a horizontal position by a vertical helical spring attached to B, AB being 1·8 m. A mass of 2 kg is allowed to fall on to the beam with a striking velocity of 3·5 m/s and the point of impact is 1·2 m from A. Assuming the mass and beam move together, determine the angular velocity of the beam immediately after impact. (*U. Lond.*)
(*Ans.*: 0·2013 rad/s)

25. A uniform rectangular beam, 3 m long, 0·2 m deep, of mass 50 kg, is supported by a transverse hinge, whose centre line is halfway along the beam and coincides with the upper surface, so that it can swing in a vertical plane. When the beam is in its equilibrium position, a 6-kg mass falls on to its upper surface, the line of fall being 1·2 m from the hinge, and the height of the fall 0·75 m. The impact is cushioned by a spring of stiffness 35 kN/m compression. Assuming no loss of energy at impact, find the height of rebound of the mass, the angle through which the beam swings and the maximum force in the spring. (*U. Lond.*)   (*Ans.*: 0·297 m ; 62° 52′ ; 1·587 kN)

26. One end of a thin uniform rod 0·45 m long is connected by a hinge to a rigid support. The rod which has a mass of 2 kg and initially hangs downward is raised through 60° from the vertical and is then released. As the rod approaches the end of its travel it is restrained by a horizontal compression spring situated 0·38 m below the hinge. The stiffness of this spring is 35 kN/m and it is arranged so that the rod just reaches the vertical position.

Neglecting any loss of energy during the impact, calculate the greatest reaction in the hinge.

Where should the spring be placed for this value to be a minimum ? (*U. Lond.*)
(*Ans.*: Horizontal force = 104·5 N ; 0·3 m from hinge)

27. In Fig. 1.26, OA is a uniform rod 0·6 m long, of mass 2·5 kg which is pivoted on a fixed support at O and hangs vertically. A compression spring S is just in contact with the lower end A of the rod and resists any movement of A to the left by a force of 18 kN/m movement of A. If the rod is turned until it makes an angle of 45° with the vertical and is then allowed to fall, what are the greatest forces exerted on the spring and on the pivot at O ? Neglect any loss of energy in the impact. (*U. Lond.*)   (*Ans.*: 278·2 N ; 139·3 N)

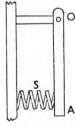

FIG. 1.26

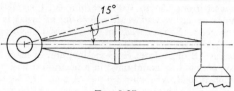

FIG. 1.27

28. To drive a pile, inaccessible for a direct drop, a 'tilt hammer', Fig. 1.27, was used. The total mass of the hammer was 90 kg, its centre of gravity was 2·5 m from the hinge pin and the radius of gyration about the centre of gravity was 1·1 m. The force on the hammer due to the blow can be considered to act 3 m from the hinge. When the hammer had fallen through an angle of 15°, it was brought to rest in 11·5 mm. Find the average force on the hinge if the hammer shaft is horizontal during this period. (*U. Lond.*)

(*Ans.*: 684 N (excluding dead weight))

29. A uniform rod AB, 0·45 m long and of mass 11 kg, can rotate in a vertical plane about a pin through the end A. Initially it is held in a horizontal position. The end B is then released so that it swings downwards in a circular arc. When the rod reaches the vertical position the end B strikes a stop pin which is elastic and yields 3 mm under the impact.

Find (*a*) the maximum force on the stop pin, (*b*) the maximum horizontal force on the hinge A, (*c*) the maximum bending moment at the mid-point of the rod. (*U. Lond.*)   (*Ans.*: 16·2 kN ; 8·11 kN ; 1·365 kN m)

30. The tailboard of a lorry is 1·5 m long and 0·75 m high. It is hinged along the bottom edge to the floor of the lorry. Chains are attached to the top corners of the board and to the sides of the lorry so that when the board is in a horizontal position the chains are parallel and inclined at 45° to the horizontal. A tension spring is inserted in each chain so as to reduce the shock and these are adjusted to prevent the board from dropping below the horizontal. Each spring exerts a force of 52 kN/m of extension.

Find the greatest force in each spring and the resultant force at the hinges when the board falls freely from the vertical position. Assume that the tailboard is a uniform body of mass 27 kg. (*U. Lond.*) (*Ans.*: 2·27 kN ; 3·59 kN)

31. A uniform rod AB, 0·75 m long and of mass 20 kg, is hinged at A and held in a horizontal position. It is allowed to fall, rotating about A in a vertical plane, until it strikes a horizontal spring C of stiffness 35 kN/m, whose centre line is 0·35 m below the level of A. When the spring force is a maximum, the rod is vertical. Losses at impact may be neglected.

Find (*a*) the maximum spring force,
     (*b*) the maximum horizontal force on the hinge A,
     (*c*) the maximum bending moment at the mid-point of the rod.
(*U. Lond.*)   (*Ans.*: 2270 N ; 682 N ; 254·5 N m)

32. The pendulum of an Izod impact testing machine has a mass of 32 kg. Its c.g. is 1·05 m from the axis of suspension and the striking knife is 0·15 m below the c.g. The time for 20 small free oscillations is 43·5 s. In making a test the pendulum is released from an angle of 60° to the vertical. Determine:

(*a*) the striking velocity and the position of the centre of percussion relative to the striking knife,

(*b*) the impulse on the pendulum and the sudden change of axis reaction when a specimen giving an impact value of 55 J is broken. (*U. Lond.*)

(*Ans.*: 3·47 m/s ; 23 mm ; 17·55 N s ; 94 N)

33. In an impact testing machine of the Izod type, the pendulum has a mass of 27 kg and its centre of gravity is 1·2 m from the fixed pivot of the pendulum. The radius of gyration of the pendulum about its centre of gravity is 0·25 m. In making a test, the pendulum is pulled away from its lowest position and is then released. At the lowest part of the swing a knife-edge in the bob of the pendulum strikes and breaks a metal test-piece.

Working from first principles, find the distance of the knife-edge from the pivot so that there shall be no impulsive force on the pivot bearing.

In a test, the initial position of the pendulum was 60° from the lowest position and the amount of the swing after the impact was 50° from the lowest position. Find the loss of energy from the pendulum and, assuming that the duration of the impact was 0·004 s, find the average force exerted on the knife-edge during that time. (*U. Lond.*)   (*Ans.*: 1·226 N ; 45·4 J ; 3·59 kN)

34. A hammer B suspended from pin C, and an anvil A suspended from pin D, are just touching each other at E when both hang freely, as shown in Fig. 1.28. B has a mass of 6·5 kg, its centre of gravity is 250 mm below C and its radius of gyration about C is 270 mm. A has a mass of 22 kg, its centre of gravity is 175 mm below D and its radius of gyration about D is 190 mm.

The hammer B is rotated 20° to the position shown dotted and released. Assume that the points of contact move horizontally at the instant of impact and that their local relative linear velocity of recoil is 0·8 times their relative linear velocity of impact.

Find the angular velocities of the hammer and of the anvil immediately after impact. (*U. Lond.*)   (*Ans.* : −0·768 rad/s ; +1·185 rad/s)

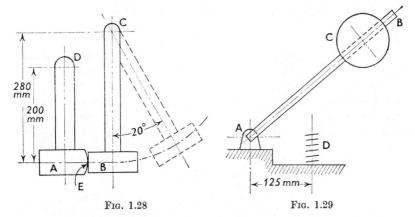

FIG. 1.28   FIG. 1.29

35. A uniform rod AB, 350 mm long and of mass 10 kg, carries a small bob C, of mass 12 kg, at a point 75 mm from B. The rod is freely hinged at A, and released from the position shown in Fig. 1.29, in which it is inclined at 40° to the horizontal. It strikes a spring stop D, of stiffness 35 kN/m, compressing it until in its lowest position the rod is horizontal.

Find (*a*) the maximum force in the spring,
　　　(*b*) the maximum vertical force on the rod at A,
　　　(*c*) the maximum bending moment at the point on the rod 125 mm
　　　　　from A.

(*U. Lond.*)   (*Ans.:* 1493 N; 752 N; 97 N m)

36. A railway carriage is standing in a siding, a door being opened at an angle θ with the train. Another carriage is shunted into it causing it to start moving with velocity v, in the direction which causes the door to shut.

The mass of the door is m, its centre of gravity is a from the hinge and the radius of gyration round a vertical line through the centre of gravity is k.

Obtain expressions for (a) the angular velocity with which the door starts to move and (b) the impulsive reaction at the hinge.

Taking the door as a uniform rectangle 0·75 m wide, of mass 70 kg, v as 5 km/h and θ as 60°, calculate the angular velocity and the total kinetic energy of the door immediately after impact. ($U.\ Lond.$)

$$\left(Ans.: \ \frac{va \sin \theta}{k^2 + a^2}; \ \frac{mv}{k^2 + a^2} \sqrt{(k^2 + a^2)^2 \cos^2 \theta + k^4 \sin^2 \theta}; \ 2 \cdot 405 \,\text{rad/s}; \ 105 \cdot 6 \,\text{J} \right)$$

(b) Hoists

37. *In an overhead travelling crane the winding drum is driven by an electric motor through a triple reduction gear. The diameter of the drum, measured to the centre of the lifting rope, is 630 mm. The load is raised one-half of the distance the rope is wound on the drum.*

Using the data given below and assuming that the overall efficiency is 75%, determine the speed and power of the motor when the crane lifts a load of 15 t at a speed of 0·12 m/s. If the motor exerts a constant torque of 750 N m, find the time taken for this speed to be attained from rest.

	Mass of shaft, with wheels, etc. (kg)	*Radius of gyration* (mm)	*Speed*
Motor	*180*	*160*	*48·0 N*
1st intermediate shaft . . .	*135*	*200*	*12·5 N*
2nd intermediate shaft . .	*230*	*300*	*5·0 N*
Drum and shaft . . .	*1800*	*600*	*N*

(U. Lond.)

When the speed of the load is uniform at 0·12 m/s,

$$\text{motor power} = \frac{15 \times 10^3 \times 9 \cdot 81 \times 0 \cdot 12}{10^3} \times \frac{100}{75} = 23 \cdot 5 \,\text{kW}$$

Peripheral speed of drum $= 2 \times 0 \cdot 12 = 0 \cdot 24$ m/s

$$\therefore \ \text{motor speed} = \frac{0 \cdot 24}{0 \cdot 315} \times 48 = 36 \cdot 5 \,\text{rad/s} = \underline{349 \,\text{rev/min}}$$

Let f be the acceleration of the load in m/s², Fig. 1.30.

Then tension in cable at load $= 15 \times 10^3 (g + f)$ N

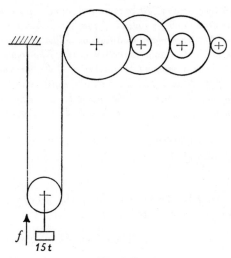

f

15 t

Fig. 1.30

$$\therefore \text{ tension in cable at drum} = \frac{15 \times 10^3}{2}(g+f) \text{ N}$$

$$\therefore \text{ torque on drum} = 7 \cdot 5 \times 10^3(g+f) \times 0 \cdot 315$$
$$= 2370(g+f) \text{ N m}$$

Equivalent moment of inertia at drum

$$= 1800 \times 0 \cdot 6^2 + 230 \times 0 \cdot 3^2 \times 5^2 + 135 \times 0 \cdot 2^2 \times 12 \cdot 5^2$$
$$+ 180 \times 0 \cdot 16^2 \times 48^2$$
$$= 12\ 630 \text{ kg m}^2$$

Torque available at drum $= 750 \times 48 \times \dfrac{75}{100} = 27\ 000$ N m

Linear acceleration of drum periphery $= 2f$ m/s^2

$$\therefore \text{ angular acceleration of drum} = \frac{2f}{0 \cdot 315} = 6 \cdot 34f \text{ rad/s}^2$$

The equation of motion of the drum is therefore

$$27\ 000 = 12\ 630 \times 6 \cdot 34f + 2370(9 \cdot 81 + f)$$

from which $\qquad f = 0 \cdot 0455$ m/s^2

$$\therefore \text{ time taken} = \frac{v}{f} = \frac{0 \cdot 12}{0 \cdot 0455} = \underline{2 \cdot 64 \text{ s}}$$

38. *A rope lifting a load of 240 kg is wound round a 1·2-m diameter drum. The drum mass is 40 kg and its radius of gyration is 520 mm. The bearing friction torque on the drum shaft is 80 N m. The drum is driven by an electric motor through a two-stage reduction gear. The rotating parts of the intermediate shaft have a moment of inertia of 3 kg m² and experience a frictional torque of 25 N m. The intermediate shaft runs at four times the drum speed. The rotating parts of the motor have a moment of inertia of 0·6 kg m², and a frictional torque of 4 N m. The motor exerts a torque of 70 N m.*

Find (a) the gear ratio between the motor and the intermediate shaft for maximum acceleration of the load, and (b) this acceleration.

(U. Lond.)

In Fig. 1.31, let f be the acceleration of the load in m/s² and n the gear ratio of the motor shaft to the intermediate shaft.

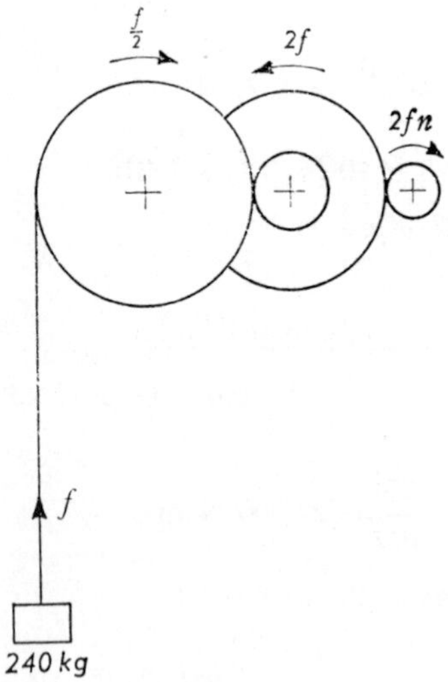

Fig. 1.31

Then   tension in rope $= 240(g + f)$ N

∴ torque on motor shaft

$$= 240(g + f) \times \frac{0·6}{4n}$$

$$= \frac{36}{n}(g + f) \text{ N m}$$

Equivalent moment of inertia of system at motor shaft

$$= 0·6 + \frac{3}{n^2} + \frac{40 \times 0·52^2}{(4n)^2}$$

$$= 0·6 + \frac{3·676}{n^2} \text{ kg m}^2$$

Equivalent friction torque at motor shaft

$$= 4 + \frac{25}{n} + \frac{80}{4n}$$

$$= 4 + \frac{45}{n} \text{ N m}$$

Angular acceleration of motor shaft $= \dfrac{f}{0·6} \times 4n = 6·67nf$ rad/s²

The equation of motion of the motor shaft is therefore

$$70 = \left(0.6 + \frac{3.676}{n^2}\right) + 6.67nf + \frac{36}{n}(9.81 + f) + \left(4 + \frac{45}{n}\right)$$

from which

$$f = \frac{66n - 398}{4n^2 + 60.52} \text{ m/s}^2$$

For maximum acceleration,

$$\frac{df}{dn} = 0$$

i.e.

$$(4n^2 + 60.52) \times 66 = (66n - 398) \times 8n$$

from which

$$n = 13.24$$

$$f_{max} = \frac{66 \times 13.24 - 398}{4 \times 13.24^2 + 60.52} = \underline{0.624 \text{ m/s}^2}$$

39. *The winding drum of a hoisting gear is driven by a motor through double reduction gearing. The pinion on the motor shaft has 18 teeth, and the spur wheel on the drum has 90 teeth; the intermediate wheels have 75 and 25 teeth. The moments of inertia of the shafts with gear wheels and other attached masses are, in kg m², motor shaft 2, intermediate shaft, 9, and drum shaft, 65. The frictional torques on these shafts are 20, 43 and 250 N m respectively. The drum diameter is 620 mm and the rope runs on to the drum at four times the speed of the load. The load is 12 t.*

(a) Find the torque required on the motor shaft to lift the load with an acceleration of 0·75 m/s².

(b) If the driving torque ceases, but no brakes are applied, find the deceleration of the load which continues to move upward.

(c) If there is neither driving torque nor braking, and the load is moving downwards, find its downward acceleration. (U. Lond.)

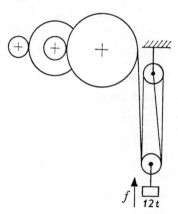

Fig. 1.32

In Fig. 1.32, let f be the acceleration of the load in m/s², upward acceleration being considered positive.

Then   tension in cable at load $= 12 \times 10^3(g+f)$ N

$$\therefore \text{ tension in cable at drum} = \frac{12 \times 10^3}{4}(g+f)$$

$$= 3 \times 10^3(g+f) \text{ N}$$

$$\therefore \text{ torque on motor shaft} = 3 \times 10^3(g+f) \times 0.31 \times \frac{25}{90} \times \frac{18}{75}$$

$$= 62(g+f) \text{ N m}$$

Equivalent moment of inertia of system at motor shaft

$$= 2 + 9\left(\frac{18}{75}\right)^2 + 65\left(\frac{18}{75} \times \frac{25}{90}\right)^2$$

$$= 2.807 \text{ kg m}^2$$

Angular acceleration of motor shaft

$$= \frac{f \times 4}{0.31} \times \frac{90}{25} \times \frac{75}{18} = 193.5f \text{ rad/s}^2$$

$\therefore$ motor torque required for acceleration of motor, intermediate shaft and drum

$$= 2.807 \times 193.5f = 543f \text{ N m}$$

Equivalent friction torque at motor shaft

$$= 20 + 43 \times \frac{18}{75} + 250 \times \frac{18}{75} \times \frac{25}{90}$$

$$= 47 \text{ N m}$$

(a) $f = 0.75$ m/s², so that

$$T = 543 \times 0.75 + 47 + 62(9.81 + 0.75)$$

$$= \underline{1109 \text{ N m}}$$

(b) When driving torque ceases,

$$0 = 543f + 47 + 62(9.81 + f)$$

from which $f = \underline{-1.084 \text{ m/s}^2}$

(c) When the load moves downwards, the direction of the friction torque is reversed, hence

$$0 = 543f - 47 + 62(9.81 + f)$$

from which $f = \underline{-0.928 \text{ m/s}^2}$

40. *A loaded mine cage is raised up a vertical shaft by a rope which passes over a pulley of 1·5 m effective diameter, and on to a winding drum 3·6 m diameter ; the masses of the cage (including load), pulley and drum are 1800 kg, 230 kg and 1400 kg respectively, and the radii of gyration of the pulley and drum 0·6 m and 1·2 m.*

The load is raised with a constant acceleration of 3·6 m/s² at first, then at uniform velocity, finally the power is cut off and the load brought to rest at the top of the shaft by the action of gravity. If the shaft length is 270 m and the total time taken is 40 s, determine the maximum power exerted at the drum, and the durations of the periods of acceleration and deceleration.

(U. Lond.)

Moment of inertia of drum $= 1400 \times 1\cdot2^2 = 2016$ kg m²

Moment of inertia of pulley $= 230 \times 0\cdot6^2 = 82\cdot8$ kg m²

$$\therefore \text{ equivalent mass to be accelerated} = 1800 + \frac{2016}{1\cdot8^2} + \frac{82\cdot8}{0\cdot75^2}$$

$$= 2570 \text{ kg}$$

Let suffices 1, 2 and 3 refer to the periods of acceleration, uniform velocity and retardation, respectively ; let f be the retardation at the top and v the uniform velocity, Fig. 1.33.

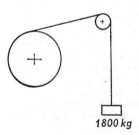

Then, during retardation, the equation of motion of the load is

$$1800 \times 9\cdot81 = 2570f$$

from which $\qquad f = 6\cdot87$ m/s²

$$s_1 = \frac{v^2}{2 \times 3\cdot6} = \frac{v^2}{7\cdot2} \text{ m} \quad . \quad (1)$$

$$s_2 = vt_2 \text{ m} \quad . \qquad . \qquad . \quad (2)$$

FIG. 1.33

$$s_3 = \frac{v^2}{2 \times 6\cdot87} = \frac{v^2}{13\cdot74} \text{ m} \quad (3)$$

1800 kg

$$s_1 + s_2 + s_3 = 270 \text{ m} \quad . \qquad . \qquad . \qquad . \qquad . \quad (4)$$

$$t_1 + t_2 + t_3 = 40 \text{ s} \quad . \qquad . \qquad . \qquad . \qquad . \quad (5)$$

$$v = 3\cdot6t_1 = 6\cdot87t_3 \qquad . \qquad . \qquad . \quad (6)$$

From equation (5), $\qquad t_2 = 40 - t_1 - t_3$

$$= 40 - \frac{v}{3\cdot6} - \frac{v}{6\cdot87} = 40 - 0\cdot424v$$

$\therefore$ in equation (4), $\dfrac{v^2}{7\cdot2} + v(40 - 0\cdot424v) + \dfrac{v^2}{13\cdot74} = 270$

i.e. $\qquad\qquad\qquad 0\cdot212v^2 - 40v + 270 = 0$

from which $\qquad\qquad\qquad\qquad\qquad v = 7\cdot1$ m/s

$$\therefore\ t_1 = \frac{7\cdot1}{3\cdot6} = \underline{1\cdot96\ \text{s}}$$

and $\qquad\qquad\qquad\qquad t_2 = \frac{7\cdot1}{6\cdot87} = \underline{1\cdot03\ \text{s}}$

The maximum power occurs immediately before the end of the acceleration period,

i.e. $\qquad$ maximum power $= \dfrac{2570 \times 9\cdot81 \times 7\cdot1}{10^3} = \underline{179\ \text{kW}}$

41. In a double reduction lifting gear the moments of inertia of the motor pinion assembly, intermediate and drum shafts are 3·5, 45 and 1000 kg m² respectively. The motor rotates at 6 times the speed of the intermediate shaft and at 6G times the speed of the drum shaft. The drum radius is 0·9 m and the load lifted is 1 t. For a constant motor torque of 550 N m find the value of G for maximum acceleration of the load and determine the value of this acceleration. (*U. Lond.*) (*Ans.:* 6·88 ; 1·26 m/s²)

42. In a double reduction lifting gear driven by an electric motor the gear ratio between the motor shaft and the intermediate shaft is 3·5 and between the intermediate shaft and drum shaft it is 4·5. The moments of inertia of the three shafts are 5, 40 and 500 kg m² respectively. On the drum, 1·2 m diameter, is suspended a loaded cage of mass 6 t and a balance mass of 4·5 t which descends as the loaded cage is lifted. Neglecting frictional losses, determine the motor torque required to give an acceleration of 0·4 m/s² to the cage when being raised. (*U. Lond.*) (*Ans.:* 830 N m)

43. The winding drum of a crane is driven by an electric motor through a triple reduction gear. The pinion of the motor shaft has 21 teeth, while the teeth on the remaining gears are, in order, 83, 29, 120, 23, 70, the last gear being rigidly attached to the winding drum shaft. The effective diameter of the drum is 0·75 m. The hoisting rope passes over a system of light pulleys which raise the load through a distance equal to a sixth of the rope travel. The moments of inertia of the various shafts, and associated wheels are, in kg m² units, motor shaft 10, first intermediate shaft 8, second intermediate shaft 16, drum shaft 100. The effective load to be hoisted is 25 t, there being an additional frictional resistance to the lifting of this load of 10 kN. Determine the motor torque necessary to accelerate this load upwards at 0·008 m/s². (*U. Lond.*) (*Ans.:* 388 N m)

44. A motor drives the 1·8-m diameter drum of a direct-acting haulage gear through two gear reduction steps, each of 5 : 1 ratio. The moments of inertia of the masses of the motor, intermediate and drum shafts are, respectively, 3, 100 and 2500 kg m². The gear and bearing friction effects may be taken as equivalent to torques of 170 N m at the intermediate shaft and 900 N m at the drum shaft. The haulage rope has a mass of 1150 kg and has to pull a load of 10 t up an incline of 1 in 50. The rope friction may be allowed at 5% of the rope weight and the rail friction at 180 N/t. If the starting torque of the motor is 800 N m, estimate the acceleration of the load. (*U. Lond.*) (*Ans.:* 0·812 m/s²)

45. In a mine hoist a loaded cage is raised and an empty cage is lowered by means of a single rope. This rope passes from one cage, over a guide pulley of 1·2 m diameter, on to the winding drum of 2·4 m diameter, and thence over a second guide pulley, also of 1·2 m diameter, to the other cage. This drum is driven by an electric motor through a double reduction gear.

Determine the motor torque required, at an instant when the loaded cage has an upward acceleration of 0·6 m/s², given the following data :

Item	Speed (rev/min)	Mass (kg)	Radius of gyration (mm)	Frictional resistance
Motor and pinion . .	N	450	150	—
First intermediate gear shaft and attached wheel .	N/5	540	225	40 N m
Drum and attached gear .	N/20	2700	900	135 N m
Guide pulley, each . .	—	110	450	30 N m
Rising rope and cage .	—	9000	—	2200 N
Falling rope and cage .	—	4500	—	1300 N

(*U. Lond.*) (*Ans. :* 3·53 kN m)

46. A loaded cage of mass 5·5 t is raised by a rope on a winding drum of 0·9 m effective diameter. A counterbalance mass of 3·5 t operates so that as the cage is lifted the counterbalance mass falls. The drum is driven by a motor through double reduction gearing, the first reduction being 3·5 to 1 and the second reduction 4·5 to 1. The moments of inertia of the three shafts in kg m² units are : motor shaft 4, intermediate shaft 13 and the drum shaft 80.

Determine the motor torque required to accelerate the system when lifting the cage with an acceleration of 1·2 m/s². (*U. Lond.*) (*Ans. :* 1·099 kN m)

47. A lift cage of mass 6 t is raised by means of a rope which passes over a driving drum and carries a counterbalance mass of 3·15 t. The driving drum is 0·6 m diameter and is driven by an electric motor through a double reduction gear. The motor shaft carries a pinion with 25 teeth. The numbers of teeth on the remaining gear wheels are, in order, 84, 21, 100, the last wheel being mounted on the drum shaft. The moments of inertia of the shafts with wheels, etc., in kg m² units, are : motor shaft 8·4 ; intermediate shaft 8 ; drum shaft 67.

If the motor develops a constant torque of 630 N m, determine the time and distance required to accelerate the cage from rest to a steady upward speed of 1·2 m/s. Neglect friction. (*U. Lond.*) (*Ans. :* 7·57 s ; 4·54 m)

48. The winding drum for an overhead crane is driven by an electric motor through triple reduction gearing, some particulars of which are tabulated below. The effective drum radius to the centre of the rope is 340 mm, and the speed of the load is one-quarter that of the rope at the drum. The overall efficiency of the gearing is 80%.

Determine :

(*a*) the speed and power of the motor when a load of 30 t is being lifted at 4 m/min ;

(*b*) the constant torque required at the motor if the load of 30 t reaches the speed of 4 m/min in 10 s from rest.

	Mass of Shaft, Wheels, etc. (*kg*)	Radius of Gyration (*mm*)	Speed Ratio
Motor and first pinion . . .	200	160	
			3·2
First intermediate shaft . . .	145	225	
			3·6
Second intermediate shaft . . .	250	290	
			4·5
Winding drum and last spur wheel .	2000	310	

(*U. Lond.*)   (*Ans.* : 388·5 rev/min ; 24·5 kW ; 632·7 N m)

49. A loaded cage (mass 7 t) is raised, and an empty cage (mass 2 t) is lowered in a vertical shaft 120 m deep by a single rope which passes from one cage over a head-pulley, down to the winding drum, and back over a second pulley to the other cage.

When the loaded cage is at the bottom of the shaft a constant accelerating torque is applied to the winding drum for 5 s ; the speed of the cages is then maintained constant until the loaded cage is 6 m from the top, when a constant torque is applied to the winding drum bringing the cage to rest at the top, in 25 s from time of starting. Find the torques applied to the winding drum during the periods of acceleration and deceleration.

	Mass	Effective Diameter	Radius of Gyration
Winding drum . .	2500 kg	2·4 m	1·0 m
Head-pulley (each) .	700 kg	4·0 m	1·6 m
Rope . . .	1400 kg	—	—

(*U. Lond.*)   (*Ans.*:   76·35 kN m ; 18 kN m)

50. The winding drum of a crane is driven by an electric motor through a triple reduction gear. The motor shaft carries a pinion having 22 teeth. The numbers of teeth on the remaining gears, in order, are 77, 23, 78, 22, 89, the last wheel being mounted on the winding drum shaft. The effective drum diameter is 0·6 m. The hoisting rope passes over a system of pulleys so that the lift rate of the load is a quarter of the speed of the rope at the winding drum. The moments of inertia of the shafts, with gear wheels, etc., in kg m² units, are : motor shaft 8 ; first intermediate shaft 5 ; second intermediate shaft 9 ; drum shaft 80.

Assuming that the efficiency of each pair of gears is 95%, determine the length of time before a load of 40 t can attain a speed of 4·5 m/min upwards from rest if the motor exerts a constant torque of 800 N m.

Determine also the greatest power required from the motor during this period.
(*U. Lond.*)   (*Ans.* : 6·03 s ; 38·5 kW)

51. In an electric hoist the driving motor is geared to rotate at 20 times the speed of the winding drum, which has an effective diameter of 0·35 m. The motor shaft is fitted with a brake drum of 0·35 m diameter, the brake band having a total contact arc of 240°. The moment of inertia of the motor and the brake drum is 4 kg m², while that of the winding drum and shaft is 2 kg m². A load of 2 t is being lowered by the hoist at a speed of 45 m/min. The coefficient of friction between the brake drum and the brake bands is 0·5.

Determine the pulls at the ends of the brake band to stop the descent of the load in a distance of 0·3 m. (*U. Lond.*)   (*Ans.* : 495 and 4025 N)

52. In a colliery winding gear a loaded cage of mass 7 t is raised, and an empty cage of mass 2 t is lowered in a vertical shaft, by a single rope which passes from one cage over a head-pulley on to a winding drum and back over a second head-pulley to the other cage. The winding drum has a diameter of 3 m, a radius of gyration of 1·2 m and a mass of 3 t. Each pulley has a mass of 2·75 t, and is 4·2 m in diameter with radius of gyration of 2 m. When the loaded cage is at the bottom of the shaft the length of rope between it and the head-pulley for that cage has a mass of 1½ t. The weight and inertia of the other lengths of rope may be neglected.

Find the torque necessary on the drum when starting to lift the loaded cage from the bottom of the shaft with an acceleration of 1·2 m/s².

What must be the minimum coefficient of friction between the rope and the head-pulleys if the rope is not to slip on either of them ? The angle of wrap is 130°. (*U. Lond.*) (*Ans. :* 127·2 kN m ; 0·084)

53. In a double reduction lifting gear, the speed of the driving motor shaft is G times the speed of the intermediate shaft, and G^2 times the speed of the drum shaft. The moments of inertia of these three shafts with their gears, etc., are 3, 35 and 850 kg m² respectively. A load of 800 kg at 0·6 m radius is suspended from the drum, and the driving motor exerts a constant torque of 570 N m.

Neglecting frictional effects, determine the value of G for maximum acceleration of the load when being raised, and evaluate this maximum acceleration. (*U. Lond.*) (*Ans. :* 5·59 ; 1·535 m/s²)

(c) Vehicle Dynamics

54. *The total resistance of a locomotive and train, of mass 550 t, on a horizontal track is $R = 3800 + 250v$, where R is in N and the velocity v in km/h. If the tractive force is kept constant at 50 kN and the train enters on an up gradient of 1 in 200 at a speed of 32 km/h, find the distance travelled and the time taken to reach 48 km/h.* (U. Lond.)

Tractive force = rail resistance + inertia force
$$+ \text{ component of weight down slope}$$

Therefore at v km/h, in N km hour units,

$$50 \times 10^3 = (3800 + 250v) + 550 \times 10^3 \times \frac{1000}{3600^2} \frac{dv}{dt} + \frac{550 \times 10^3 \times 9\cdot81}{200}$$

from which

$$\frac{dv}{dt} = 453 - 5\cdot89v \qquad . \qquad . \qquad . \qquad . \qquad . \qquad (1)$$

$$\therefore t = \int_{32}^{48} \frac{dv}{453 - 5\cdot89v}$$

$$= \frac{-1}{5\cdot89}\left[\log_e (453 - 5\cdot89v)\right]_{32}^{48}$$

$$= -0\cdot17 \log_e 0\cdot6425 = 0\cdot0752 \text{ hour} = \underline{4\cdot52 \text{ min}}$$

Equation (1) may be written

$$v \frac{dv}{ds} = 453 - 5\cdot89v$$

from which

$$s = \int_{32}^{48} \frac{v\,dv}{453 - 5\cdot89v}$$

$$= \int_{32}^{48} \left\{ -\frac{1}{5\cdot89} + \frac{453}{5\cdot89(453 - 5\cdot89v)} \right\} dv$$

$$= \left[-\frac{v}{5\cdot89} - \frac{453}{5\cdot89^2} \log_e (453 - 5\cdot89v) \right]_{32}^{48}$$

$$= -2\cdot715 - 13\cdot05 \log_e 0\cdot6425 = \underline{3\cdot06 \text{ km}}$$

55. *An engine fitted to a motor car develops a maximum power of 60 kW at 4000 rev/min and a maximum torque of 200 N m at 1600 rev/min. The vehicle has an all up mass of 1150 kg, and tests indicate that its rolling resistance is given by $155 + 0\cdot55v$, and its air resistance by $0\cdot08v^2$, the resistance being in N and the speed v in km/h in each case.*

On test the vehicle just reaches a speed of 128 km/h on the level and in still air at an engine speed of 4000 rev/min.

Calculate the transmission efficiency in top gear and the acceleration in m/s^2 at the condition of maximum engine torque in top gear on the level.

(I. Mech. E.)

At a uniform speed of 128 km/h, resistance to motion = tractive force

$$= 155 + 0\cdot55 \times 128 + 0\cdot08 \times 128^2 = 1537 \text{ N}$$

$$\therefore \text{ output power} = \frac{1537 \times 128 \times \dfrac{10^3}{3600}}{10^3} = 54\cdot6 \text{ kW}$$

$$\therefore \text{ transmission efficiency} = \frac{54\cdot6}{60} \times 100 = \underline{91\%}$$

At 1600 rev/min,   speed $= \dfrac{1600}{4000} \times 128 = 51\cdot2$ km/h

$\therefore$ resistance to motion $= 155 + 0\cdot55 \times 51\cdot2 + 0\cdot08 \times 51\cdot2^2 = 393$ N

At 128 km/h,   engine torque $= \dfrac{60 \times 10^3 \times 60}{2\pi \times 4000} = 143$ N m

At 51·2 km/h,   engine torque $= 200$ N m

$\therefore$ at 51·2 km/h,   tractive force $= \dfrac{200}{143} \times 1537 = 2150$ N

from which $\therefore 2150 = 393 + 1150f$ $f = \underline{1\cdot526 \text{ m/s}^2}$

56. *A motor vehicle, total mass 1350 kg, has road wheels of 0·6 m effective diameter. The effective moment of inertia of the four road wheels and of the rear axle together is 6·8 kg m², while that of the engine and flywheel is 0·84 kg m². The transmission efficiency is 85% and the tractive resistance at a speed of 24 km/h is 270 N. The total available engine torque is 200 N m.*

(a) Determine the gear ratio, engine to back axle, to provide maximum acceleration on an upgrade whose sine is 0·25, when travelling at 24 km/h.

(b) What is this maximum acceleration ?

(c) Determine the engine rev/min and power under these conditions.

(U. Lond.)

(a) Let n be the gear ratio and f the acceleration in m/s².

Then tractive force = road resistance + inertia force

$$+ \text{ component of weight down slope}$$

$$= 270 + 1350f + 1350 \times 9\cdot81 \times \tfrac{1}{4}$$

$$= 3580 + 1350f \text{ N}$$

Torque to accelerate engine parts $= 0\cdot84\alpha_e = 0\cdot84 \times \dfrac{f}{0\cdot3} \times n$

Torque to accelerate wheels $\quad = 6\cdot8\alpha_w = 6\cdot8 \times \dfrac{f}{0\cdot3}$

$\therefore$ net torque on wheels $= \left\{ 200 - \dfrac{0\cdot84}{0\cdot3}nf \right\} \times n \times 0\cdot85 - \dfrac{6\cdot8}{0\cdot3}f$

$$= 170n - (22\cdot68 + 2\cdot38n^2)f$$

$\therefore \dfrac{\{170n - (22\cdot68 + 2\cdot38n^2)f\}}{0\cdot3} = 3580 + 1350f$

from which $\qquad\qquad f = \dfrac{170n - 1075}{2\cdot38n^2 + 427\cdot7}$

For maximum acceleration, $\dfrac{df}{dn} = 0$

i.e. $\qquad (2\cdot38n^2 + 427\cdot7) \times 170 = (170n - 1075) \times 4\cdot76n$

from which $\qquad\qquad n = \underline{21\cdot14}$

(b) Maximum acceleration $\quad = \dfrac{170 \times 21\cdot14 - 1075}{2\cdot38 \times 21\cdot14^2 + 427\cdot7}$

$$= \underline{1\cdot69 \text{ m/s}^2}$$

(c) At 24 km/h, wheel speed $= \dfrac{24}{0\cdot3} \times \dfrac{10^3}{3600} = 22\cdot25$ rad/s

$\therefore$ engine speed $= 22\cdot25 \times 21\cdot14$

$$= 470 \text{ rad/s} = \underline{4490 \text{ rev/min}}$$

$$\text{Engine power} = \frac{200 \times 470}{10^3} = \underline{94 \text{ kW}}$$

Note that the transmission efficiency does not affect the torque required to accelerate the engine parts.

57. *A motor car has a mass of 800 kg; each of the four road wheels has an effective diameter of 0·6 m and a moment of inertia of 1·7 kg m². The rotating parts of the engine have a moment of inertia of 0·34 kg m².*

The car is coasting at 24 km/h in bottom gear of 20 to 1 ratio and the clutch pedal depressed.

If the clutch pedal is suddenly released, find the change in speed of the car when the engine is (a) initially at rest, (b) idling at a speed of 700 rev/min.

(U. Lond.)

Equivalent moment of inertia of car at wheels
$$= 4 \times 1{\cdot}7 + 800 \times 0{\cdot}3^2 = 78{\cdot}8 \text{ kg m}^2$$

$\therefore$ equivalent moment of inertia at clutch,

$$I_1 = \frac{78{\cdot}8}{20^2} = 0{\cdot}197 \text{ kg m}^2$$

Moment of inertia of engine, $I_2 = 0{\cdot}34$ kg m²

Initial speed of driven side of clutch,

$$\omega_1 = \frac{24}{0{\cdot}3} \times \frac{10^3}{3600} \times 20 = 444 \text{ rad/s}$$

When the clutch is engaged, the total angular momentum before and after engagement remains constant,

i.e. $$I_1\omega_1 + I_2\omega_2 = (I_1 + I_2)\Omega$$

where Ω is the final common speed of the two sides of the clutch.

(a) $\omega_2 = 0$ $\therefore 0{\cdot}197 \times 444 + 0{\cdot}34 \times 0 = 0{\cdot}537\Omega$

$$\therefore \Omega = 163 \text{ rad/s}$$

$$\therefore \text{ new car speed} = 24 \times \frac{163}{444} = 8{\cdot}8 \text{ km/h}$$

$$\therefore \text{ change in speed} = 24 - 8{\cdot}8 = \underline{15{\cdot}2 \text{ km/h}}$$

(b) $\omega_2 = 700 \times \dfrac{2\pi}{60} = 73{\cdot}25$ rad/s

$$\therefore 0{\cdot}197 \times 444 + 0{\cdot}34 \times 73{\cdot}25 = 0{\cdot}537\Omega$$

$$\therefore \Omega = 209 \text{ rad/s}$$

$$\therefore \text{ new car speed} = 24 \times \frac{209}{444} = 11{\cdot}3 \text{ km/h}$$

$$\therefore \text{ change in speed} = 24 - 11{\cdot}3 = \underline{12{\cdot}7 \text{ km/h}}$$

58. *In a motor cycle the axles of the road wheels are 1·35 m apart. The centre of gravity of the cycle and rider is 0·6 m in front of the rear axle and 0·68 m above the ground level.*

If only the rear wheel is braked, determine the maximum possible retardation of the cycle on a level road when (a) the cycle is travelling in a straight path and (b) when the cycle is moving in a curve of 60 m radius at 65 km/h.

Neglect air resistance and assume the coefficient of friction between the tyre and the ground to be 0·7. (U. Lond.)

(a) Resolving forces vertically and horizontally, Fig. 1.34,

$$N_r + N_f = mg \quad . \qquad . \qquad . \qquad . \qquad (1)$$

$$0·7N_r = mf \quad . \qquad . \qquad . \qquad . \qquad (2)$$

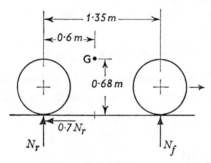

Fig. 1.34

Taking moments about G,

$$0·7N_r \times 0·68 + 0·6N_r = 0·75N_f$$

i.e.
$$N_f = 1·435N_r \quad . \qquad . \qquad . \qquad . \qquad (3)$$

Substituting in equation (1),

$$2·435N_r = mg$$

Substituting in equation (2),

$$0·7 \times \frac{mg}{2·435} = mf$$

$$\therefore f = \frac{0·7 \times 9·81}{2·435} = \underline{2·82 \text{ m/s}^2}$$

(b) Centrifugal force at 65 km/h $= m \times \dfrac{\left(65 \times \dfrac{10^3}{3600}\right)^2}{60} = 5·44m$

$\therefore$ radially inward force on rear wheel $= \dfrac{0·75}{1·35} \times 5·44m = 3·03m$

If P is the force at the rear wheel in the direction of motion, the friction force, $0{\cdot}7N_r$, is the resultant of P and $3{\cdot}03m$. Therefore, from Fig. 1.35,

$$P = \sqrt{(0{\cdot}7N_r)^2 - (3{\cdot}03m)^2} \quad . \quad . \quad . \quad (4)$$

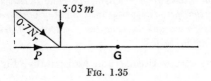

FIG. 1.35

Neglecting the obliquity of the motor cycle while cornering, equations (1), (2) and (3), become

$$N_r + N_f = mg \quad . \quad . \quad . \quad (5)$$

$$P = mf \quad . \quad . \quad . \quad (6)$$

$$0{\cdot}68P + 0{\cdot}6N_r = 0{\cdot}75N_f \quad . \quad . \quad . \quad (7)$$

Eliminating N_f between equation (5) and equation (7),

$$P = 10{\cdot}83m - 1{\cdot}985N_r = mf \quad \text{from equation (6)}$$

$$\therefore \; N_r = \frac{10{\cdot}83m - mf}{1{\cdot}985}$$

$\therefore$ in equation (4),

$$mf = \sqrt{\left\{\frac{0{\cdot}7}{1{\cdot}985}(10{\cdot}83 - f)m\right\}^2 - (3{\cdot}03m)^2}$$

i.e. $f^2 + 3{\cdot}075f - 6{\cdot}13 = 0$

from which $f = \underline{1{\cdot}376} \text{ m/s}^2$

59. A motor vehicle, of mass 1300 kg, with the engine at full throttle, can travel at a speed of 160 km/h on a level road with the engine developing 75 kW. The resistance to motion due to windage and road drag varies as the square of the road speed.

Determine the time taken for the speed of the vehicle to rise from 72 km/h to 120 km/h at full throttle, on an upgrade of 1 in 20, assuming that the engine torque remains constant. (*U. Lond.*)   (*Ans.:* 51·7 s)

60. A car has a total mass of 2000 kg. It has four wheels, each 0·75 diameter, 0·32 m radius of gyration and of mass 18 kg. The rotating parts of the engine have moment of inertia 0·47 kg m². The engine develops a torque of 115 N m. The car is going up an incline of 1 in 120 against a wind resistance of 180 N acting parallel to the road.

Calculate the ratio of engine speed to back-axle speed for maximum car acceleration and calculate this acceleration. (*U. Lond.*)

(*Ans.:* 25·9 ; 1·79 m/s²)

61. To maintain a uniform speed of 72 km/h on the level, a car of mass 1·5 t requires 9 kW. The total resistance to motion is given by $P = p + qv^2$, where P is the resistance in N, p and q are constants, and v is the speed in km/h.

If while running at 72 km/h the car starts to climb a uniform slope of 1 in 20, how soon will the speed drop to 40 km/h, assuming that the torque applied to the wheels is maintained unchanged, and that the constant p has a value of 160 N ? (*U. Lond.*) (*Ans.:* 21·45 s)

62. A motor vehicle has a mass of 900 kg and road wheels of 0·6 m rolling diameter. The total inèrtia of all four road wheels together with the half shafts is 8·5 kg m² while that of the engine and clutch is 0·8 kg m². The engine torque is 135 N m, the transmission efficiency is 90% and the tractive resistance is constant at 450 N.

Determine (*a*) the gear ratio between the engine and the road wheels to give maximum acceleration on an upgrade of 1 in 20, and (*b*) this acceleration rate in m/s². (*U. Lond.*) (*Ans.:* 13·0; 1·75 m/s²)

63. A car travelling down an incline of 1 in 13 at 32 km/h experiences a frictional resistance to motion of 320 N in addition to a friction torque on the engine of $0·0016N$ N m, where N is the engine speed in rev/min. The total mass of the car is 1350 kg. The wheels are 650 mm diameter and have a total moment of inertia of 3 kg m². The rotating parts of the engine and flywheel have moment of inertia 0·4 kg m². The engine torque is 67 N m. There is a four-speed gearbox, giving ratios of 20, 13, 8 and 5 between the engine speed and axle speed. What gear should be selected for maximum acceleration ? How much will the acceleration be ? (*U. Lond.*) (*Ans.:* 13 ; 1·548 m/s²)

64. A motor vehicle of mass 1100 kg has wheels 0·65 m diameter and a gear ratio between the engine and the back axle of 6·3 to 1. The total moment of inertia of the four wheels is 5 kg m² and the moment of inertia of the rotating parts of the engine is 0·35 kg m². When the vehicle moves with a speed of v m/s the linear resistance to its motion is $(250 + 1·4v^2)$ N and the torque developed by the engine is $(290 - 0·06v^2)$ N m. Losses of energy in transmission between the engine and road wheels may be neglected.

Calculate (*a*) the acceleration of the vehicle when moving at 12 m/s on a level road, (*b*) the time required for the vehicle to accelerate from 12 to 20 m/s on a level road. (*U. Lond.*) (*Ans.:* 0·862 m/s² ; 13·83 s)

65. For the motion of a train on the level the resistance per tonne mass is $a + 0·4v$ N, where a is a constant and v is the speed in km/h. If the tractive effort is constant at all speeds and the limiting steady speed is 145 km/h, find the distance ' lost ' in accelerating from rest to this speed, i.e. the difference between the actual distance and that corresponding to constant maximum speed for the same time. (*U. Lond.*) (*Ans.:* 28 km)

66. A motor vehicle of mass 1150 kg absorbs 75 kW when travelling at a steady speed of 160 km/h on a level road. The resistance due to windage and road drag may be assumed to vary as the square of the speed. At full throttle the engine is capable of delivering 95 kW at 160 km/h and the torque output from the engine at full throttle may be assumed to be constant over a wide range of speed. Each of the four road wheels has a mass of 55 kg and a radiuç of gyration of 230 mm and a rolling radius of 350 mm. The rotating parts of the engine have a mass of 160 kg and a radius of gyration of 80 mm. The gear ratio, engine to back axle, is 4·8 to 1.

Determine the time necessary for this vehicle to accelerate from 100 km/h to 150 km/h at full throttle on an upgrade of 1 in 50. (*U. Lond.*) (*Ans.:* 25·7 s)

67. The resistance to motion of a motor car of mass 800 kg is given by $R = (54 + 0.8v^2)$ N, where v is the speed of the car in m/s. The gear ratio between the engine and rear axle is 5·3/1 ; the moment of inertia of the rotating parts of the engine is 0·3 kg m^2 and that of the road wheels and rear axle 1·0 kg m^2. The rear wheels have an effective diameter of 0·52 m.

Find the distance travelled by the car up an incline of 1 in 40 whilst accelerating from 25 to 50 km/h assuming the engine to develop a constant torque of 40 N m during this period. (*U. Lond.*) (*Ans. :* 145 m)

68. In a performance test, a motor vehicle attained a maximum speed of 80 km/h in third gear when travelling up a gradient of 1 in 12 against a head wind of 16 km/h, with the following test conditions : mass of vehicle as tested, 1250 kg ; rolling radius of driving wheels, 0·33 m ; engine torque, 113 N m ; transmission efficiency in third gear, 85% ; back-axle ratio, 4·79 to 1 ; third gear ratio in gearbox, 1·49 to 1 ; rolling resistance per tonne at 80 km/h, 140 N.

Calculate the total air drag resistance and, given that the projected frontal area is 2 m^2 and that the air drag resistance is given by the expression $C_D A V^2/20$, where A is in m^2 and V is air speed relative to vehicle in km/h, calculate the value of C_D for this vehicle.

Assuming that the rolling resistance is independent of the speed and a transmission efficiency in top gear of 88%, calculate the power required to maintain a speed of 128 km/h in top gear on the level in still air. (*I. Mech. E.*)

(*Ans. :* 882 N ; 0·957 ; 70·4 kW)

69. A motor car has a mass of 1·6 t. Its engine develops 70 kW when the car is travelling on a level road at a steady speed of 140 km/h in top gear. Under these conditions the engine speed is 4000 rev/min, there is a direct drive through the gearbox and the transmission efficiency is 95%.

The resistance to motion of the car is $A + BV^{1·7}$ N, where V is the speed of the car in km/h, A is a constant and $B = 0·24$.

The car now climbs a gradient which rises 1 m in every 6·5 m of road. It travels at 65 km/h in second gear. The gearbox ratio in second gear is 2·13 and the transmission efficiency is 92%. Find the engine speed and power. (*U. Lond.*) (*Ans. :* 3955 rev/min : 65·6 kW)

70. A motor vehicle has a maximum speed on a level road of 250 km/h, at which speed the engine develops 180 kW, running at 6000 rev/min. The four road wheels have a rolling radius of 0·3 m, a radius of gyration of 0·2 m and the mass of each wheel is 23 kg. The rotating parts of the engine have a moment of inertia of 1 kg m^2. The total mass of the vehicle is 1350 kg. The torque output of the engine may be regarded as sensibly constant over a wide speed range.

If the resistance to motion varies as the square of the road speed, determine the time taken for the speed to rise from 160 km/h to 240 km/h on a down grade whose sine is 0·05. (*U. Lond.*) (*Ans. :* 22·35 s)

71. A supercharged road racing automobile has an engine capable of giving an output torque of 950 N m, this torque being reasonably constant over a speed range from 100 to 250 km/h in top gear. The road wheels are of 0·75 m effective diameter, and the back-axle ratio is 3·3 to 1. When travelling at a steady speed of 160 km/h in top gear on a level road, the power absorbed is 56 kW. The vehicle has a mass of 1 t ; the four road wheels each have a mass of 40 kg and a radius of gyration of 0·25 m ; the moment of inertia of the engine and all parts forward of the differential is 1·7 kg m^2.

Assuming that the resistance caused by windage and road drag varies as the square of the speed, determine the time taken for the speed to rise from 100 km/h to 250 km/h in top gear, at full throttle, on an upgrade of 1 in 30. (*U. Lond.*) (*Ans. :* 7·84 s)

72. A motor vehicle is driven by the rear wheels only. When the vehicle is stationary, 0·55 of the total mass is supported on the rear wheels. The height of the centre of gravity above the ground is one-fifth of the wheel base. In a test on a level road it was found that the greatest acceleration obtainable without skidding the rear wheels was 3 m/s². Calculate the coefficient of friction between the tyres and the ground. Using the same coefficient of friction, find the steepest gradient which the vehicle could climb. (*U. Lond.*)

$$(Ans.:\ 0·5;\ \tan^{-1} 0·3055)$$

73. (*a*) A vehicle has a mass of 9 t and accelerates at the rate of 1·5 m/s from rest upon a level road. The wheel base is 3·6 m. The centre of gravity is 1·5 m in front of the rear axle and 1·2 m above the surface of the road. The drive is through the rear wheels only. What is the smallest coefficient of friction between the wheels and the road which will enable the desired acceleration to be obtained ?

(*b*) If the vehicle maintains the given acceleration until the speed reaches 16 km/h, and thereafter the power output remains constant, how far will the vehicle have travelled from its starting point when the speed reaches 32 km/h ? Neglect road and air resistance. (*U. Lond.*)   (*Ans.:* 0·241 ; 38 m)

74. A vehicle of total mass 1 t has a wheel base of 2 m. The co-ordinates of c.g. are 0·82 m from the front wheel axis and 0·75 m above the road surface. Each wheel is 0·65 m diameter. Each pair of wheels has a mass of 36 kg with polar radius of gyration of 0·28 m. Determine the necessary coefficient of friction between the rear wheel tyres and the road surface to produce a maximum acceleration of 1·5 m/s² up an incline of 1 in 15. Neglect windage.
(*U. Lond.*)   (*Ans.:* 0·461)

CHAPTER 2

MECHANISMS: VELOCITY, ACCELERATION AND INERTIA FORCES

2.1 Velocities in mechanisms. If the ends A and B of a rigid link, Fig. 2.1, are moving with velocities v_a and v_b respectively, the velocity of A relative to B is given by ba in the relative velocity diagram and is denoted by v_{ab}, Fig. 2.2.

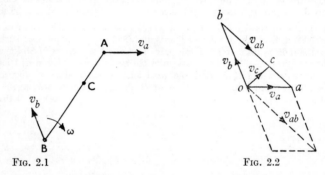

FIG. 2.1   FIG. 2.2

v_{ab} may also be obtained by imparting a velocity equal and opposite to v_b to the link, thereby bringing B to rest, in which case the velocity of A relative to B is as shown dotted. The link may then be regarded as rotating about the point B and the angular velocity of the link is given by

$$\omega = \frac{v_{ab}}{AB} \qquad . \qquad . \qquad . \qquad (2.1)$$

It is evident that v_{ab} must be perpendicular to AB, otherwise the link would have to extend or contract along its length. Thus, if the magnitude and direction of v_b are known but only the direction of v_a is known, its magnitude may be determined by drawing from b a line perpendicular to AB to intersect the line of action of A at a. Hence, in a mechanism consisting of a number of rigid links, it is only necessary to know the velocity of one point in magnitude and direction in order to find the velocities of all other points for a given configuration.

The velocity of any point C on AB may be obtained by dividing ab so that $bc : ba = BC : BA$. The line oc then represents the velocity of C.

2.2 Velocities of a block sliding on a rotating link. Let ω be the angular velocity of the link about the fixed point O, Fig. 2.3, and v_a the velocity of the block, assumed known in magnitude and direction. If A' is the point on the link coincident with the block, then the velocity of A' relative to O is perpendicular to OA' and the velocity of A relative

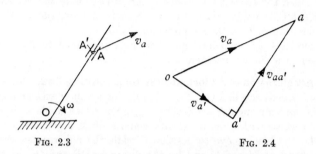

FIG. 2.3   FIG. 2.4

to A' is parallel to OA'. Therefore, if a line perpendicular to OA' is drawn from the point o, Fig. 2.4, and a line parallel to OA' is drawn from the point a, the intersection gives the point a'. The velocity of sliding of A relative to A' is then given by $a'a$.

2.3 Accelerations in mechanisms. Consider a link AB, Fig. 2.5, rotating with angular velocity and acceleration ω and α respectively. Let f_a and f_b be the accelerations of A and B respectively. Then if oa and ob are drawn from the point o, Fig. 2.6, to represent f_a and f_b, the acceleration of A relative to B is given by ba in the relative acceleration diagram. This relative acceleration has two components: (a) a centripetal acceleration, $\omega^2 AB$ or $\dfrac{v_{ab}^2}{AB}$, acting in the direction of A to B, and (b) a tangential acceleration, αAB, perpendicular to AB. These are represented respectively by ba_1 and a_1a in Fig. 2.6.

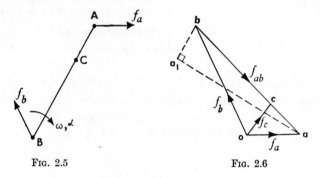

FIG. 2.5   FIG. 2.6

Normally, only one acceleration, such as f_b, will be known completely and the other, f_a, in direction only. If f_b is drawn as before, the centripetal acceleration ba_1, can be set off from b, parallel to AB. The intersection of a line through o, parallel to f_a, and a line through a_1, perpendicular to AB, will then give the required point a.

ab is called the acceleration image of AB, and to find the acceleration of any point C, a point c is taken on ab such that $ac : ab = AC : AB$. The acceleration of C is then represented by oc.

The accelerations of points on other links connected to AB can be determined in a similar manner.

2.4 Acceleration of a block sliding on a rotating link. (*a*) *Vector derivation.* Consider a link OA, Fig. 2.7, rotating about O with an angular velocity ω and an angular acceleration α, upon which a block B slides radially outwards with a linear velocity v and a linear acceleration f. In time dt, let the block move to position C while the link rotates through an angle $d\theta$ to position OA' and let the distances OB and OC be r and $r + dr$, respectively.

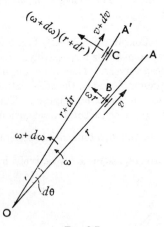

FIG. 2.7

The radial velocities of B and C are represented by ob_1 and oc_1, in the vector diagram, Fig. 2.8 (*a*), the change in velocity being represented by b_1c_1.

Radial component of
$b_1c_1 = d_1c_1 = dv$ (outwards)

Tangential component of
$b_1c_1 = b_1d_1 = v\,d\theta$ (to the left)

The tangential velocities of B and C are represented by ob_2 and oc_2 in the vector diagram, Fig. 2.8(*b*), the change in velocity being represented by b_2c_2.

Radial component of
$b_2c_2 = b_2d_2 = \omega r\,d\theta$ (inwards)

Tangential component of $b_2c_2 = d_2c_2 = \omega\,dr + r\,d\omega$ (to the left)
Total change in radial velocity $= dv - \omega r\,d\theta$ (outwards)

$$\therefore \text{ radial acceleration} = \frac{dv}{dt} - \omega r \frac{d\theta}{dt}$$

$$= f - \omega^2 r \qquad . \qquad . \qquad . \qquad (2.1)$$

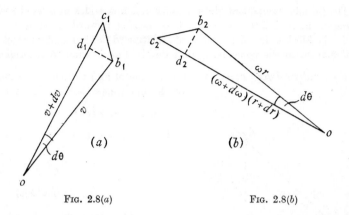

FIG. 2.8(a)   FIG. 2.8(b)

Total change in tangential velocity $= v\,d\theta + \omega\,dr + r\,d\omega$ (to the left)

$$\therefore \text{ tangential acceleration} = \frac{v\,d\theta}{dt} + \frac{\omega\,dr}{dt} + r\frac{d\omega}{dt}$$

$$= v\omega + \omega v + r\alpha$$

$$= \alpha r + 2v\omega \qquad . \qquad . \qquad . \qquad (2.2)$$

The transverse acceleration $2v\omega$ of the block *relative to the link* is called the *Coriolis* component and acts perpendicular to the sliding surfaces. Its sense is such that it leads the sliding velocity vector by 90°, measured in the direction of rotation of the link.

The various accelerations of the block are shown in Fig. 2.9; the terms $\omega^2 r$ and αr represent the radial and tangential accelerations of the

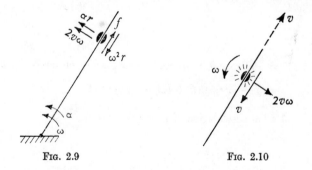

FIG. 2.9   FIG. 2.10

coincident link point relative to O, while the terms f and $2v\omega$ represent the radial and tangential accelerations of the block relative to the co-incident link point.

The Coriolis component also arises when a link slides in a fixed swivel or trunnion, Fig. 2.10. If the link is moving upwards relative to the block, the block is moving downwards relative to the link and its acceleration relative to the coincident point on the link is then to the right.

(b) *Analytical derivation.* Let the horizontal and vertical co-ordinates of the block be x and y, Fig. 2.11.

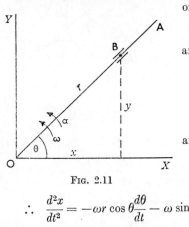

FIG. 2.11

Then $x = r \cos \theta$

and $y = r \sin \theta$

$$\therefore \frac{dx}{dt} = -r \sin \theta \frac{d\theta}{dt} + \cos \theta \frac{dr}{dt}$$

$$= -\omega r \sin \theta + v \cos \theta$$

and $$\frac{dy}{dt} = r \cos \theta \frac{d\theta}{dt} + \sin \theta \frac{dr}{dt}$$

$$= \omega r \cos \theta + v \sin \theta$$

$$\therefore \frac{d^2x}{dt^2} = -\omega r \cos \theta \frac{d\theta}{dt} - \omega \sin \theta \frac{dr}{dt} - r \sin \theta \frac{d\omega}{dt}$$

$$- v \sin \theta \frac{d\theta}{dt} + \cos \theta \frac{dv}{dt}$$

$$= -\omega^2 r \cos \theta - \omega v \sin \theta - \alpha r \sin \theta - \omega v \sin \theta + f \cos \theta$$

$$= (f - \omega^2 r) \cos \theta - (\alpha r + 2v\omega) \sin \theta$$

and $$\frac{d^2y}{dt^2} = -\omega r \sin \theta \frac{d\theta}{dt} + \omega \cos \theta \frac{dr}{dt} + r \cos \theta \frac{d\omega}{dt}$$

$$+ v \cos \theta \frac{d\theta}{dt} + \sin \theta \frac{dv}{dt}$$

$$= -\omega^2 r \sin \theta + \omega v \cos \theta + \alpha r \cos \theta + \omega v \cos \theta + f \sin \theta$$

$$= (f - \omega^2 r) \sin \theta + (\alpha r + 2v\omega) \cos \theta$$

$$\text{Radial acceleration of B} = \frac{d^2x}{dt^2} \cos \theta + \frac{d^2y}{dt^2} \sin \theta$$

$$= f - \omega^2 r$$

$$\text{Tangential acceleration of B} = \frac{d^2y}{dt^2} \cos \theta - \frac{d^2x}{dt^2} \sin \theta$$

$$= \alpha r + 2v\omega$$

2.5 Crank and connecting rod: graphical constructions for velocity and acceleration. **(i) Velocity.** Let ω be the angular velocity

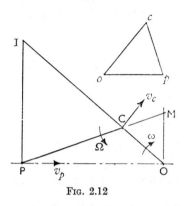

FIG. 2.12

of the crank OC, Fig. 2.12, and Ω the angular velocity of the connecting rod, PC. Then, since I is the instantaneous centre for PC,

$$\frac{v_c}{IC} = \frac{v_p}{IP} = \frac{v_{pc}}{PC} = \Omega$$

If PC is produced to intersect a vertical through O at M, then triangles PIC and OCM are similar.

Hence

$$v_p = v_c \times \frac{IP}{IC} = v_c \times \frac{OM}{OC} = \omega \times OM \qquad . \qquad . \qquad (2.3)$$

Also

$$v_{pc} = v_c \times \frac{PC}{IC} = v_c \times \frac{CM}{OC} = \omega \times CM \qquad . \qquad . \qquad (2.4)$$

and

$$\Omega = \frac{v_{pc}}{PC} = \omega \times \frac{CM}{PC} \qquad . \qquad . \qquad . \qquad . \qquad (2.5)$$

The velocity diagram for the mechanism is shown by triangle *ocp*.

(ii) Acceleration—Klein's Construction.* Draw a circle with PC as diameter, Fig. 2.13, and produce PC to cut the vertical through O at M. With centre C and radius CM, describe another circle, intersecting the first at H and K. Draw the common chord HK, intersecting OP at N and PC at L. Then the quadrilateral OCLN represents the acceleration diagram to the same scale that OC represents the centripetal acceleration of C.

---

* This construction can only be used when the crank has a uniform angular velocity.

c

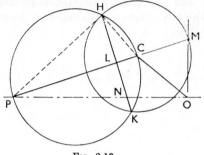

FIG. 2.13   FIG. 2.14

Thus $$f_p = \omega^2 ON \quad . \quad . \quad . \quad . \quad (2.6)$$
Also the centripetal acceleration of P relative to $C = \omega^2 LC$. (2.7)
and the tangential acceleration of P relative to $C = \omega^2 LN$. (2.8)

$$\text{The angular acceleration of PC, } \alpha = \omega^2 \frac{LN}{PC} \quad . \quad (2.9)$$

Proof. Fig. 2.14 shows the acceleration diagram for the mechanism. *oc* represents the centripetal acceleration of C relative to O $(= \omega^2 OC)$ and cp_1 represents the centripetal acceleration of P relative to $C\left(= \dfrac{v_{cp}^2}{CP}\right)$. The tangential acceleration of P relative to C is perpendicular to PC and the acceleration of P relative to O is horizontal, thus giving the point *p*.

From equation (2.4), $v_{cp} = \omega CM \quad . \quad . \quad . \quad . \quad (1)$
also $\qquad oc = \omega^2 OC$
and $\qquad cp_1 = \dfrac{v_{cp}^2}{CP} \qquad = \dfrac{(\omega CM)^2}{CP}$

$$\therefore \frac{cp_1}{oc} = \frac{\omega^2 CM^2}{\omega^2 OC . CP} = \frac{CM^2}{OC . CP} \quad . \quad . \quad (2)$$

If P and C are joined to H, triangles CLH and CHP are similar,

$$\therefore \frac{CL}{CH} = \frac{CH}{CP}$$

$$\therefore CL = \frac{CH^2}{CP} = \frac{CM^2}{CP}$$

$$\therefore \frac{CL}{OC} = \frac{CM^2}{OC . CP} \quad . \quad . \quad . \quad (3)$$

$\therefore$ from equations (2) and (3), OCLN is similar to ocp_1p,
$\therefore$ since $f_c = \omega^2 OC$ then $f_p = \omega^2 ON$.

2.6 Crank and connecting rod: analytical determination of velocity and acceleration. Let x be the displacement of the piston from the inner dead centre position, Fig. 2.15.

Then $\qquad x = (r + l) - (r \cos \theta + l \cos \phi)$

Also $\qquad r \sin \theta = l \sin \phi$

$$\therefore \sin \phi = \frac{r}{l} \sin \theta$$

$$= \frac{\sin \theta}{n}$$

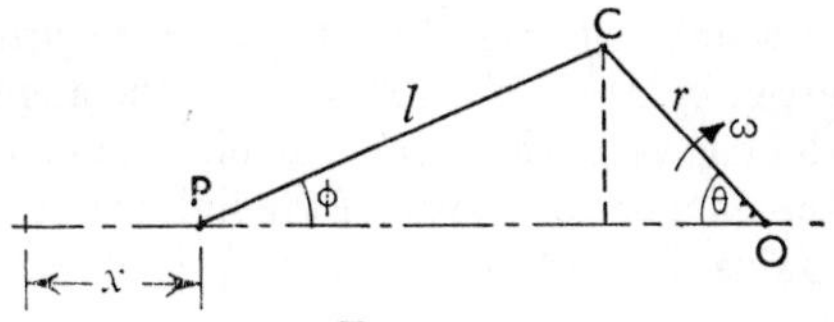

where $\qquad n = \dfrac{l}{r}$

FIG. 2.15

$$\therefore \cos \phi = \sqrt{\left\{1 - \left(\frac{\sin \theta}{n}\right)^2\right\}} \simeq 1 - \frac{\sin^2 \theta}{2n^2} \quad \text{since } \frac{1}{n} \text{ is small*}$$

$$\therefore x = r(1 - \cos \theta) + l \frac{\sin^2 \theta}{2n^2}$$

$$\therefore v_p = \frac{dx}{dt} = \left(r \sin \theta + l \frac{\sin 2\theta}{2n^2}\right) \frac{d\theta}{dt}$$

$$= \omega r \left(\sin \theta + \frac{\sin 2\theta}{2n}\right) \quad . \qquad . \qquad . \qquad . \qquad . \quad (2.10)$$

$$\therefore f_p = \frac{d^2 x}{dt^2} = \omega r \left(\cos \theta + \frac{\cos 2\theta}{n}\right) \frac{d\theta}{dt}$$

$$= \omega^2 r \left(\cos \theta + \frac{\cos 2\theta}{n}\right) \quad . \qquad . \qquad . \qquad . \quad (2.11)$$

Let Ω and α be the angular velocity and acceleration of the connecting rod respectively.

$$\sin \phi = \frac{\sin \theta}{n}$$

$$\therefore \cos \phi \frac{d\phi}{dt} = \frac{\cos \theta}{n} \frac{d\theta}{dt}$$

---

* The complete expansion is of the form

$$\cos \phi = 1 - \frac{\sin^2 \theta}{2n^2} - \frac{\sin^4 \theta}{8n^4} - \frac{\sin^6 \theta}{16n^6} - \ldots$$

from which $\qquad v_p = \omega r(\sin \theta + B_2 \sin 2\theta + B_4 \sin 4\theta + \ldots)$

and $\qquad f_p = \omega^2 r(\cos \theta + A_2 \cos 2\theta + A_4 \cos 4\theta + \ldots)$

The constants A_2, A_4, B_2, B_4, etc., are functions of n but equations (2.10) and (2.11) are sufficiently accurate for most purposes. Further terms may, however, be needed in some engine balancing problems; see Art. 7.4 and Example 5, p.172.

i.e.

$$\Omega = \frac{d\phi}{dt} = \frac{\cos\theta}{\cos\phi}\frac{\omega}{n} \simeq \omega\frac{\cos\theta}{n} \qquad . \qquad (2.12)$$

$$\alpha = \frac{d\Omega}{dt} \simeq -\omega^2 \frac{\sin\theta}{n} . \qquad . \qquad . \qquad (2.13)$$

2.7 Inertia force on a link. If the linear acceleration of the c.g. of a link is f, Fig. 2.16, the force required to produce this acceleration is given by $P = mf$. Similarly, if the angular acceleration of the link about the c.g. is α, the torque required is given by $T = I\alpha = mk^2\alpha$ where I is the moment of inertia of the link about the c.g. and k is the corresponding radius of gyration.

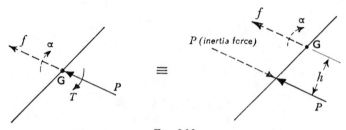

FIG. 2.16

The resultant effect of P and T can be obtained from the action of a single force, of magnitude P, acting parallel to the direction of f and at a perpendicular distance h from G such that

$$h = \frac{T}{P} = \frac{I\alpha}{mf} = \frac{k^2\alpha}{f} \qquad . \qquad . \qquad . \qquad (2.14)$$

The sense of the angular acceleration α determines to which side of G the distance h is measured.

The inertia force on the link is equal and opposite to this single force P, as shown dotted, and is transmitted to adjacent links through the pin-joints at the ends.

2.8 Use of equivalent dynamical system. If a rigid link of mass m and radius of gyration k is to be replaced by a dynamically equivalent system of two concentrated masses, m_1 and m_2, situated at distances a and b on either side of G, Fig. 2.17, then

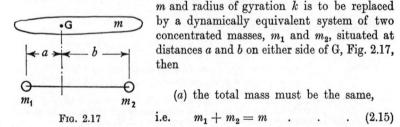

FIG. 2.17

(a) the total mass must be the same,

i.e. $m_1 + m_2 = m$ . . . (2.15)

(b) the position of the c.g. must be the same,

i.e. $$m_1 a = m_2 b \qquad . \qquad . \qquad . \qquad . \qquad (2.16)$$

(c) the moment of inertia about an axis through the c.g. must be the same,

i.e. $$m_1 a^2 + m_2 b^2 = m k^2 . \qquad . \qquad . \qquad . \qquad (2.17)$$

From equations (2.15) and (2.16)

$$m_1 = \frac{b}{a+b} m \quad \text{and} \quad m_2 = \frac{a}{a+b} m$$

Substituting in equation (2.17) gives the essential condition for the placing of the masses, i.e. $ab = k^2$. Either a or b can be chosen arbitrarily and the other term obtained from this relation.

If one of the masses, m_1, is to be placed at the end A of the link AB,

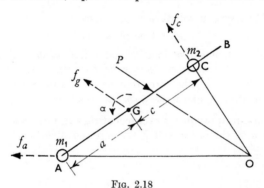

FIG. 2.18

Fig. 2.18, the other mass, m_2, must be placed at C such that $ac = k^2$. The resultant accelerating force on the link is of magnitude mf_g and must pass through the intersection of the lines of action of the accelerating forces on the two masses, m_1 and m_2. It must also be parallel to the acceleration of G and in the direction of f_g, so that its line of action is thereby completely defined. The inertia force, P, is then equal and opposite to this resultant accelerating force.

If, as an approximation, it is required to place one of the masses at each of the ends A and B, Fig. 2.19, then

$$m_1 = \frac{b}{a+b} m$$

and $$m_2 = \frac{a}{a+b} m$$

but the condition $ab = k^2$ will not, in general, be fulfilled. The moment

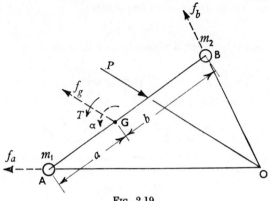

FIG. 2.19

of inertia of the system will then be given by

$$m_1a^2 + m_2b^2 = mab$$

The error in the moment of inertia is therefore $m(ab - k^2)$; to allow for this, however, a correction couple $m(ab - k^2)\alpha$ may be applied to the system. If $ab > k^2$, the correction couple has the same sense as the angular acceleration α, and if $ab < k^2$, the couple has the opposite sense.

2.9 Crankshaft torque in reciprocating engine mechanism. Let F be the force on the piston and X the corresponding force at the crank-

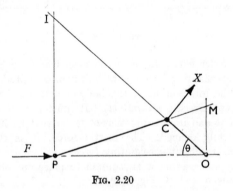

FIG. 2.20

pin, perpendicular to the crank, Fig. 2.20. Then, neglecting the inertia and gravity effects of the connecting rod,

$$F \times v_p = X \times v_c$$

i.e.

$$X = F \times \frac{v_p}{v_c}$$

$$= F \times \frac{IP}{IC} = F \times \frac{OM}{OC}$$

$$\therefore \text{ crankshaft torque, } T = X \times OC = F \times OM \qquad . \qquad (2.18)$$

If p is the pressure in the cylinder and a the area of the piston, the force on the piston due to gas pressure is pa. If R is the mass of reciprocating parts, an inertia force Rf_p must be subtracted from the gas force during the acceleration period and added to it during the retardation period.

Thus
$$F = pa - R\omega^2 r\left(\cos\theta + \frac{\cos 2\theta}{n} \right) \qquad . \qquad . \qquad (2.19)$$

the sign of the inertia force changing at the point of maximum velocity.

In vertical engines, the dead weight Rg assists the piston effort on the downstroke and opposes it on the upstroke.

2.10 Effect of mass and inertia of connecting rod. For the calculation of inertia forces, the connecting rod may be replaced by two concentrated masses at P and D such that $PG \times GD = k^2$, where k is the radius of gyration of the connecting rod about G, Fig. 2.21.

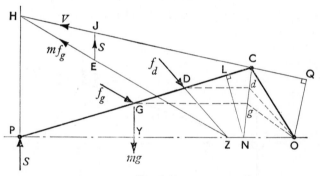

Fig. 2.21

Using Klein's construction, the acceleration diagram OCLN is obtained, CN being the acceleration image for the rod PC. The points d and g on CN are obtained by drawing lines parallel to PO through D and G respectively, dO and gO then giving the respective directions of the accelerations of D and G; the magnitudes of these accelerations are given by $\omega^2 dO$ and $\omega^2 gO$ respectively.

The inertia force due to the mass at P acts along PO and that due to the mass at D acts through D parallel to dO. The resultant inertia force of the rod must pass through the intersection, Z, of these lines of action and must also be parallel to the direction of the acceleration of G, given by gO.

The three forces acting on the rod are the inertia force, the side thrust, S, on the piston (neglecting friction) and the force, V, at the crankpin, C. The lines of action of these forces must be concurrent, so that the force through C must pass through the intersection, H, of the inertia force and the vertical through P. The magnitude of the inertia force is mf_g, and so resolving this force into components parallel to CH and PH, JH represents the force at the crankpin and EJ represents the side thrust at the piston. The crankshaft torque is then the product of the crankpin force and the perpendicular distance, OQ, of its line of action from O, this being in addition to that due to the gas force and inertia of the reciprocating parts.

The vertical force at C due to the dead weight of the rod is $mg \times \dfrac{\text{PG}}{\text{PC}}$,

so that the crankshaft torque is then this force multiplied by the distance of its line of action from O.

If, as an approximation, the rod is replaced by masses m_1 and m_2 placed at P and C, respectively, Fig. 2.22, a correction couple

$$m \,(\text{PG.GC} - k^2)\alpha$$

must be applied since equation (2.17) will not be satisfied.

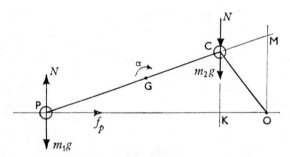

Fig. 2.22

The inertia force due to m_1 at P is $m_1 f_p$ and produces an inertia torque of magnitude $m_1 f_p \times \text{OM}$. The inertia force due to m_2 at C has no effect on the crankshaft torque but the dead weight of this mass produces a torque about O of magnitude $m_2 g \times \text{KO}$.

The correction couple may be applied by the action of two vertical forces N applied at P and C, such that

$$N \times \text{PK} = m(\text{PG.GC} - k^2)\alpha$$

the couple being in the same direction as α if $\text{PG.GC} > k^2$ and in the opposite direction if $\text{PG.GC} < k^2$.

The force N at P has no effect on the crankshaft torque but the torque due to the force N at C is $N \times$ KO.

Thus total crankshaft torque $= -m_1 f_p \times \text{OM} - (m_2 g + N) \times$ KO, this being in addition to that due to the gas force and inertia of the reciprocating parts.

2.11 Effect of friction. At sliding surfaces, the reaction between the surfaces becomes inclined at the friction angle ϕ to the normal at the point of contact, the component perpendicular to this normal being such as to oppose the relative motion.

At rotating surfaces, the reaction R between the surfaces is again inclined at the friction angle ϕ to the normal (radius) and hence is tangential to a circle of radius $r \sin \phi$ ($\eqsim r \tan \phi \eqsim \mu r$), where r is the radius of the surface of contact; this circle is termed the *friction circle*, Fig. 2.23.

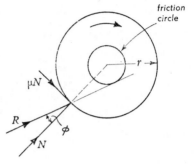

FIG. 2.23

In the case of a link connected to adjacent links by pin-joints at which friction is present, the force in the link is tangential to the friction circles, its line of action being termed the *friction axis*. The points of tangency at the two ends are such that the moments of the force about the pin centres oppose the relative rotation at the joints.

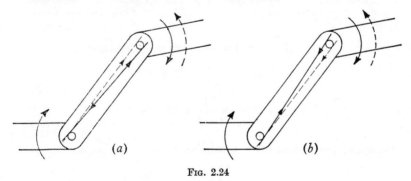

(a) (b)

FIG. 2.24

Fig. 2.24 shows such a link, the arrows on the adjacent links showing the direction of their motion relative to this link. Fig. 2.24(a) shows the friction axis if the link is in compression and Fig. 2.24(b) shows this axis if the link is in tension. The dotted axes correspond to the relative motion of the adjacent links shown dotted.

(a) Acceleration Diagrams

1. *In Fig. 2.25, a rod PR is constrained by guides to move horizontally and is driven by a crank OA and a sliding block at P. For the given*

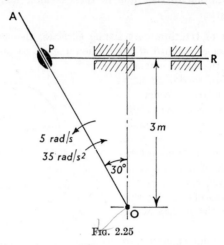

Fig. 2.25

configuration determine graphically or otherwise the acceleration of PR when OA has an angular velocity of 5 rad/s in a counter-clockwise direction and an angular acceleration of −35 rad/s², i.e. clockwise.

Let P′ be the point on the link OA which is coincident with the block P. Then, in the velocity diagram, Fig. 2.26, op' represents the absolute velocity of P′, op the absolute velocity of P and $p'p$ the relative velocity of P to P′.

$$v_p' = op' = 5 \times \text{PO} = 5 \times 2\sqrt{3} = 10\sqrt{3} \text{ m/s}$$

$$\therefore v_p = op = 10\sqrt{3} \times \frac{2}{\sqrt{3}} = 20 \text{ m/s}$$

and

$$v_{pp'} = p'p = 10\sqrt{3} \times \frac{1}{\sqrt{3}} = 10 \text{ m/s}$$

Centripetal acceleration of P′ relative to $0 = \dfrac{v_p'^2}{\text{OP}} = \dfrac{(10\sqrt{3})^2}{2\sqrt{3}}$

$$= 86 \cdot 6 \text{ m/s}^2$$

Tangential acceleration of P′ relative to $0 = \alpha_{op'} \times \text{OP} = 35 \times 2\sqrt{3}$

$$= 121 \cdot 2 \text{ m/s}^2$$

Tangential acceleration of P relative to $\text{P}' = 2v_{pp'} \, \omega_{op'} = 2 \times 10 \times 5$

$$= 100 \text{ m/s}^2$$

In the acceleration diagram, Fig. 2.27, op_1' and $p_1'p'$ are respectively the centripetal and tangential accelerations of P' relative to O. The

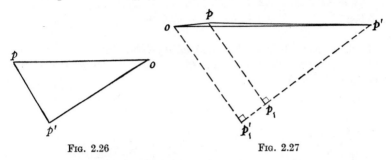

FIG. 2.26   FIG. 2.27

direction of the Coriolis component of the acceleration of P relative to P' is obtained by rotating the relative velocity vector $v_{pp'}$ through 90° in the direction of $\omega_{op'}$ and this vector is represented by $p'p_1$. The sliding acceleration of P relative to P' is parallel to OA and the diagram is completed by drawing a horizontal line through o to represent the acceleration of P relative to O, the intersection of these last two lines giving the point p.

From the diagram, $f_r = op = \underline{25 \text{ m/s}^2}$.

2. The end A of a bar AB, Fig. 2.28, is constrained to move along the vertical path AD and the bar passes through a swivel bearing pivoted at C. Draw velocity and acceleration diagrams for the given configuration when A has a velocity of 3 m/s towards D and an acceleration of 25 m/s² in the opposite direction.

Determine: (a) *the velocity and acceleration of sliding of the bar through the swivel;*

(b) *the angular velocity and angular acceleration of AB.*

(U. Lond.)

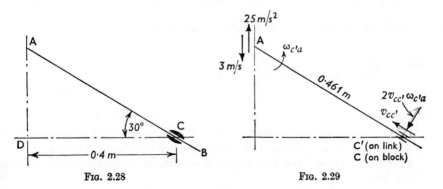

FIG. 2.28   FIG. 2.29

Let C' be the point on the link AB, Fig. 2.29, which is coincident with the swivel C.

Then $$AC' = 0.4 \times \frac{2}{\sqrt{3}} = 0.461 \text{ m}$$

In the velocity diagram, Fig. 2.30, da represents the absolute velocity of A, cc' represents the relative velocity of C' to C and ac' the relative velocity of C' to A.

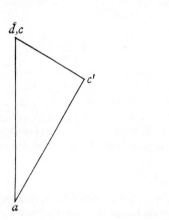

FIG. 2.30

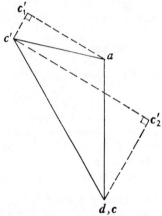

FIG. 2.31

$$v_{cc'} = c'c = 3 \times \tfrac{1}{2} = 1.5 \text{ m/s}$$

$$v_{c'a} = ac' = 3 \times \frac{\sqrt{3}}{2} = 2.6 \text{ m/s}$$

$$\omega_{ac'} = \frac{v_{c'a}}{AC'} = \frac{2.6}{0.461} = 5.63 \text{ rad/s}$$

Centripetal acceleration of C' relative to A $= \dfrac{v_{c'a}^2}{AC'} = \dfrac{2.6^2}{0.461}$

$$= 14.67 \text{ m/s}^2$$

Tangential acceleration of C relative to C' $= 2v_{cc'}\,\omega_{ac'}$

$$= 2 \times 1.5 \times 5.63$$

$$= 16.9 \text{ m/s}^2$$

In the acceleration diagram, Fig. 2.31, da represents the absolute acceleration of A. The centripetal acceleration of C' relative to A is represented by ac_1' and the tangential acceleration is perpendicular to this line.

The block C is moving towards A and AC′ is rotating about A in an anticlockwise direction, so that the direction of the Coriolis component of the acceleration of C relative to C′ is as shown in Fig. 2.29. The tangential acceleration of C′ relative to C is therefore in the opposite direction and is represented by $cc_2′$ in the acceleration diagram. The sliding acceleration of C′ relative to C is perpendicular to $cc_2′$ and the intersection of this line with that perpendicular to $ac_1′$ gives the point $c′$.

Velocity of sliding of bar through swivel $= cc′ = \underline{1\cdot5\ \text{m/s}}$

Acceleration of sliding of bar through swivel $= c_2′c′ = \underline{27\cdot1\ \text{m/s}^2}$

Angular velocity of AB $= \dfrac{v_{c′a}}{\text{AC}′} = \dfrac{2\cdot6}{0\cdot461} = \underline{5\cdot63\ \text{rad/s}}$ (anticlockwise)

Angular acceleration of AB $= \dfrac{f_{c_1′c′}}{\text{AC}′} = \dfrac{4\cdot8}{0\cdot461} = \underline{10\cdot4\ \text{rad/s}^2}$ (clockwise)

3. *In the mechanism shown in Fig. 2.32, the slide QD is driven with a uniform angular velocity of 10 rad/s and the block C has a mass of 1·5 kg.*

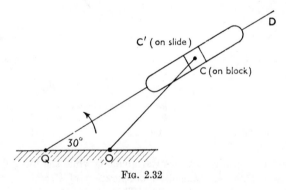

FIG. 2.32

Draw the velocity and acceleration diagrams for the mechanism in the position shown and determine the thrust which the slide exerts on the block C and the tension in the rod OC.

Indicate clearly the direction of the thrust which the slide exerts on the block. Neglect friction. OQ = 1 m; OC = 2 m. (U. Lond.)

By measurement, QC = 2·8 m

In the velocity diagram, Fig. 2.33, $qc′$ represents the absolute velocity of C′ and $c′c$ and oc are the velocities of C relative to C′ and O respectively.

$$v_{co} = oc = 28\cdot8\ \text{m/s}, \quad v_{cc′} = c′c = 7\cdot5\ \text{m/s}$$

Centripetal acceleration of C′ relative to Q $= \omega_{qd}^2 \times \text{QC}′ = 10^2 \times 2\cdot8$
$$= 280\ \text{m/s}^2$$

Centripetal acceleration of C relative to $O = \dfrac{v_{co}^2}{OC} = \dfrac{28 \cdot 8^2}{2} = 414$ m/s^2

Tangential acceleration of C relative to $C' = 2v_{cc'}\,\omega_{qd} = 2 \times 7 \cdot 5 \times 10$
$$= 150 \text{ m/s}^2$$

In the acceleration diagram, Fig. 2.34, qc' represents the absolute acceleration of C'. The centripetal acceleration of C relative to O is represented by oc_1 and the tangential acceleration is perpendicular to this line.

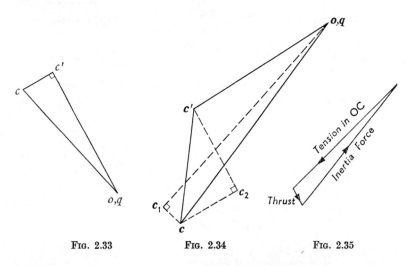

FIG. 2.33   FIG. 2.34   FIG. 2.35

The link is rotating anticlockwise about Q while the block C is sliding towards Q so that the tangential acceleration of C relative to C' (the Coriolis component) is to the right. This is represented by $c'c_2$ and the sliding acceleration of C relative to C' cuts the perpendicular to qc_1 at the point c.

Thus,   acceleration of $C = oc = 417$ m/s^2

The inertia force on the block acts in the opposite direction to the acceleration, i.e. in the direction co, and is of magnitude $1 \cdot 5 \times 417 = 625 \cdot 5$ N. This force is opposed by the side thrust on the block, which is perpendicular to QD since friction is neglected, and the tension in the link OC. Therefore, from the triangle of forces, Fig. 2.35,

thrust exerted by $QD = \underline{72 \cdot 5 \text{ N}}$ and tension in $OC = \underline{580 \text{ N}}$

The direction of the thrust which the slide exerts on the block is indicated in Fig. 2.35.

4. *A crank OA, 100 mm long, rotates clockwise at 100 rev/min as shown in Fig. 2.36. Rod AC, 500 mm long, slides in a swivelling pin at B. The end C slides on a swinging link DE. When the angle BOA is 120°, find the angular velocity and angular acceleration of DE.* (U. Lond.)

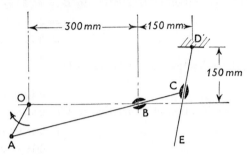

FIG. 2.36

If B′ and C′ are the points on AC and DE which are coincident with the blocks B and C respectively, then in the velocity diagram, Fig. 2.37, *oa* represents the absolute velocity of A, *ab′* the velocity of B′ relative to A and *bb′* the velocity of B′ relative to B. C is obtained by extending *ab′* to *c* such that *ab′* : *b′c* = AB′ : B′C and *dc′* and *cc′* are then the vectors representing the velocities of C′ relative to D and C respectively.

$$v_a = oa = 100 \times \frac{2\pi}{60} \times 0.1 = 1.048 \text{ m/s}$$

$$v_{b'a} = ab' = 0.75 \text{ m/s}$$

$$v_{ca} = ac = 1.062 \text{ m/s}$$

$$v_{b'b} = bb' = 0.75 \text{ m/s}$$

$$v_{c'd} = dc' = 0.588 \text{ m/s}$$

$$v_{c'c} = cc' = 0.576 \text{ m/s}$$

$$\omega_{ac} = \frac{v_{ca}}{AC} = \frac{1.062}{0.5}$$

$$= 2.124 \text{ rad/s}$$

$$\omega_{de} = \frac{v_{cd}}{DC'} = \frac{0.588}{0.11}$$

$$= 5.35 \text{ rad/s}$$

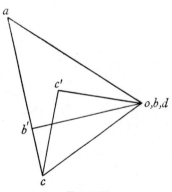

FIG. 2.37

Centripetal acceleration of A relative to $O = \dfrac{v_a{}^2}{OA} = \dfrac{1.046^2}{0.1}$

$$= 10.95 \text{ m/s}^2$$

$$\text{Centripetal acceleration of B' relative to A} = \frac{v_{b'a}{}^2}{AB'} = \frac{0 \cdot 75^2}{0 \cdot 35}$$
$$= 1 \cdot 61 \text{ m/s}^2$$
$$\text{Tangential acceleration of B' relative to B} = 2v_{b'b}\,\omega_{ac}$$
$$= 2 \times 0 \cdot 75 \times 2 \cdot 124$$
$$= 3 \cdot 19 \text{ m/s}^2$$
$$\text{Centripetal acceleration of C' relative to D} = \frac{v_{c'd}{}^2}{DC'} = \frac{0 \cdot 588^2}{0 \cdot 11}$$
$$= 3 \cdot 15 \text{ m/s}^2$$
$$\text{Tangential acceleration of C' relative to C} = 2v_{c'c}\,\omega_{de}$$
$$= 2 \times 0 \cdot 576 \times 5 \cdot 35$$
$$= 6 \cdot 16 \text{ m/s}^2$$

In the acceleration diagram, Fig. 2.38, **oa** represents the absolute acceleration of A. **ab₁′** and **b₁′b′** represent the centripetal and tangential accelerations of B' relative to A and **bb₂′** and **b₂′b′** the tangential and sliding accelerations of B' relative to B. **c** is then obtained by extending **ab′** to **c** such that **ab′** : **b′c** = AB' : B'C. The tangential and sliding accelerations of C' relative to C are represented by **cc₁′** and **c₁′c′** and the centripetal and tangential accelerations of C' relative to D by **dc₂′** and **c₂′c′,** the directions of the tangential (or Coriolis) accelerations of B' and C' relative to B and C respectively being determined as in Example 3.

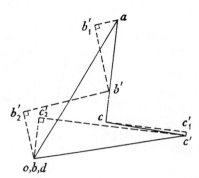

FIG. 2.38

$$\text{Angular velocity of DE} = \frac{v_{c'd}}{DC'} = \frac{0 \cdot 588}{0 \cdot 11} = \underline{5 \cdot 35 \text{ rad/s}} \text{ (clockwise)}$$
$$\text{Angular acceleration of DE} = \frac{f_{c_2'c'}}{DC'} = \frac{10 \cdot 2}{0 \cdot 11} = \underline{92 \cdot 7 \text{ rad/s}^2} \text{ (anticlockwise)}$$

5. *The quick return mechanism of a shaping machine is shown in Fig. 2.39. The crank rotates in an anticlockwise direction at a speed of 90 rev/min. The length QP is 800 mm. Find (a) the maximum velocity and the maximum acceleration of P, (b) the acceleration of P when θ = 45°.* (U. Lond.)

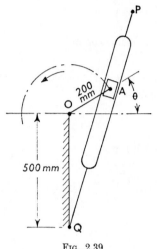

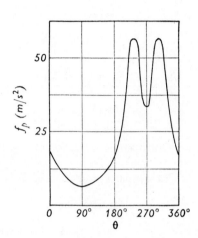

FIG. 2.39   FIG. 2.40

(*a*) Let $\qquad \omega$ = angular velocity of link OA

$\qquad\qquad\qquad \Omega$ = angular velocity of link QP

and $\qquad\qquad \phi$ = angle OQA.

Then, in triangle OAQ,

$$\frac{0{\cdot}2}{\sin \phi} = \frac{0{\cdot}5}{\sin \{180 - (90 + \theta + \phi)\}} = \frac{0{\cdot}5}{\cos (\theta + \phi)}$$

$$\therefore \; \sin \phi = 0{\cdot}4 \cos (\theta + \phi)$$

$$= 0{\cdot}4 \, (\cos \theta \cos \phi - \sin \theta \sin \phi)$$

i.e. $\qquad\qquad 2{\cdot}5 = \cos \theta \cot \phi - \sin \theta$

from which

$$\tan \phi = \frac{\cos \theta}{2{\cdot}5 + \sin \theta} \qquad . \qquad . \qquad . \qquad . \qquad . \qquad . \qquad (1)$$

$$\therefore \; \sec^2 \phi \frac{d\phi}{dt} = \frac{-(2{\cdot}5 + \sin \theta) \sin \theta - \cos \theta \cos \theta}{(2{\cdot}5 + \sin \theta)^2} \frac{d\theta}{dt}$$

i.e. $\qquad\qquad (1 + \tan^2 \phi)\Omega = -\dfrac{2{\cdot}5 \, \sin \theta + 1}{(2{\cdot}5 + \sin \theta)^2} \, \omega$

$$\therefore \Omega = - \frac{2 \cdot 5 \sin \theta + 1}{5 \sin \theta + 7 \cdot 25} \omega$$

substituting for $\tan \phi$ from equation (1).

This is a maximum when $\theta = 270°$, i.e. when A is vertically below O.

$$\therefore \Omega_{max} = - \frac{-2 \cdot 5 + 1}{-5 + 7 \cdot 25} \times \left(\frac{2\pi}{60} \times 90 \right) = 2\pi \text{ rad/s}$$

$$\therefore \text{ maximum velocity of P} = 2\pi \times 0 \cdot 8 = \underline{5 \cdot 03 \text{ m/s}}$$

$$\frac{d\Omega}{dt} = - \left\{ \frac{(5 \sin \theta + 7 \cdot 25) \times 2 \cdot 5 \cos \theta - (2 \cdot 5 \sin \theta + 1) \times 5 \cos \theta}{(5 \sin \theta + 7 \cdot 25)^2} \right\} \omega^2$$

$$= - \frac{13 \cdot 125 \cos \theta}{(5 \sin \theta + 7 \cdot 25)^2} \omega^2$$

Centripetal acceleration of P $= \Omega^2 QP = 0 \cdot 8\Omega^2 \text{ m/s}^2$

Tangential acceleration of P $= \dfrac{d\Omega}{dt} QP = 0 \cdot 8 \dfrac{d\Omega}{dt} \text{ m/s}^2$

$\therefore$ resultant acceleration of P,

$$f_p = \sqrt{\left\{ (0 \cdot 8\Omega^2)^2 + \left(0 \cdot 8 \frac{d\Omega}{dt} \right)^2 \right\}}$$

$$= 0 \cdot 8 \sqrt{\left\{ \Omega^4 + \left(\frac{d\Omega}{dt} \right)^2 \right\}}$$

$$= 0 \cdot 8\omega^2 \sqrt{\left\{ \frac{(2 \cdot 5 \sin \theta + 1)^4 + (13 \cdot 125 \cos \theta)^2}{(5 \sin \theta + 7 \cdot 25)^4} \right\}}$$

By differentiation or plotting, this has a maximum value when

$$\theta = 241°45' \quad \text{or} \quad 298°15'$$

Hence,

$$f_{p\,max} = 0 \cdot 8 \times (3\pi)^2 \sqrt{\left\{ \frac{2 \cdot 091 + 38 \cdot 5}{65 \cdot 5} \right\}}$$

$$= \underline{55 \cdot 95 \text{ m/s}^2}$$

(b) When $\theta = 45°$

$$f_p = 0 \cdot 8 \times (3\pi)^2 \sqrt{\left\{ \frac{58 \cdot 5 + 86}{13,500} \right\}}$$

$$= \underline{7 \cdot 33 \text{ m/s}^2}$$

The graph of f against θ is shown in Fig. 2.40.

6. A part of a mechanism operating a sliding block is shown in Fig. 2.41. In the given configuration the link OB swings about O and has an angular velocity of 11·2 rad/s and an angular acceleration of 56·5 rad/s², both in a clockwise direction. The reciprocating mass at E is 32 kg. The mass of the rod CE is 23 kg; its centre of gravity is at its mid-point and its moment of inertia about this point is 0·18 kg m².

Determine the kinetic energy of the reciprocating mass at E and of the rod CE and also the value of the force P necessary to accelerate the mass at E only. (*U. Lond.*) (*Ans.* : 37·5 J; 30·4 J; 433 N)

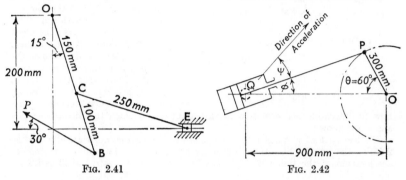

FIG. 2.41   FIG. 2.42

7. In the mechanism of a trunnion engine shown in Fig. 2.42, O is the axis of rotation of the crank OP and Q is the axis of rotation of the trunnions which support the cylinder. If the crank rotates at a uniform speed, prove that the direction of the acceleration of the point on the piston rod at Q is such that $\tan \psi = 2 \tan \phi$, where ψ and ϕ are as shown. Hence draw the velocity and acceleration images for the piston rod for $\theta = 60°$ and a crank speed of 90 rev/min, and determine the angular velocity and angular acceleration of the cylinder. (*U. Lond.*) (*Ans.* : 0·688 rad/s; 37·5 rad/s²)

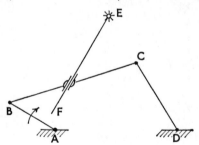

FIG. 2.43

8. As shown in Fig. 2.43, a four-bar chain ABCD drives an arm EF through a slide carried on a swivel at the mid-point of BC. The dimensions are AB = 30 mm, BC = 90 mm, CD = 60 mm and AD = 90 mm. E lies on the perpendicular bisector of AD at a distance 75 mm from AD. If AB rotates clockwise at a constant speed of 10 rad/s, determine the angular acceleration of EF when B, A and D lie on a straight line in the order given. (*U. Lond.*)

(*Ans.* : 19·3 rad/s²)

9. Fig. 2.44 shows a quick-return motion in which the driving crank OA rotates at 120 rev/min in a clockwise direction. For the position shown determine the magnitude and direction of: (*a*) the acceleration of the block D; (*b*) the angular acceleration of the slotted bar QB. (*U. Lond.*)

(*Ans.:* 6·75 m/s²; 16·8 rad/s²)

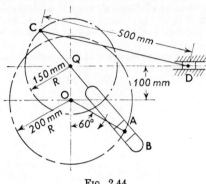

Fig. 2.44

10. In Fig. 2.45(a) the bar OA is rotating about O with a uniform angular velocity ω rad/s; the block B is sliding along OA with a linear velocity v m/s; C is the point on OA which is instantaneously coincident with B. Show that the value of the Coriolis component acceleration of B relative to C is $2v\omega$ and explain how to determine the direction of this acceleration when the directions of ω and v are known.

In the part of a quick-return mechanism, shown diagrammatically in Fig. 2.45(b), the crank OA rotates uniformly at 2·5 rad/s. Determine the angular acceleration of the link CB. (*U. Lond.*)   (*Ans.:* 0·33 rad/s²)

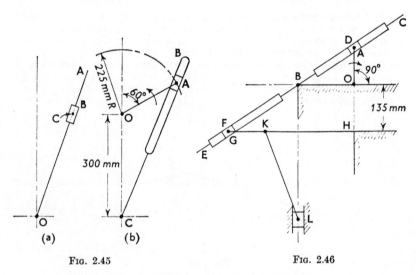

Fig. 2.45   Fig. 2.46

11. If, in the mechanism shown in Fig. 2.46, the link OA rotates at 3·183 rev/s, determine the velocity and acceleration of the slider L for the position shown.

OA = 90 mm, OB = 150 mm, HG = 375 mm, HK = 240 mm, KL = 300 mm. D and F are points on CBE instantaneously coincident with A and G respectively. (*U. Lond.*)   (*Ans.:* 1·035 m/s; 10·62 m/s²)

12. A straight rod PQ, 180 mm long, forms part of a mechanism. The end P of the rod is constrained to move in a straight vertical path with simple harmonic motion, making 5 complete oscillations per second. The travel of P between extreme positions is 60 mm. The rod PQ slides in a small block pivoted at a fixed point O. O is situated 60 mm to the right, and 60 mm below the mean position of the point P.

Determine the velocity and acceleration of Q at an instant when P is 15 mm below the centre of the line of stroke, and is moving upwards. (*U. Lond.*)

(*Ans.*: 1·035 m/s; 22·8 m/s²)

13. Fig. 2.47 shows a link mechanism in which the link OA rotates uniformly in an anticlockwise direction at 10 rad/s. The lengths of the various links are OA = 75 mm, OB = 150 mm, BC = 150 mm, CD = 300 mm. Determine, for the position shown, the instantaneous acceleration of D. (*U. Lond.*)

(*Ans.*: 0·8 m/s²)

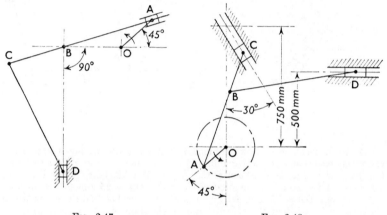

FIG. 2.47   FIG. 2.48

14. In the mechanism shown in Fig. 2.48, the crank OA rotates anticlockwise at 240 rev/min and moves the crosshead C in fixed guides by means of the connecting rod AC; the crosshead D is moved in fixed guides by the rod BD pinned to AC at B. OA = 0·2 m, AB = 0·5 m, AC = 0·75 m, BD = 1·25 m.

Find the accelerations of C and D for the position shown. Find also the magnitude and direction of the forces acting on the bar AC at A, B and C, due to the inertia of a mass of 20 kg at D. (Neglect friction and the mass of all other components.) (*U. Lond.*)

(*Ans.*: f_c = 166 m/s²; f_d = 15·6 m/s²; F_a = 114 N in direction 58° to horizontal; F_b = 306 N in direction of BD; F_c = 278 N perpendicular to guides)

15. In a Whitworth quick-return motion a crank AB rotates about a fixed centre A. The end B operates a slider reciprocating in a slotted link rotating about a fixed centre D 50 mm vertically above A. The crank AB, which is 100 mm long, rotates in a clockwise direction at a speed of 100 rev/min. Find the angular acceleration of the slotted link for the configuration in which AB has turned through an angle of 45° past its lowest position. (*U. Lond.*)

(*Ans.*: 8 rad/s²)

16. In the quick-return mechanism represented by Fig. 2.49, the crank QC rotates uniformly at 150 rev/min in a clockwise direction about the fixed centre Q, which is 100 mm from the other fixed centre O. OB = 200 mm, QC = 150 mm, BD = 600 mm. Determine, for the position in which the angle OQC is 135°, the magnitude and direction of : (a) the acceleration of D, (b) the angular acceleration of the slotted link BOA. (*U. Lond.*)

(*Ans.:* 10·5 m/s² ; 11·7 rad/s²)

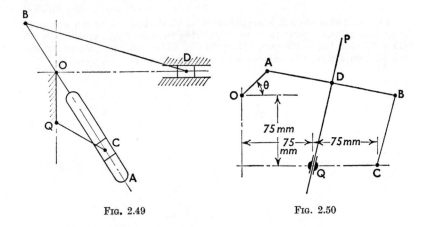

FIG. 2.49 FIG. 2.50

17. In the mechanism shown in Fig. 2.50, OA is a crank rotating at 60 rev/min about the fixed centre O. The crankpin A is connected by a rod AB to a swinging link BC, pivoted at the fixed centre C. A rod PD is hinged to AB at D and an extension of PD slides in a pivoted block at Q. OA = 25 mm, AD = 75 mm, BC = 75 mm, DB = 75 mm, DP = 50 mm. For the position shown, in which the angle θ is 45°, find the velocity and acceleration of P. (*U. Lond.*)

(*Ans.:* 0·208 m/s ; 1·06 m/s²)

18. In Fig. 2.51, the crank OA is 100 mm long and rotates in a clockwise direction at a speed of 100 rev/min. The straight rod BCD rocks on a fixed

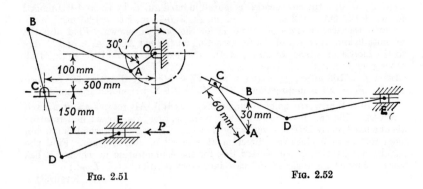

FIG. 2.51 FIG. 2.52

pivot at C. BC and CD are each 200 mm long and the link AB is 300 mm long. The slider E, which has a mass of 10 kg, is driven from D by the rod DE which is 250 mm long. The coefficient of friction between the slider and the horizontal guides is 0·1.

When OA is 30° below the horizontal as shown, find: (a) the velocity and acceleration of E, (b) the torque required at the driving shaft through O if the force P opposing the motion of the slider is 320 N. (U. Lond.)

(Ans.: 1·24 m/s; 10·62 m/s²; 26·5 N m)

19. In the mechanism shown in Fig. 2.52, the fixed bearings A and B are 30 mm apart. The crank AC of length 60 mm rotates at a uniform speed of 60 rev/min about A, and carries a slider C which moves along BC, causing CBD to rotate about B. BD is 45 mm and drives the head E through the link DE which is 180 mm long. BE is perpendicular to BA. Find the velocity and acceleration of E when the angle BAC = 45°; find also the angular acceleration of CD. (U. Lond.)   (Ans.: 0·118 m/s; 3·57 m/s²; 36·5 rad/s²)

20. In the mechanism shown in Fig. 2.53, the crank AB, 70 mm long, rotates clockwise at 110 rev/min. CD is 140 mm long. Link BD is 260 mm long and slides through a swivelling pin E at the lower end of rod EF. EF slides in vertical guides as shown.

For the position shown, find the angular velocity and angular acceleration of DB and the linear velocity and linear acceleration of F. (U. Lond.)

(Ans.: 1·395 rad/s; 31·4 rad/s²; 0·452 m/s; 2·36 m/s²)

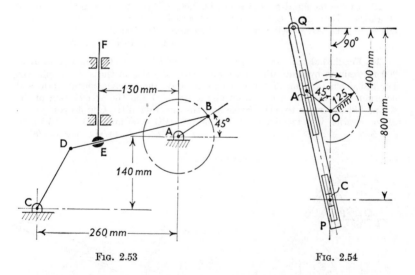

FIG. 2.53   FIG. 2.54

21. Fig. 2.54 shows a quick-return mechanism for a small shaping machine. The link PQ is driven by the crank OA which rotates clockwise at 25 rev/min. The end P of the link is slotted and guided by the block C which pivots about a fixed axis. The end Q of the link moves horizontally and operates the tool head. For the position shown, find the velocity and acceleration of Q. (U. Lond.)

(Ans.: 0·447 m/s; 0·388 m/s²)

22. In the mechanism shown in Fig. 2.55, the slide A_2S is driven with a uniform angular velocity of 10 rad/s and the block B has a mass of 1·5 kg. Draw the velocity and acceleration diagrams for the mechanism in the position shown and determine the thrust which the slide exerts on the block B and the tension in the rod A_1B. Indicate clearly the direction of the thrust which the slide exerts upon the block. Neglect friction. (*U. Lond.*)

(*Ans.:* 21·8 N; 174 N)

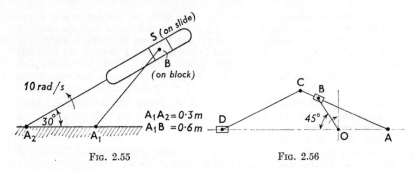

FIG. 2.55   FIG. 2.56

23. In the mechanism shown in Fig. 2.56, OB rotates at a uniform rate of 20 rad/s. Determine the velocity and acceleration of D. OA = 150 mm, OB = 75 mm, AC = 300 mm, CD = 250 mm (*U. Lond.*)

(*Ans.:* 1·04 m/s; 20·5 m/s²)

24. Fig. 2.57 shows part of a mechanism in which cranks OA and CB rotate at 200 rev/min in opposite directions, and block B slides in the slotted link AD. For the given position of the mechanism, draw the acceleration diagram and determine the angular acceleration of the link AD and the acceleration of the sliding block relative to AD. State how the direction of the Coriolis component of acceleration is determined. OA = 50 mm, BC = 50 mm, AB = 83·75 mm. (*U. Glas.*)

(*Ans.:* 30 m/s²; 470 rad/s²)

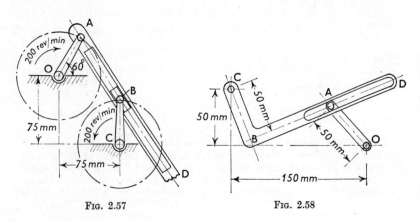

FIG. 2.57   FIG. 2.58

25. In Fig. 2.58, a bent lever CBD is pivoted at the fixed point C. The angle CBD is 90°. The crank OA rotates anticlockwise at 180 rev/min about the fixed point O. The crankpin A slides in a closely-fitting slot in the lever, as shown. For the given position, in which the crank is at an angle of 45° from the horizontal, find the angular velocity and the angular acceleration of the lever and also the acceleration of the point B. (*U. Lond.*)

(Ans.: 2·03 rad/s; 53 rad/s²; 2·68 m/s²)

26. A link XY, 200 mm long, is arranged so that the end X moves along a vertical path with simple harmonic motion, the frequency of oscillation being 10 Hz. The travel of X, between extreme positions, is 50 mm. The link slides through a small block pivoted at P, where P is a fixed point 75 mm below and 75 mm to the right of the centre of the line of stroke of X.

Determine the magnitude of the velocity and of the acceleration of Y at an instant when X is 12·5 mm below the upper limit of its travel and is moving downwards. (*U. Lond.*)   *(Ans.:* 1·26 m/s; 44·4 m/s²)

27. The cylinders of a rotary engine rotate at a uniform angular speed of 900 rev/min about the lower end of a fixed vertical crank 50 mm long. The connecting rods, 170 mm long, rotate about the upper end of the crank and reciprocate the pistons in the cylinders. Determine the acceleration of the piston relative to the cylinder and the angular acceleration of the connecting rod for a cylinder which has turned through an angle of 45° past the outer dead centre position. (*U. Lond.*)   *(Ans.:* 582 m/s²; 1775 rad/s²)

(b) Inertia Forces

28. *A single-cylinder horizontal steam engine has a stroke of 0·75 m and a connecting rod 1·8 m long. The mass of the reciprocating parts is 520 kg and that of the connecting rod is 230 kg. The c.g. of the rod is 0·8 m from the crankpin and the moment of inertia about an axis through the c.g. perpendicular to the plane of motion is 100 kg m². For an engine speed of 90 rev/min and a crank position of 45° to the inner dead centre, determine the torque on the crankshaft and the force on the crankshaft bearings due to the inertia of these parts.* (U. Lond.)

(*a*) *Exact two-mass system.* The connecting rod is replaced by two masses, placed at P and D, such that

$$PG.GD = k^2 = \frac{100}{230} = 0·435 \text{ m}^2$$

But $$PG = 1 \text{ m}$$

$$\therefore GD = \frac{0·435}{1} = 0·435 \text{ m}$$

In Fig. 2·59, OCLN is the acceleration diagram, obtained by Klein's Construction. G*g* and D*d* are drawn parallel to the line of stroke and

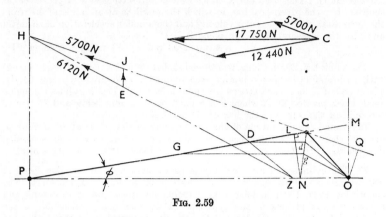

Fig. 2.59

*g*O and *d*O then represent the magnitudes and directions of the accelerations of G and D respectively.

$$\omega = \frac{2\pi}{60} \times 90 = 3\pi \text{ rad/s}$$

$$\therefore f_g = \omega^2 . gO = (3\pi)^2 \times 0\cdot3 = 26\cdot6 \text{ m/s}^2$$

$$\therefore \text{ inertia force on rod} = mf_g = 230 \times 26\cdot6 = 6120 \text{ N}$$

From the triangle of forces for the rod, HJE,

force at crankpin = HJ = 5700 N

$$\therefore \text{ crankshaft torque} = 5700 \times OQ$$

$$= 5700 \times 0\cdot15 = 855 \text{ N m}$$

Inertia force due to mass, R, of reciprocating parts,

$$= Rf_p = R\omega^2ON$$

$$= 520 \times (3\pi)^2 \times 0\cdot265 = 12\ 250 \text{ N}$$

$\therefore$ crankshaft torque due to the inertia force on reciprocating parts

$$= 12\ 250 \times OM$$

$$= 12\ 250 \times 0\cdot305 = 3740 \text{ N m}$$

$\therefore$ total crankshaft torque = <u>4595 N m</u>

The forces acting on the crankpin are:

(*a*) the component of the inertia force on the connecting rod along HC, of magnitude 5700 N and acting in the direction C to H;

(*b*) the force along the connecting rod due to the inertia of the piston, of magnitude 12 250 sec ϕ = 12 440 N and acting in the direction C to P.

The resultant force on the crankpin, of magnitude 17 750 N, is obtained from the parallelogram of forces and the reaction at the crankshaft is equal and opposite to this force.

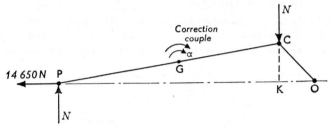

FIG. 2.60

(b) *Approximate two-mass system.* If m_1 is the mass to be placed at P and m_2 the mass at C, Fig. 2.60, then

$$m_1 + m_2 = 230$$

and

$$m_1 \times 1 = m_2 \times 0.8$$

$$\therefore\ m_1 = 102.2 \text{ kg} \quad \text{and} \quad m_2 = 127.8 \text{ kg}$$

∴ total reciprocating mass at P $= 520 + 102.2 = 622.2$ kg

∴ inertia force due to this mass $= 622.2 \times (3\pi)^2 \times 0.265$

$$= 14\ 650 \text{ N}$$

∴ inertia torque due to this mass $= 14\ 650 \times$ OM

$$= 14\ 650 \times 0.305 = 4470 \text{ N m}$$

Tangential acceleration of P relative to C

$$= \omega^2 \text{LN}$$

$$= (3\pi)^2 \times 0.265 = 23.5 \text{ m/s}^2$$

∴ angular acceleration of PC, $\alpha = \dfrac{23.5}{0.18} = 13.1 \text{ rad/s}^2$ (clockwise)

Correction couple to be applied to system

$$= m\,(\text{PG}.\text{GC} - k^2)\alpha$$

$$= 230\,(1 \times 0.8 - 0.435) \times 13.1$$

$$= 1100 \text{ N m}$$

This couple is positive and is therefore in the same direction as α. It is provided by two forces, N, at P and C, perpendicular to the line of stroke, such that

$$N \times \text{PK} = 1100 \text{ N m}$$

$$\therefore\ N = \frac{1100}{1.775} = 620 \text{ N}$$

$\therefore$ crankshaft torque due to N at $C = 620 \times KO$

$$= 620 \times 0{\cdot}263 = 163 \text{ N m}$$

$\therefore$ total crankshaft torque $= \underline{4633 \text{ N m}}$

29. *In the mechanism shown in Fig. 2.61, AB is rotated about the fixed centre A with a uniform speed of 10 rad/s, and CD oscillates about the fixed centre D. For the position shown, find the angular velocity and angular*

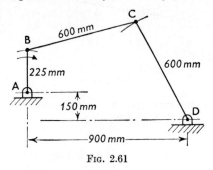

FIG. 2.61

acceleration of CD. If CD is equivalent to a uniform thin rod of mass 16 kg with its centre of gravity lying midway between the two points C and D, find the magnitude and direction of the turning moment which must then be applied to the shaft at A to accelerate CD. (I. Mech. E.)

$$v_b = 10 \times 0{\cdot}225 \text{ m/s} = 2{\cdot}25 \text{ m/s}$$

From the velocity diagram, Fig. 2.62,

$$v_c = 2{\cdot}325 \text{ m/s}$$

and

$$v_{bc} = 1{\cdot}274 \text{ m/s}$$

$$\therefore \omega_{cd} = \frac{2{\cdot}325}{0{\cdot}6} = \underline{3{\cdot}875 \text{ rad/s}}$$

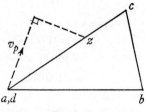

FIG. 2.62

Centripetal acceleration of B relative to $A = 10^2 \times 0{\cdot}225 = 22{\cdot}5 \text{ m/s}^2$

Centripetal acceleration of C relative to $B = \dfrac{1{\cdot}274^2}{0{\cdot}6} = 2{\cdot}7 \text{ m/s}^2$

Centripetal acceleration of C relative to $D = \dfrac{2 \cdot 325^2}{0 \cdot 6} = 9$ m/s^2

From the acceleration diagram, Fig. 2.63, tangential acceleration of C relative to D $= c_2 c$

$$= 11 \cdot 4 \text{ m/s}^2$$

$$\therefore \; \alpha_{cd} = \frac{11 \cdot 4}{0 \cdot 6} = \underline{19 \text{ rad/s}^2} \text{ (anticlockwise)}$$

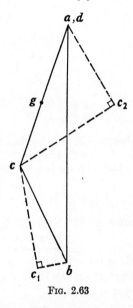

FIG. 2.63

If the centre of gravity of CD is denoted by G, then

$$\text{acceleration of } G = dg = 7 \cdot 2 \text{ m/s}^2$$

$$\therefore \; \text{accelerating force at } G = mf_g = 16 \times 7 \cdot 2$$

$$= 115 \cdot 3 \text{ N}$$

Accelerating torque on CD $= I_g \alpha_{cd}$

$$= 16 \times \frac{0 \cdot 6^2}{12} \times 19$$

$$= 9 \cdot 12 \text{ N m}$$

The resultant accelerating force on CD is therefore parallel to f_g, its perpendicular distance, h, from G being such that

$$h = \frac{9 \cdot 12}{115 \cdot 3} = 0 \cdot 0792 \text{ m}$$

The inertia force, P, on CD is equal and opposite to this resultant accelerating force. The line of action of P is such as to produce a clockwise inertia couple.

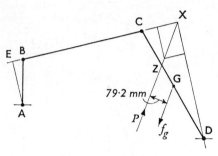

FIG. 2.64

The three forces acting on CD are (a) the inertia force, P, (b) the reaction at D, (c) the reaction at C, which acts along BC. These forces must be concurrent and hence, from the parallelogram of forces at X, Fig. 2.64,

$$\text{reaction at C} = 63.4 \text{ N}$$

$$\therefore \text{ torque at A} = 63.4 \times \text{AE} = 63.4 \times 0.2715$$

$$= \underline{13.8 \text{ N m}}$$

NOTE. If the link CD is replaced by an equivalent two-mass system with one of the two masses placed at D, then

$$\text{DG} \times \text{GZ} = k^2$$

i.e.
$$\text{GZ} = \frac{0.6^2}{12 \times 0.3} = 0.1 \text{ m}$$

Since the inertia force on the mass at D is zero, the resultant force on CD must pass through the point Z and its direction is that of the acceleration of Z. The line of action of this force is thus determined without consideration of the magnitude or direction of the angular acceleration of CD.

Alternative Solution.

$$\text{Work done by } P = \text{work done at A}$$

i.e.
$$P \times v_p = T_{ab} \times \omega_{ab}$$

i.e.
$$115.3 \times 1.2 = T_{ab} \times 10$$

$$\therefore T_{ab} = \underline{13.8 \text{ N m}}$$

30. *In the plane mechanism shown in Fig. 2.65, the driving crank OA, which is 50 mm between centres, oscillates about the fixed axis O. The rod AC, which is uniform, 100 mm long and of total mass 0·05 kg, slides through the trunnion B. The distance OB is 75 mm. When the crank is at 20° from the downward vertical through O it has an anticlockwise angular velocity of 40 rad/s and zero angular acceleration. For this configuration of the mechanism determine (a) the angular acceleration of the rod AC and the acceleration of its centre of mass, (b) the force between the rod AC and the trunnion, (c) the bending moment at section D of the rod AC, the distance CD being 25 mm. The masses of all moving parts other than the rod AC and the effects of friction and gravity are to be neglected.* (U. Lond.)

$$v_a = 40 \times 0·05 = 2 \text{ m/s}$$

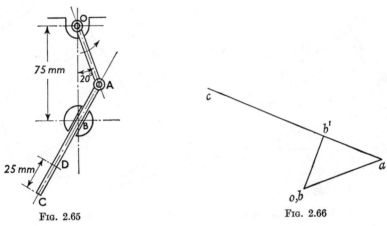

FIG. 2.65 · · · · FIG. 2.66

From the velocity diagram, Fig. 2.66,

$$v_{b'a} = 1·275 \text{ m/s} \quad \text{and} \quad v_{b'b} = 1·53 \text{ m/s}$$

Centripetal acceleration of A relative to O $= \omega^2 OA = 40^2 \times 0·05$
$$= 80 \text{ m/s}^2$$

Centripetal acceleration of B' relative to A $= \dfrac{v_{b'a}^2}{\text{B'A}} = \dfrac{1·275^2}{0·033} = 49·2 \text{ m/s}^2$

Tangential acceleration of B' relative to B $= 2 \times v_{b'b} \times \omega_{ac}$
$$= 2 \times 1·53 \times \frac{1·275}{0·033}$$
$$= 118 \text{ m/s}^2$$

(a) From the acceleration diagram, Fig. 2.67,

acceleration of centre of AC $= \boldsymbol{og} = \underline{248 \text{ m/s}^2}$

Angular acceleration of AC $= \dfrac{b_1'b'}{AB'} = \dfrac{180}{0\cdot033} = \underline{5450 \text{ rad/s}^2}$

(b) If P is the force between the rod AC and the trunnion, then

$$P \times AB = I_{ac}\alpha_{ac} = \frac{ml^2}{3}\alpha_{ac}$$

$$\therefore P = \frac{0\cdot05 \times 0\cdot1^2}{3} \times \frac{5450}{0\cdot03} = \underline{27\cdot5 \text{ N}}$$

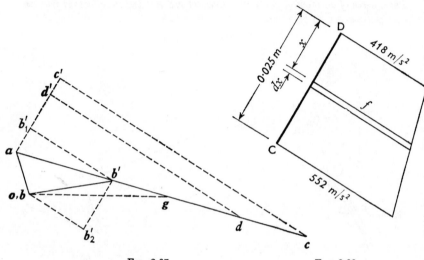

FIG. 2.67   FIG. 2.68

(c) The tangential accelerations of C and D are given by cc' and dd' respectively, Fig. 2.67. These are of magnitude 552 and 418 m/s² respectively and are set out on the part CD of the rod, Fig. 2.68.

Therefore, at a distance x from D, the tangential acceleration,

$$f = 418 + \frac{x}{0\cdot025}(552 - 418)$$

$$= 418 + 5360x$$

Thus the inertia force on an element of the rod, of length dx, is given by

$$dP = dm \times f$$

$$= \frac{dx}{0\cdot025} \times 0\cdot05 \times (418 + 5360x)$$

$$= 2(418 + 5360x)\,dx$$

$$\therefore \text{ bending moment at D} = dP \times x$$
$$= 2x(418 + 5360x) \, dx$$
$$\therefore \text{ total bending moment at D} = \int_0^{0 \cdot 025} 2(418x + 5360x^2) \, dx$$
$$= \underline{0 \cdot 3175 \text{ N m}}$$

31. *In the mechanism shown in Fig. 2.69, two blocks A and B of negligible mass slide between guides having centre-lines which are in the same vertical plane and which intersect at right angles at O. The blocks are connected by a light rigid link CD, 150 mm long, which is pinned to the blocks at C*

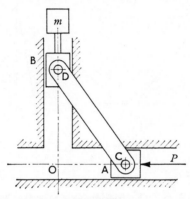

FIG. 2.69

and D. The pins at C and D each have a diameter of 15 mm and the coefficient of friction for all sliding surfaces at the guides and the pins is 0·2. The block B supports a mass m of 14 kg.

If when OC = 90 mm, A has a velocity towards O of 0·6 m/s and an acceleration towards O of 0·75 m/s², find the value of the force P required to drive the mechanism at this instant. (U. Lond.)

Either analytically* or graphically, the acceleration of D is found to be 4·125 m/s², downwards.

Thus

$$\text{upward inertia force acting on mass } m = 14 \times 4\cdot125$$
$$= 57\cdot75 \text{ N}$$
$$\therefore \text{ resultant downward force on B} = 14 \times 9\cdot81 - 57\cdot75$$
$$= 79\cdot65 \text{ N}$$

---

* If the acceleration of D is determined analytically, it must be remembered that $\frac{dx}{dt}$ and $\frac{d^2x}{dt^2}$ are both negative since they are directed towards O.

D

Radius of friction circles at C and D = 7·5 × sin (tan⁻¹ 0·2)

$$= 1·47 \text{ mm}$$

CD is in compression and the forces on the system are as shown in Fig. 2.70; the reactions R_1 and R_2 at the guides tend to oppose the

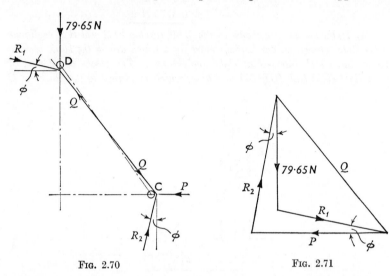

FIG. 2.70   FIG. 2.71

motions of A and B and the friction axis is such that the moments of the thrust, Q, in CD about the pin centres tend to oppose the relative rotations at the pins.*

The three forces R_1, Q and 79·65 N at the top block must be concurrent and similarly for the three forces R_2, Q and P at the lower block.

Fig. 2.71 shows the combined force diagrams at C and D, from which P is found to be 85·6 N.

---

* As a check on the correct position for the friction axis, it must be such as to lead to an *increase* in the value of P, compared with the value required without friction.

32. In a horizontal engine the crank OA of length 125 mm rotates at 240 rev/min. The connecting rod AB is 600 mm long, its mass is 11·25 kg, its centre of gravity is 250 mm from A and its radius of gyration about its centre of gravity is 200 mm. When the crank is vertical, determine the vertical force exerted by the guides on the cross-head due to the inertia of the connecting rod. (*U. Lond.*)   (*Ans.:* 106 N)

33. In a single-cylinder reciprocating engine the line of stroke of the piston passes through the axis of the crankshaft. The mass of the piston is m, the crank radius is r, the ratio of the length of the connecting rod to the crank radius is n and the angular velocity of the crankshaft (assumed constant) is ω rad/s.

Show that if n is so large that terms involving $(1/n)^2$ and higher powers of $(1/n)$ can be neglected—

(*a*) the acceleration of the piston is given by

$$\omega^2 r(A \cos \theta + B \cos 2\theta)$$

(*b*) the torque required at the crankshaft to accelerate the piston is given by

$$m\omega^2 r^2(C \sin \theta + D \sin 2\theta + E \sin 3\theta)$$

where θ is the angle of rotation of the crankshaft from the inner dead-centre position and A, B, C, D and E are constants depending on the value of n. Determine these constants. (*U. Lond.*) $\left(Ans.: \ 1; \ \dfrac{1}{n}; \ -\dfrac{1}{4n}; \ \dfrac{1}{2}; \ \dfrac{3}{4n}\right)$

34. A connecting rod has a mass of 1·125 kg and the distance between the centres of the end bearings is 250 mm. The centre of gravity is 162·5 mm from the centre of the small end bearing and the relevant moment of inertia is 0·0118 kg m² about an axis through the centre of gravity and perpendicular to the centre line of the rod. The crank radius is 62·5 mm and the speed is 200 rad/s. For a crank angle of 30° past the inner dead centre position, determine : (*a*) the torque required at the crankshaft to accelerate the rod, assuming the rod to be equivalent to two mass particles, one at each end of the rod, (*b*) the percentage error in the torque on this assumption. (*U. Lond.*)

(*Ans.:* 37·6 N m ; 11·8%)

35. A vertical engine of 120 mm stroke has a connecting rod 260 mm long between centres and of mass 1·25 kg. The mass centre is 80 mm from the big end centre and when suspended as a pendulum from the gudgeon-pin axis the rod makes 21 complete oscillations in 20 seconds.

Determine (*a*) the radius of gyration of the rod about an axis through the mass centre, and (*b*) the inertia torque exerted on the crankshaft when the crank is 40° from the top dead centre and is rotating at 1500 rev/min. (*U. Lond.*)

(*Ans. :* 89·4 mm ; 24·35 N m)

36. A connecting rod is 1·2 m long and 75 mm diameter, assumed uniform throughout its length. The crank is 0·3 m long and the engine speed is 240 rev/min. Draw the inertia-load diagram for the connecting rod when the crank is at 60° from the inner dead centre position. Determine the value of the maximum bending moment and state its position. Density of material, 8 Mg/m³. (*U. Lond.*)

(*Ans.:* 600 N m ; 0·795 m from piston)

37. In the mechanism shown in Fig. 2.72, crank O_1A rotates at 300 rev/min clockwise. Draw the acceleration diagram and determine the torque required at the crankshaft to overcome the inertia of links AB and O_2B. $O_1A = 60$ mm, $O_2B = 120$ mm, AB = 180 mm, $AG_1 = 75$ mm, $BG_2 = 75$ mm. For AB, mass = 0·9 kg, $k_G = 63$ mm ; for O_2B, mass = 0·65 kg, $K_G = 51$ mm. (*U. Glas.*)

(*Ans.:* 0·835 N m)

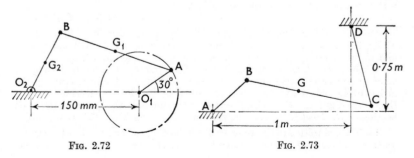

Fig. 2.72   Fig. 2.73

38. Fig. 2.73 represents a mechanism in which a crank AB rotates anticlockwise about A at 70 rev/min. The link CD swings about D and is connected to the crankpin B by the link BC. The lengths are : AB, 0·25 m, BC, 1·0 m, CD, 0·75 m. The connecting link BC has a mass of 14 kg, its centre of gravity G is at the centre of its length and its radius of gyration about a transverse axis through G is 0·3 m. The joints at B, C and D are frictionless and the link CD is of negligible mass.

For the given position, in which AB is at 45° from the horizontal, find the forces exerted at the joints B and C resulting from the inertia of the rod BC. (*U. Lond.*)   (*Ans.:* 150·5 N ; 29·4 N)

39. Fig. 2.74 shows the drive of a bell-crank lever that operates a pump rod at D. The connecting rod AB has a mass of 36 kg. Its centre of gravity is at the mid-length and the radius of gyration about the c.g. is 175 mm. The bell-crank has a mass of 90 kg and a moment of inertia about the fulcrum C of 1·8 kg m². When in the position shown, there is a vertical force, P, at D of 10·85 kN. The crank speed is 30 rad/s anticlockwise. Determine, for the configuration shown, the necessary driving torque on the crank to overcome inertia and resistance effects. (*U. Lond.*)   (*Ans.:* 1·565 kN m)

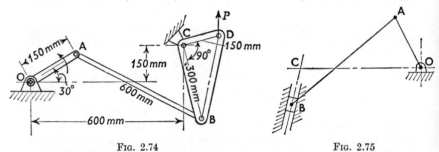

Fig. 2.74   Fig. 2.75

40. In the mechanism shown in Fig. 2.75, OA is a crank, rotating about a fixed centre O. The connecting rod AB moves the slider B, which slides in fixed guides. OA = 75 mm, AB = 175 mm, OC = 137·5 mm, ∠OCB = 110°. The slider has a mass of 6·3 kg and the coefficient of friction between it and its guides is 0·1. In the position shown, ∠COA = 65°, OA has an anticlockwise angular acceleration of 40 rad/s² and an angular velocity, which may be either clockwise or anticlockwise, of 7·5 rad/s.

Taking into account the weight, inertia and friction of the slider, determine the external torque required on the crank OA, (*a*) if the angular velocity of OA is clockwise, (*b*) if it is anticlockwise. (*U. Lond.*)

(*Ans.:* 2·94 N m ; 2·55 N m)

41. In a vertical engine the connecting rod is 800 mm long. Its mass is 45 kg, the centre of gravity is 525 mm from the axis of the small end bearing and the radius of gyration is 275 mm. The crank is 200 mm long.

When the crank has turned through 45° from the top dead centre, its angular velocity and acceleration are 25 rad/s and 250 rad/s². Find the forces at the gudgeon pin and at the main bearing caused by gravity and inertia forces on the connecting rod. Give the direction of the force on the frame at the main bearing. (*U. Lond.*)   (*Ans.:* 622 N ; 5·22 kN ; 10° to vertical)

42. The connecting rod of an internal combustion engine has a length between centres of 450 mm and a total mass of 3·25 kg. The centre of gravity is 325 mm from the small end and the radius of gyration about an axis through the centre of gravity and at right angles to the plane of swing of the connecting rod is 187·5 mm. The piston and gudgeon pin have a mass 2¼ kg, the cylinder is 150 mm diameter, the crank radius 137·5 mm and the engine speed 3200 rev/min.

The pressure in the cylinder when the crank is 30° after T.D.C. is 1·55 MN/m² above atmospheric. Determine the magnitude and direction of the force which the crankpin exerts on the big end bearing in this position. The direction of the force must be completely specified. (*U. Lond.*)

(*Ans.:* 54·1 kN at 14° to horizontal towards crankshaft)

43. In a four-bar chain ABCD, A and D are fixed centres 0·9 m apart in a horizontal line. The driving crank AB = 0·3 m, the driven crank DC = 0·6 m and the coupler BC = 0·6 m. When AB is perpendicular to AD, B and C are both above AD. If the angular velocity of AB is then uniform at 100 rev/min anticlockwise, find the angular velocity and angular acceleration of CD. If the moment of inertia of CD about D is equivalent to 40 kg at a radius of 0·3 m, find the turning moment required at A to produce the rotation of CD. (*I. Mech. E.*) (*Ans.:* 5·1 rad/s; 35 rad/s²; 61·3 N m)

44. Fig. 2.76 shows a mechanism in which a crank OP revolves about O at 180 rev/min. The crank carries a pin P which slides in the slotted lever CQ. The lever, its shaft and associated parts are counterbalanced so that the centre of gravity lies at the centre of the shaft Q. The total mass is 2·7 kg and the radius of gyration about Q is 100 mm. The length of OQ is 125 mm and OP is 50 mm.

For the position where the angle POQ is 30°, find the angular acceleration of the lever and the torque required on the crankshaft at O to overcome the inertia of the lever and attached parts. (*U. Lond.*) (*Ans.:* 274 rad/s²; 3 N m)

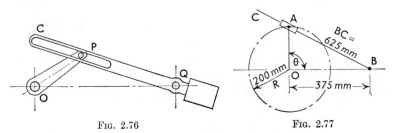

FIG. 2.76   FIG. 2.77

45. Part of a quick-return motion is shown in Fig. 2.77, where the driving crank OA rotates at 150 rev/min in an anticlockwise direction. The swinging link BC has a mass of 30 kg and its mass may be considered to be uniformly distributed along its length. Find the angular acceleration of the link BC when the angle $\theta = 105°$ and the transverse forces and reactions acting on BC. Neglect friction and gravitational forces. (*U. Lond.*) (*Ans.:* 36·2 rad/s²; 303 N; 174 N)

46. In a horizontal combusion engine, liquid for cooling the piston head is fed in through the 'walking pipe' system shown diagrammatically in Fig. 2.78, where the pipes AB and BC have swivelling joints at their ends.

(*a*) Determine the maximum inertia bending moment in the section AB when the crank is at 45° from the inner dead centre, the engine speed being 250 rev/min. The mass of AB may be taken as 18 kg per metre run, including the cooling liquid.

(*b*) Sketch and explain the form of the acceleration diagram for the mechanism when the crankpin is passing through the inner dead centre position. (*U. Lond.*)

(*Ans.*: 27·4 N m)

47. In the mechanism shown in Fig. 2.79, the side shaft O_2 is oscillated by the link CD, pin-jointed to the connecting rod at C and to the lever O_2D at D. $O_1B = 75$ mm, $BA = 225$ mm, $BC = 75$ mm, $CD = 150$ mm, $O_2D = 112$ mm. If the crankshaft runs at 15 rev/s anticlockwise, find, for the position shown, the velocity of the piston A and the angular velocity of the side shaft.

If the piston has a mass of 5·4 kg with its speed decreasing at the rate of 455 m/s² and the inertia of the side shaft is equivalent to 4·5 kg at a radius of 100 mm with its angular velocity increasing at the rate of 1040 rad/s², find the turning moment required at the crank to overcome the inertia of these parts. (*I. Mech. E.*) (*Ans.*: 2·475 m/s; 48·7 rad/s; 90 N m)

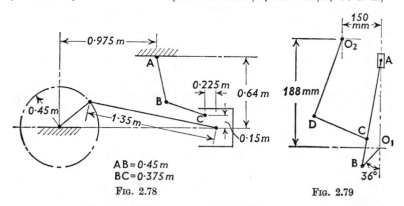

AB = 0·45 m
BC = 0·375 m

FIG. 2.78

FIG. 2.79

48. A light slider A slides between horizontal guides, as shown in Fig. 2.80. It is connected by a light rod AB, 200 mm long, to a light slider B, which slides between vertical guides. A concentrated body C, of mass 3 kg, is fixed to the rod at a point 125 mm from A. The coefficient of friction between each slider and the guides is 0·1, and friction at turning pairs may be neglected.

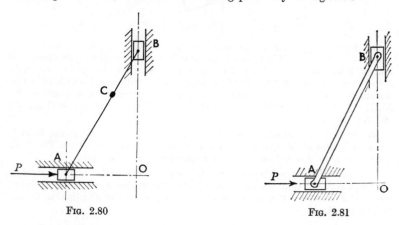

FIG. 2.80

FIG. 2.81

At the instant when the angle BAO = 60°, A has a velocity to the right of 1·5 m/s and an acceleration to the right of 12 m/s².

Determine the horizontal force P required under these conditions, taking into account the weight and inertia of C and the sliding friction at A and B. (*U. Lond.*) (*Ans.:* 8·82 N)

49. A slider A, Fig. 2.81, of mass 1·8 kg, moves along horizontal guides, whilst another slider B, of mass 2·7 kg, moves in vertical guides. The two sliders are joined by a light connecting rod 200 mm long. The coefficient of friction at the sliding surfaces is 0·08, whilst friction at turning pairs may be neglected.

Find the horizontal force P required to drive the mechanism at an instant when angle BAO is 60°, the velocity of A then being 0·9 m/s to the right, and the acceleration of A 1·5 m/s² to the right. (*U. Lond.*) (*Ans.:* 12·05 N)

50. Fig. 2.82 shows a uniform link AB, 500 mm in length and of mass 9 kg, the ends of which are constrained to move along perpendicular straight lines. For the position shown a force $P = 305$ N applied at A gives A a constant velocity of 1·8 m/s.

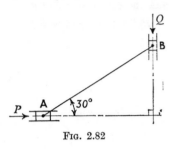

Taking into account the inertia of the link, (*a*) determine the magnitude of the resisting force Q applied at B; (*b*) make a scale diagram showing the inertia loading on AB. Neglect gravitational effects and any effects of friction at A and B. (*U. Lond.*) (*Ans.:* 335 N)

Fig. 2.82

51. In the mechanism shown in Fig. 2.83 a small block can slide along the link OE which rotates about the fixed axis O. The crank CB rotates about the fixed axis C and is pinned to the block at B. The link OE has a mass of 4 kg per metre run.

$$OC = 100 \text{ mm}; \quad CB = 50 \text{ mm}; \quad OE = 250 \text{ mm}$$

At the instant when the angle θ is 90° the crank CB has an *anticlockwise* angular

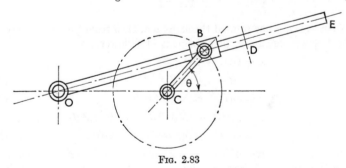

Fig. 2.83

velocity of 40 rad/s and a *clockwise* angular acceleration of 800 rad/s². For this configuration determine (*a*) the angular acceleration of the link OE, and (*b*) the bending moment at D in the link OE, where OD = 175 mm, neglecting the effect of gravity. (*U. Lond.*) (*Ans.:* 99·4 N m)

CHAPTER 3

CAMS

3.1 Cam with straight flanks; roller-ended follower. Let the base circle and nose radii be R and r respectively, the follower radius r_0 and the centre distance d.

Roller in contact with flank AB, Fig. 3.1.

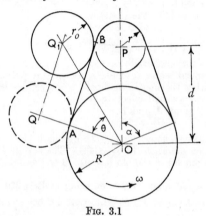

FIG. 3.1

When the cam has rotated through an angle θ from the lowest position of the follower centre Q, displacement of follower,

$$x = OQ_1 - OQ$$
$$= (R + r_0) \sec \theta - (R + r_0) \quad . \quad . \quad . \quad (3.1)$$
$$\therefore v = \omega(R + r_0) \sec \theta \tan \theta \quad . \quad . \quad . \quad (3.2)$$
$$\therefore f = \omega^2(R + r_0)(\sec^3 \theta + \sec \theta \tan^2 \theta)$$
$$= \omega^2(R + r_0)(2 \sec^3 \theta - \sec \theta) \quad . \quad . \quad (3.3)$$

The velocity increases from zero at A to a maximum at B while the acceleration increases from a minimum at A to a maximum at B, as shown in Fig. 3.5.

If β is the angle turned through by the cam while the roller moves from A to B, then

$$\tan \beta = \frac{d \sin \alpha}{R + r_0} \quad \text{where } \alpha \text{ is the angle of lift.}$$

96

The maximum velocity and acceleration on the flank are then obtained by substituting $\theta = \beta$ in equations (3.2) and (3.3).

Roller in contact with nose BC, Fig. 3.2.

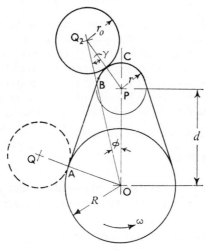

FIG. 3.2

When the cam has rotated through an angle ϕ from the highest position of the follower, displacement of follower,

$$x = OQ_2 - OQ$$
$$= \{d \cos \phi + (r + r_0) \cos \gamma\} - (R + r_0)$$
$$d \sin \phi = (r + r_0) \sin \gamma$$
$$\therefore \sin \gamma = \frac{d}{r + r_0} \sin \phi = \frac{\sin \phi}{n} \quad \text{where } n = \frac{r + r_0}{d}$$
$$\therefore \cos \gamma = \sqrt{\left\{1 - \left(\frac{\sin \phi}{n}\right)^2\right\}}$$
$$\therefore x = d \cos \phi + (r + r_0)\sqrt{\left\{1 - \left(\frac{\sin \phi}{n}\right)^2\right\}} - (R + r_0)$$
$$= d\{\cos \phi + \sqrt{(n^2 - \sin^2 \phi)}\} - (R + r_0) \qquad . \qquad . \qquad (3.4)$$
$$\therefore v = -\omega d\left\{\sin \phi + \frac{\sin 2\phi}{2\sqrt{(n^2 - \sin^2 \phi)}}\right\} \qquad . \qquad . \qquad (3.5)$$
$$\therefore f = -\omega^2 d\left\{\cos \phi + \frac{\sin^4 \phi + n^2 \cos 2\phi}{(n^2 - \sin^2 \phi)^{3/2}}\right\} \qquad . \qquad . \qquad (3.6)$$

The velocity decreases from a maximum at B to zero at C while the acceleration falls from a maximum at B to a minimum at C, as shown in Fig. 3.5. The acceleration is negative and at the highest point,

$$f = -\omega^2 d\left(1 + \frac{1}{n}\right)$$

Substituting $\phi = \alpha - \beta$ in equation (3.6) gives the deceleration when the roller makes contact with the nose at B. The maximum deceleration on the nose may occur at either B or C, depending on the values of d and n. The possible deceleration curves are shown in Fig. 3.5.

Equations (3.4), (3.5) and (3.6) are the exact analytical expressions for the displacement, velocity and acceleration of the piston in the simple slider-crank mechanism. The approximate expressions given in Art. 2.6 are not always applicable since the ratio $\dfrac{\text{connecting rod length}}{\text{crank length}}$ may be small or even fractional.

3.2   Cam with curved flanks ; flat-ended follower. Let the base circle and nose radii be R and r respectively, the flank radius ρ and the centre distance d.

Follower in contact with flank AB, Fig. 3.3.

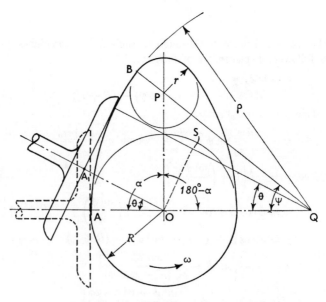

FIG. 3.3

When the cam has rotated through an angle θ from the lowest position of the follower, A, displacement of follower,

$$x = OA_1 - OA = \{\rho - (\rho - R)\cos\theta\} - R$$
$$= (\rho - R)(1 - \cos\theta) \qquad . \qquad . \qquad . \qquad . \qquad (3.7)$$
$$\therefore v = \omega(\rho - R)\sin\theta \qquad . \qquad . \qquad . \qquad . \qquad . \qquad (3.8)$$
$$\therefore f = \omega^2(\rho - R)\cos\theta \qquad . \qquad . \qquad . \qquad . \qquad (3.9)$$

The velocity increases from zero at A to a maximum at B while the acceleration decreases from a maximum at A to a minimum at B, as shown in Fig. 3.6.

To determine the flank radius ρ for a given angle of lift, α:

From triangle OPQ,

$$(\rho - r)^2 = (\rho - R)^2 + d^2 - 2(\rho - R)d\cos(180° - \alpha)$$

from which $\qquad \rho = \dfrac{R^2 - r^2 + d^2 - 2\,Rd\cos\alpha}{2(R - r - d\cos\alpha)} \qquad . \qquad . \qquad . \qquad (3.10)$

Also $\qquad \dfrac{d}{\sin\psi} = \dfrac{\rho - r}{\sin(180° - \alpha)} = \dfrac{\rho - r}{\sin\alpha}$

$$\therefore \sin\psi = \dfrac{d\sin\alpha}{\rho - r} \qquad . \qquad . \qquad . \qquad . \qquad . \qquad (3.11)$$

Follower in contact with nose BC, Fig. 3.4.

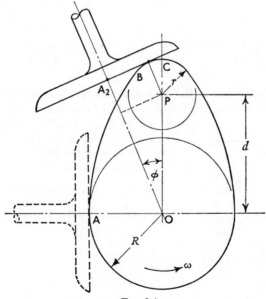

FIG. 3.4

When the cam has rotated through an angle ϕ from the highest position of the follower, displacement of follower,

$$x = OA_2 - OA$$
$$= (d \cos \phi + r) - R \qquad . \qquad . \qquad . \qquad (3.12)$$
$$\therefore \ v = -\omega d \sin \phi \qquad . \qquad . \qquad . \qquad (3.13)$$
$$\therefore \ f = -\omega^2 d \cos \phi \qquad . \qquad . \qquad . \qquad (3.14)$$

The velocity decreases from a maximum at B to zero at C while the acceleration increases from a minimum at B to a maximum at C, as shown in Fig. 3.6. The acceleration is negative and at the highest point,

$$f = -\omega^2 d$$

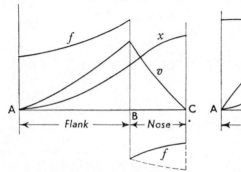

Tangent cam, roller-ended follower
FIG. 3.5

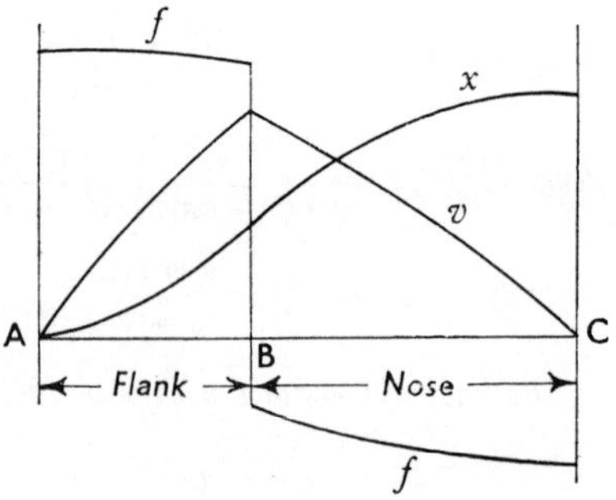

Circular arc cam, flat-ended follower
FIG. 3.6

3.3 Circular cam; flat-ended follower.

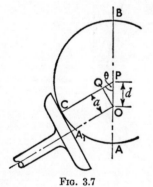

FIG. 3.7

Let the radius of the cam be R and the eccentricity of the axis of rotation O from the centre of the cam P be d, Fig. 3.7. When the cam has rotated through an angle θ from the lowest position of the follower, A, displacement of follower,

$$x = OA_1 - OA = QC - OA$$
$$= (R - d \cos \theta) - (R - d)$$
$$= d(1 - \cos \theta) \qquad . \qquad . \qquad . \qquad (3.15)$$
$$\therefore \ v = \omega d \sin \theta \ . \qquad . \qquad . \qquad . \qquad (3.16)$$
$$\therefore \ f = \omega^2 d \cos \theta \ . \qquad . \qquad . \qquad . \qquad (3.17)$$

The follower therefore moves with S.H.M. and there are no periods of dwell.

The velocity increases from zero at A to a maximum when $\theta = 90°$ and then falls again to zero at B. The acceleration has its maximum positive value at A and its maximum negative value at B, being zero at $\theta = 90°$, as shown in Fig. 3.8.

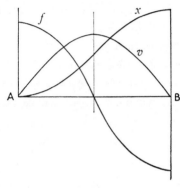

FIG. 3.8

3.4 Circular cam; roller-ended follower. This case is similar to that of the tangent cam, Art. 3.1, when the roller is in contact with the nose. The displacement, velocity and acceleration of the follower are therefore as given in equations (3.4), (3.5) and (3.6).

3.5 Spring force. While on the flank, the follower is either driven by the cam or retarded by it, but when it is in contact with the nose, the acceleration of the follower towards the cam axis must be provided by a spring (if the follower is vertically above the cam, gravity will assist in producing this force, but in high-speed engines or machines, this will have a negligible effect). In calculating the spring force required, it is necessary to determine the maximum acceleration *towards the cam axis* (i.e. negative acceleration) to which the follower is subjected. The force required is then given by

$$P = mf$$

where m is the effective mass of the follower and attached parts and possibly an allowance for the mass of the spring.

In the tangent cam, the maximum negative acceleration may occur either at B or C (depending on the dimensions of the cam); in the cam with curved flanks or in the circular cam, the maximum negative acceleration occurs at the highest point of the follower travel.

3.6 Reaction torques. Let m be the effective mass of the follower, S the spring stiffness and y the initial compression of the spring (i.e. at the commencement of the lift). When the follower has moved through a distance x, spring force $= S(x + y)$. If, at this instant, the acceleration of the follower is f, the total force on the follower is

$$mf + S(x + y)$$

In the case of the tangent cam with a roller-ended follower, Fig. 3.9, the normal force, F, between the roller and flank is given by

$$F \cos \theta = mf + S(x + y)$$

and the reaction torque is then

$$F \times a = F \times (R + r_0) \tan \theta$$
$$= \{mf + S(x + y)\}(R + r_0) \sec \theta \tan \theta \qquad . \quad (3.18)$$

This torque reaches its maximum value when the roller is in contact at the end of the flank.

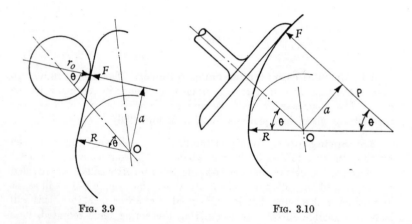

FIG. 3.9   FIG. 3.10

In the case of the cam with curved flanks and flat-ended follower, Fig. 3.10, the reaction is parallel to the follower axis, so that

$$F = mf + S(x + y)$$

and the reaction torque is then

$$F \times a = F \times (\rho - R) \sin \theta$$
$$= \{mf + S(x + y)\}(\rho - R) \sin \theta \qquad . \qquad . \quad (3.19)$$

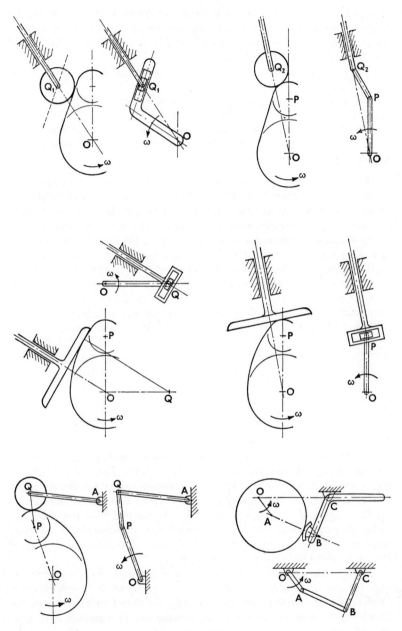

Fig. 3.11

The position of the cam for maximum torque must be determined by differentiation. If there is no spring the torque reaches its maximum value when $\theta = 45°$ or when the follower is in contact with the end of the flank if $\psi < 45°$ (see Ex. 3).

The circular cam, Fig. 3.7, is similar to the preceding case, except that $a = d \sin \theta$.

The appropriate expressions for x and f in terms of θ for the above cases are given in Arts. 3.1, 3.2 and 3.3.

The reaction torque when the follower is in contact with the nose is usually of little interest, since the acceleration of the follower is then negative.

3.7   Equivalent mechanisms. Certain types of cam problem lend themselves to solution by graphical methods and particularly by the use of an equivalent mechanism, such as a four-bar chain or slider-crank mechanism. Having ascertained the equivalent mechanism from a study of the cam and follower in a particular position, the velocity and acceleration diagrams are drawn for the mechanism in the usual way. This method gives the velocity and acceleration of the follower for one position only, whereas the analytical method gives a general solution. To obtain velocity and acceleration curves covering all positions of the cam would necessitate repeating the diagrams for various cam positions.

Fig. 3.11 shows some examples of equivalent mechanisms.

1. *The camshaft of a petrol engine operates an overhead valve through a vertical tappet and push-rod, and a horizontal rocker, Fig. 3.12. A spring acting against a collar on the valve stem maintains contact between the tappet and cam. The line of action of the tappet passes through the cam axis.*

The masses of the moving parts are: tappet and push-rod, 0·16 kg; rocker, 0·18 kg; valve, 0·07 kg. The arm length of the rocker on the push-rod side is 25 mm and on the valve side 40 mm. The radius of gyration of the rocker about its axis of rotation is 20 mm.

The cam has a base circle of 25 mm diameter, straight sides, a nose radius of 6 mm and a lift of 6 mm. The roller on the tappet is 12 mm diameter.

Find the spring force necessary to keep the tappet in contact with the cam when the valve has its maximum opening and the camshaft is rotating at 2000 rev/min. (U. Lond.)

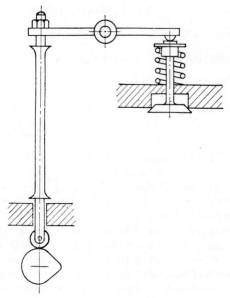

Fɪɢ. 3.12

Acceleration of valve at maximum opening,

$$f = \omega^2 d\left\{\cos\phi + \frac{\sin^4\phi + n^2\cos 2\phi)}{(n^2 - \sin^2\phi)^{3/2}}\right\} \quad \text{from equation (3.6)}$$

$$= \omega^2 d\left(1 + \frac{1}{n}\right) \quad \text{since } \phi = 0$$

$$\omega = 2000 \times \frac{2\pi}{60} = 209.5 \text{ rad/s}$$

$$d = l + R - r = 6 + 12 - 6 = 12 \text{ mm}$$

$$n = \frac{r + r_o}{d} = \frac{6 + 6}{12} = 1.0$$

$$\therefore f = 209.5^2 \times 0.012\,(1 + 1) = 1050 \text{ m/s}^2$$

Effective mass of system at tappet*

$$= 0.16 + 0.18\left(\frac{0.02}{0.025}\right)^2 + 0.07\left(\frac{0.04}{0.025}\right)^2 = 0.455 \text{ kg}$$

If P is the force in the spring,

$$\text{effective spring force at tappet} = P \times \frac{0.04}{0.025} = 1.6P$$

* See Art. 1.15.

$$\therefore \ 1 \cdot 6P = 0 \cdot 455 \times 1050$$

$$\therefore \ P = \underline{299 \text{ N}}$$

More exactly, allowing for the dead weight of the valve and tappet,

$$1 \cdot 6(P - 0 \cdot 07 \times 9 \cdot 81) + 0 \cdot 18 \times 9 \cdot 81 = 0 \cdot 455 \times 1050$$

assuming the rocker arm to be pivoted at its c.g., from which

$$P = \underline{298 \cdot 2 \text{ N}}$$

2. *In Fig. 3.13 a cam rotating with uniform speed about a fixed centre O moves a lever PQ which is pivoted at Q. The base circle radius of the cam is 25 mm. The nose of the cam is an arc of a circle of 10 mm radius with centre E, the distance OE being 30 mm. The flanks are straight lines, tangential to the base and nose circles.*

The lever carries at P a roller of 20 mm radius. The length PQ is 160 mm. The centre of gravity of the lever (including the roller) is at G, 100 mm from Q, and the radius of gyration about G is 60 mm. The lever moves through equal angles above and below the horizontal position, and the path of P may be taken as approximating to a vertical straight line passing through O. The lever rests upon the cam only by its own weight.

Find the greatest speed of the camshaft at which the roller will always be in contact with the cam. For this speed of the camshaft, find:

(a) the angular acceleration of the lever when the roller is just leaving its lowest position;

(b) the angular acceleration of the lever when the roller is just about to leave the straight flank of the cam on the upward stroke. (U. Lond.)

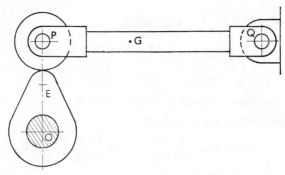

Fig. 3.13

The maximum downward acceleration of P occurs either when the roller is in contact with the nose at the point B or at the point C, Fig. 3.14.

$$\cos \alpha = \frac{R - r}{d}$$

$$= \frac{25 - 10}{30} = 0.5$$

$$\therefore \ \alpha = 60° = \theta + \phi$$

$$EF = 10 + 20 = 30 \text{ mm}$$

Also $OE = 30 \text{ mm}$

$$\therefore \ \phi = 30°$$

$$n = \frac{r + r_0}{d}$$

$$= \frac{10 + 20}{30} = 1$$

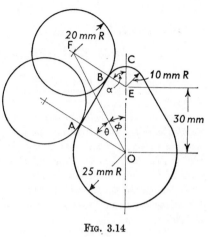

Fig. 3.14

At B, $f = \omega^2 \times 0.03 \left\{ \cos 30° + \dfrac{\sin^4 30° + 1 \times \cos 60°}{(1 - \sin^2 30°)^{3/2}} \right\}$

.   . from equation (3.6)

$$= 0.052\omega^2 \text{ m/s}^2$$

At C, $f = \omega^2 \times 0.03(1 + 1)$ since $\phi = 0$

$$= 0.09\omega^2 \text{ m/s}^2$$

The maximum downward acceleration therefore occurs at the point C.
The equation of motion of PQ is given by

$$mg \times 0.1 = I_Q \alpha = m(0.06^2 + 0.1^2) \times \frac{f}{0.16}$$

$$\therefore f = 11.55 \text{ m/s}^2$$

Therefore, for contact to be maintained,

$$11.55 > 0.09\omega^2$$

$$\therefore \ \omega^2 < 128.3$$

$$\therefore \ \omega = 11.33 \text{ rad/s} \quad \text{or} \quad N < \underline{108.3 \text{ rev/min}}$$

(a) While in contact with the flank AB,

$$f = \omega^2(R + r_0)(2 \sec^3 \theta - \sec \theta) \quad . \quad . \quad \text{from equation (3.3)}$$

At A, $\theta = 0$

$$\therefore f = 128.3(0.025 + 0.02)(2 - 1) = 5.77 \text{ m/s}^2$$

$$\therefore \ \alpha = \frac{5.77}{0.16} = \underline{36.1 \text{ rad/s}^2}$$

(b) At B,

$$\theta = 30°$$

$$\therefore f = 128{\cdot}3(0{\cdot}025 + 0{\cdot}02)(2 \sec^3 30° - \sec 30°) = 11{\cdot}1 \text{ m/s}^2$$

$$\therefore \alpha = \frac{11{\cdot}1}{0{\cdot}16} = \underline{69{\cdot}4 \text{ rad/s}^2}$$

3. *A flat-ended valve tappet is operated by a symmetrical cam having circular arcs for flank and nose profiles; the base circle diameter is 30 mm, the nose radius is 4 mm and the lift is 10 mm; the total angle of action is 180°. The mass effect of the valve, spring and tappet is equivalent to a mass of 0·8 kg concentrated at the tappet.*

Calculate the flanks radius of the cam and the maximum reaction torque, due to the accelerated masses, acting on the camshaft when the cam rotates at 1200 rev/min. (U. Lond.)

Referring to Fig. 3.3,

$$\rho = \frac{R^2 - r^2 + d^2 - 2Rd \cos \alpha}{2(R - r - d \cos \alpha)} \qquad . \quad \text{from equation (3.10)}$$

But
$$d = R + l - r$$

$$= 15 + 10 - 4 = 21 \text{ mm}$$

$$\therefore \rho = \frac{15^2 - 4^2 + 21^2 - 2 \times 15 \times 21 \times \cos 90°}{2(15 - 4 - 21 \cos 90°)}$$

$$= \underline{29{\cdot}54 \text{ mm}}$$

The maximum reaction torque will occur while the follower is in contact with the flank since it is accelerating during this part of the motion.

$$\therefore f = \omega^2(\rho - R) \cos \theta \qquad . \qquad . \qquad \text{from equation (3.9)}$$

$\therefore$ inertia force
$$P = m\omega^2(\rho - R) \cos \theta$$

$\therefore$ inertia torque,

$$T = m\omega^2(\rho - R) \cos \theta \times \text{OS}$$

$$= m\omega^2(\rho - R) \cos \theta \times (\rho - R) \sin \theta$$

$$= m\omega^2(\rho - R)^2 \frac{\sin 2\theta}{2}$$

Thus the maximum reaction torque occurs when $\theta = 45°$, provided that $\psi > 45°$, when the maximum torque will occur when the follower is in contact with the flank at the point B.

In this case,

$$\sin \psi = \frac{d \sin \alpha}{\rho - r} \qquad . \qquad . \qquad . \qquad \text{from equation (3.11)}$$

$$= \frac{21 \sin 90°}{29\cdot54 - 4} = 0\cdot822$$

$$\therefore \ \psi = 55° \ 18'$$

Thus $\quad T_{\max} = \dfrac{m}{2}\omega^2(\rho - R)^2 \quad$ when $\theta = 45°$

$$= \frac{0\cdot8}{2} \times \left(\frac{2\pi}{60} \times 1200\right)^2 \times (0\cdot0295 - 0\cdot015)^2$$

$$= \underline{1\cdot32 \text{ N m}}$$

4. *In Fig. 3.15, a cam rotating about a fixed point O moves a rocking lever which is pivoted at C. The profile of the cam consists of a base circle of 20 mm radius, a nose circle of 10 mm radius and two flank circles of*

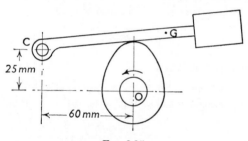

Fig. 3.15

radius 35 mm, each touching nose and base circles. The centre of the nose circle is 20 mm from the point O. The lever has a mass of 1·5 kg, its centre of gravity G is 80 mm from C, and its radius of gyration about the pivot C is 90 mm. The cam rotates at 120 rev/min.

(a) Find the time of the upward stroke of the lever and the time of the downward stroke.

(b) Find the force exerted between the cam and the lever at the beginning of each stroke and at the end of each stroke. (U. Lond.)

Referring to Fig. 3.3,

$$\rho = \frac{R^2 - r^2 + d^2 - 2Rd \cos \alpha}{2(R - r - d \cos \alpha)} \qquad . \qquad \text{from equation (3.10)}$$

i.e. $\quad 35 = \dfrac{20^2 - 10^2 + 20^2 - 2 \times 20 \times 20 \times \cos \alpha}{2(20 - 10 - 20 \times \cos \alpha)}$

from which $\quad \alpha = 90°$

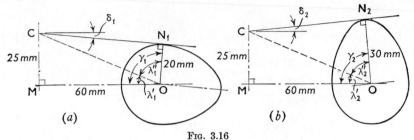

FIG. 3.16

Configuration with lever in lowest position	Configuration with lever in highest position

$$OC = \sqrt{(60^2 + 25^2)} = 65 \text{ mm}$$
$$CN_1 = \sqrt{(65^2 - 20^2)} = 61\cdot84 \text{ mm}$$
$$\lambda_1' = \tan^{-1}\frac{25}{60} \qquad = 22°\,37'$$
$$\lambda_1'' = \cos^{-1}\frac{20}{65} \qquad = 72°\,4'$$
$$\therefore \ \gamma_1 = \lambda_1' + \lambda_1'' \qquad = 94°\,41'$$
$$\text{and } \delta_1 = \gamma_1 - 90° \qquad = 4°\,41'$$

$$OC = \sqrt{(60^2 + 25^2)} = 65 \text{ mm}$$
$$CN_2 = \sqrt{(65^2 - 30^2)} = 57\cdot66 \text{ mm}$$
$$\lambda_2' = \tan^{-1}\frac{25}{60} \qquad = 22°\,37'$$
$$\lambda_2'' = \cos^{-1}\frac{30}{65} \qquad = 62°\,30'$$
$$\therefore \ \gamma_2 = \lambda_2' + \lambda_2'' \qquad = 85°\,7'$$
$$\text{and } \delta_2 = 90° - \gamma_2 \qquad = 4°\,53'$$

(a)
$$\omega = 120 \times \frac{2\pi}{60} \qquad = 4\pi \text{ rad/s}$$

$$\text{Angle of action} = 2\alpha \qquad\qquad = 180°$$

$$\text{Angle of lift} = 90° + (\gamma_1 - \gamma_2) = 99°\,34'$$

$$\therefore \ \text{angle of fall} = 180° - 99°\,34' = 80°\,26'$$

$$\therefore \ \text{time of upstroke} = \frac{99\cdot57 \times \dfrac{\pi}{180}}{4\pi} = \underline{0\cdot1383 \text{ s}}$$

$$\text{time of downstroke} = \frac{80\cdot43 \times \dfrac{\pi}{180}}{4\pi} = \underline{0\cdot1117 \text{ s}}$$

(b) At beginning of upstroke,

$$f = \omega^2(\rho - R)\cos 0 \ . \qquad . \quad \text{from equation (3.9)}$$
$$= 16\pi^2(0\cdot035 - 0\cdot02)\cos 0 = 2\cdot37 \text{ m/s}^2$$

$$\therefore \ \text{angular acceleration of arm}$$
$$= \frac{2\cdot37}{0\cdot061\,84} = 38\cdot3 \text{ rad/s}^2$$

If F_d is the dynamic reaction between the cam and the arm,

$$F_d \times 0\cdot061\,84 = I_c\alpha = 1\cdot5 \times 0\cdot09^2 \times 38\cdot3$$
$$\therefore \ F_d = 7\cdot53 \text{ N}$$

If F_s is the static reaction between the cam and the arm,

$$F_s \times 0.061\ 84 = 1.5 \times 9.81 \times 0.08 \times \cos 4° 41'$$
$$\therefore F_s = 18.96 \text{ N}$$

$\therefore$ total reaction,

$$F = 18.96 + 7.53 = \underline{26.49 \text{ N}}$$

At end of upstroke,

$$f = -\omega^2 d \cos \phi \quad . \quad . \quad \text{from equation (3.12)}$$
$$= -16\pi^2 \times 0.02 \times \cos 0 = -3.16 \text{ m/s}^2$$

$\therefore$ angular acceleration of arm

$$= \frac{-3.16}{0.057\ 66} = -54.8 \text{ rad/s}^2$$

$$\therefore F_d \times 0.057\ 66 = -1.5 \times 0.09^2 \times 54.8$$
$$\therefore F_d = -11.55 \text{ N}$$
$$F_s \times 0.057\ 66 = 1.5 \times 9.81 \times 0.08 \times \cos 4° 53'$$
$$\therefore F_s = 20.33$$
$$\therefore F = 20.33 - 11.55 = \underline{8.78 \text{ N}}$$

Conditions at the beginning of the downstroke are the same as at the end of the upstroke and conditions at the end of the downstroke are the same as at the beginning of the upstroke.

5. *A lever, shown diagrammatically in Fig. 3.17, swings in a vertical plane through 5° above and 5° below the horizontal. It is actuated by a cam with a circular profile which slides on a flat palm on the lever. The radius*

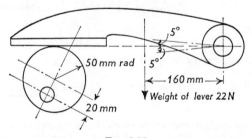

FIG. 3.17

of gyration of the lever about its mass centre is 80 mm. Find (a) the maximum normal reaction between the lever and the cam when the latter rotates at 100 rev/min; (b) the greatest possible speed of the camshaft for contact between cam and lever to be unbroken. (U. Lond.)

The vertical movement of the follower face is 20 mm when the lever turns through 5°. Let the horizontal distance between the cam axis and the lever axis be L.

Then
$$20 = L \tan 5°$$
$$\therefore L = 228 \cdot 6 \text{ mm}$$

Maximum upward acceleration occurs when lever is in lowest position, i.e.
$$f_{max} = \omega^2 d \quad \text{when } \theta = 0 \quad . \text{ from equation (3.17)}$$
$$= \left(100 \times \frac{2\pi}{60}\right)^2 \times 0 \cdot 02 = 2 \cdot 19 \text{ m/s}^2$$

$\therefore$ maximum angular acceleration of lever,
$$\alpha = \frac{2 \cdot 19}{0 \cdot 2286} = 9 \cdot 59 \text{ rad/s}^2$$

If F_d is the dynamic reaction between the cam and the lever, then
$$F_d \times L = I\alpha$$

i.e.
$$F_d \times 0 \cdot 2286 = \frac{22}{9 \cdot 81}(0 \cdot 08^2 + 0 \cdot 16^2) \times 9 \cdot 59$$
$$\therefore F_d = 3 \cdot 01 \text{ N}$$

If F_s is the static reaction between the cam and the lever, then
$$F_s \times 0 \cdot 2286 = 22 \times 0 \cdot 16$$
$$\therefore F_s = 15 \cdot 38 \text{ N}$$
$$\therefore \text{ total reaction} = 3 \cdot 01 + 15 \cdot 38 = \underline{18 \cdot 39 \text{ N}}$$

Maximum downward acceleration occurs when lever is in highest position, i.e. when $\theta = 180°$,
$$\therefore f_{max} = 2 \cdot 19 \text{ m/s}^2 \qquad \text{from equation (3.17)}$$

At N rev/min,
$$F_d = 3 \cdot 01 \times \left(\frac{N}{100}\right)^2$$

Therefore contact ceases at the top position, when
$$3 \cdot 01 \times \left(\frac{N}{100}\right)^2 = 15 \cdot 38 \quad \therefore N = \underline{226 \text{ rev/min}}$$

6. A cam has straight working faces which are tangential to a base circle of diameter D. The follower is a roller of diameter d and the centre of the roller moves along a straight line passing through the centre line of the camshaft. Show that the acceleration of the follower is given by
$$\omega^2 \left(\frac{D+d}{\cos^3 \theta}\right)\left(1 - \frac{\cos^2 \theta}{2}\right)$$
where θ is the angle turned through by the cam, measured from the start of rise of the follower, and ω is the angular velocity.

In such a cam the base circle diameter is 70 mm and the roller diameter 30 mm. The angle between the tangential faces of the cam is 90° and the faces are joined by a nose circle of 8 mm radius. The speed of rotation of the cam is 120 rev/min.

Find the acceleration of the roller centre :

(a) when, during the lift, the roller is just about to leave the straight flank,

(b) when the roller is at the outer end of its lift. (*U. Lond.*)

(*Ans.:* 14·2 m/s² ; 16·05 m/s²)

7. A straight-sided cam has both sides tangential to the base circle which is 25 mm radius and the total angle of action is 120°. A lift of 10 mm is given to a roller 20 mm diameter, the centre of which moves along a straight line passing through the axis of the cam. The camshaft has a speed of 240 rev/min.

Determine (a) the radius of the nose arc ;

(b) the speed of the roller centre when the roller is in contact with the cam at the end of one of the straight flanks adjacent to the nose ;

(c) the greatest acceleration of the roller centre. (*U. Lond.*)

(*Ans.:* 15 mm ; 0·485 m/s ; 36·2 m/s² when follower is about to leave flank [acceleration at peak = 1·9 m/s² ; acceleration on nose after leaving flank = 1·4m/s²])

8. A valve is operated by a cam which has a base circle diameter of 42 mm and a lift of 15 mm ; the cam has tangent flanks and a circular nose, and the total angle of action is 120°. The follower, which has a roller of 18 mm diameter, moves along a straight line passing through the cam axis. Find the maximum load to be exerted by the spring to maintain contact between cam and roller at all times while rotating at 1000 rev/min. The effective mass of the valve, tappet and spring is 0·6 kg. (*U. Lond.*) (*Ans.:* 947 N)

9. A symmetrical cam has a base circle 60 mm radius, arc of action 110°, straight flanks and the tip is a circular arc. The line of action of the follower passes through the centre line of the camshaft. The follower, which has a 40 mm diameter roller, has a lift of 26 mm. Calculate the velocity and acceleration of the follower when moving outwards and contact is just reaching the end of the straight flank, when the cam is rotating at 500 rev/min. (*U. Lond.*)

(*Ans.:* 3·09 m/s ; 460 m/s²)

10. A cam of base circle diameter D has tangent flanks, and operates a follower through a roller of radius R, the path of the roller centre being a straight line passing through the camshaft axis. The follower acts against a spring of stiffness S and initial compression x. The total effective mass of the follower is M and the spring mass is m.

Obtain an expression for the torque exerted on the camshaft when it is rotating at ω rad/s and the cam has turned through an angle θ from the point at which the roller makes contact with the flank. Neglect the effect of friction.

(*U. Lond.*)

$$\left(Ans.: \left\{S\left[x + \left(\frac{D}{2} + R\right)(\sec\theta - 1)\right] + \left[\left(M + \frac{m}{3}\right)\omega^2\left(\frac{D}{2} + R\right)\right.\right.\right.$$
$$\left.\left.\left.(2\sec^3\theta - \sec\theta)\right]\right\}\left(\frac{D}{2} + R\right)\sec\theta\tan\theta\right)$$

11. Fig. 3.18 shows a cam which operates a lever follower, the lever being fitted with a roller which is in continuous contact with the cam profile. The flanks are straight and tangential to the base and nose circles. The cam rotates anticlockwise at a constant speed of 400 rev/min about the centre C.

Determine, graphically or otherwise, the angular velocity and angular acceleration of the lever at the instant when the cam has rotated 100° anticlockwise from the position shown. (*U. Lond.*)

(*Ans.:* 9 rad/s ; 1330 rad/s². See Fig. 3.11 for equivalent mechanism)

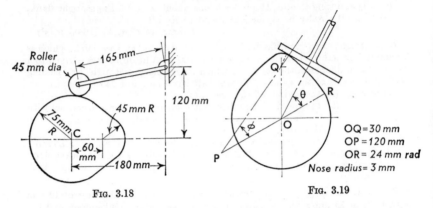

FIG. 3.18   FIG. 3.19

12. The circular arc cam shown in Fig. 3.19 rotates uniformly at 600 rev/min and operates a flat-footed reciprocating follower. The follower retaining spring is to be designed so that at the beginning of the nose radius the spring force is 10% greater than the value required to hold the follower on the cam and in the full lift position the spring force is 5% greater than the required value.

Find the spring stiffness and the spring force in the zero lift position if the total mass of the reciprocating parts is 1·25 kg.

Discuss the disadvantages associated with this type of profile when used at higher speeds and describe how the profile can be modified to overcome these. (*U. Lond.*)   (*Ans. :* 4·4 kN/m ; 116 N)

13. A flat-ended valve tappet is operated by a symmetrical cam with circular arcs for flank and nose profiles ; the straight-line path of the tappet passes through the cam axis. The total angle of action is 150°, the lift is 6 mm, the base circle diameter is 30 mm, and the period of acceleration is half that of deceleration during lift ; the cam rotates at 1250 rev/min.

Determine (*a*) the nose and flank radii ; (*b*) the maximum acceleration and deceleration while lifting. (*U. Lond.*)

(*Ans.:* 9·6 mm ; 35·55 mm ; 352·5 m/s² at start of lift ; 195 m/s² at peak)

14. A cam with convex flanks, operating a flat-ended follower whose lift is 18 mm, has a base circle radius of 36 mm and a nose radius of 9·6 mm. The cam is symmetrical about a line drawn through the centre of curvature of the nose and the centre of the camshaft. If the total angle of cam action is 120°, find the radius of the convex flanks. Determine the maximum velocity and the maximum acceleration and retardation when the camshaft speed is 500 rev/min. (*U. Lond.*)   (*Ans.:* 188 mm ; 1·716 m/s ; 414 m/s² ; 122 m/s²)

15. The profile of a cam consists of circular arcs. The base circle has a radius of 24 mm and is joined to the nose circle by two flank circles of equal radii. The follower moves vertically, has a flat horizontal face, and its rise and fall is completed during 180° of cam rotation. The lift of the follower is 21 mm.

Contact between the cam and the follower is maintained by means of a spring acting on the follower along its line of action. When the camshaft speed is 300 rev/min, the maximum force necessary in the spring is to be 80 N. The follower has a mass of 2·9 kg.

Determine (a) the radii of the nose and flank circles;
(b) the velocity and acceleration of the follower when its velocity is a maximum. (*U. Lond.*)

(*Ans. :* 7·1 mm ; 58 mm ; 0·795 m/s ; 22·5 m/s²)

16. In the mechanism shown in Fig. 3.20, a circular cam C, of 144 mm diameter and eccentricity OQ of 36 mm, rotates at a uniform speed about the axis O. A follower F, of mass 0·8 kg, is pressed against the cam by a spring of stiffness 10 kN/m.

It is found that at a certain speed the follower ceases to have contact with the cam when the latter has moved through 120° from its lowest position. Find that speed and the maximum height reached by the follower above the axis O. The initial compression of the spring (for the cam position shown) is 30 mm. The dead weight of the follower and the mass of the spring may be neglected. (*U. Lond.*)

(*Ans. :* 2305 rev/min ; 84·8 mm)

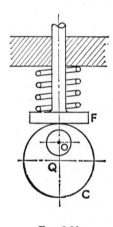

Fig. 3.20

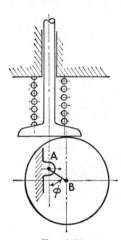

Fig. 3.21

17. Fig. 3.21 shows an eccentric circular cam and its follower. The diameter of the cam is 125 mm and the eccentricity 25 mm. A is the axis of rotation and B is the centre of the cam. The follower has a mass of 5 kg, the spring stiffness is 8·5 kN/m and the spring force is 50 N when the follower is in its lowest position. The camshaft has constant angular speed.

For a camshaft speed of 500 rev/min find the camshaft torque required to overcome inertia of the follower, spring force and gravity, when ϕ is 60°.

What is the maximum camshaft speed if separation of cam from follower is to be avoided ? (*U. Lond.*) (*Ans. :* 8·15 N m ; 619 rev/min)

18. A circular cam, 120 mm diameter, with eccentricity 45 mm, rotates about centre O, as shown in Fig. 3.22, at 100 rev/min in a clockwise direction. It operates a lever follower pivoted at B.

When the cam has rotated 120° clockwise from the position shown, find the angular velocity and angular acceleration of the follower. Graphical methods may be used. (*U. Lond.*) (*Ans.:* 1·95 rad/s ; 14 rad/s²)

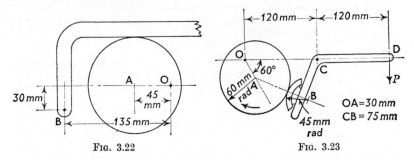

FIG. 3.22　　　　　　　　　　FIG. 3.23

19. Fig. 3.23 shows a circular eccentric, 120 mm in diameter, rotating about an axis O which is offset 30 mm from the centre of the eccentric. The follower, of mass 0·25 kg, has a radius of gyration of 45 mm about its centre of gravity situated at the pivot C. A vertical force P is applied at D. For a cam speed of 120 rev/min, determine the minimum value of P which will ensure contact between the cam and the follower when the angle AOC is 60° (*U. Lond.*)

(*Ans.:* 0·26 N. See Fig. 3.11 for equivalent mechanism)

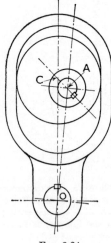

20. The lower shaft, centre O, in Fig. 3.24, is made to oscillate rotationally by means of a cam and slotted link mechanism. The camshaft, centre A, rotates clockwise at a uniform speed of 250 rev/min. OA = 100 mm. The cam is circular, with centre C, and its eccentricity AC = 12 mm. The slotted link, with other parts attached to the lower shaft, has a moment of inertia about O of 1·6 kg m². Friction and gravitational effects may be neglected.

For the position in which angle OAC = 130°, find (a) the angular acceleration of the lower shaft and (b) the driving torque required on the camshaft. (*U. Lond.*) (*Ans.:* 43·8 rad/s² ; 70 N m)

FIG. 3.24

CHAPTER 4

CRANK EFFORT DIAGRAMS

4.1 Crank effort diagrams. The torque of an engine crankshaft varies considerably throughout the working cycle, due to variations in the crank position, the pressure in the cylinder and the inertia force on the piston and connecting rod. If the crankshaft torque is plotted against the crank angle, a crank effort diagram is obtained, Fig. 4.1. The total area under the curve represents the work done by the crankshaft during the cycle.

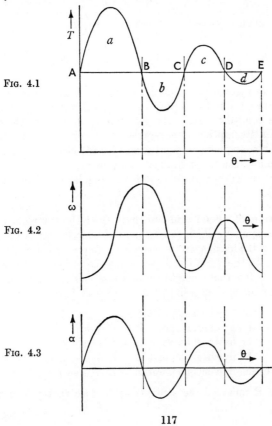

Fig. 4.1

Fig. 4.2

Fig. 4.3

If the resisting torque is constant, this is represented by the line AE, which also represents the mean engine torque. Between points A and B, the engine torque exceeds the resisting torque and the crankshaft accelerates, the area a of the loop representing the excess energy supplied during that time. Similarly, between B and C, the engine torque is less than the resisting torque and the crankshaft decelerates, the area b representing the deficit in energy available during that time. At the points of intersection, A, B, etc., the engine and load torques are equal, so that there is no acceleration or deceleration of the flywheel; hence the speed is a maximum or a minimum at these points.

Fig. 4.2 shows the variation in velocity during the cycle and Fig. 4.3 shows the variation in displacement of the flywheel relative to one rotating uniformly at the mean speed.

Since $\qquad$ angular acceleration of flywheel $= \dfrac{T}{I}$

then $\qquad$ angular velocity of flywheel $= \displaystyle\int \dfrac{T}{I}\, dt$

and $\qquad$ angular displacement of flywheel $= \displaystyle\iint \dfrac{T}{I}\, dt\, dt$

The torque is usually given as a function of θ, which must then be expressed in the form ωt, but to simplify the integration it is necessary to assume ω to be a constant. Since the variation is very small, the error introduced by this assumption is negligible.

For multi-cylinder engines, the total torque for any crankshaft position is the algebraic sum of the torques exerted by the various cranks. These are not equal since the cranks will have different angular positions relative to the crankshaft.

4.2 Fluctuation of speed and energy. Over the complete cycle, the area of the loops above the mean torque line exactly equals the area of the loops below if the mean speed is constant. To prevent large fluctuations of speed due to the variation in net crankshaft torque, however, a flywheel is fitted, its function being to act as a reservoir, absorbing energy as the speed increases and releasing it as the speed falls.

If the energy of the flywheel at point A $= U$
  the energy of the flywheel at point B $= U + a$
  the energy of the flywheel at point C $= U + a - b$
  the energy of the flywheel at point D $= U + a - b + c$
and the energy of the flywheel at point E $= U + a - b + c - d = U$

The energy at E must be the same as at A since the cycle is then repeated.

The points at which the energy is greatest and least can be determined by inspection and these correspond to the points of greatest and least speeds respectively. The difference, or fluctuation, in energy between these points must be equal to the energy stored in the flywheel due to the change in speed.

If ω is the mean speed, ω_1 and ω_2 are the maximum and minimum speeds respectively and E is the work done during the cycle, the *coefficient of fluctuation of speed* is the ratio

$$\frac{\text{greatest fluctuation of speed per cycle}}{\text{mean speed}}$$

i.e.
$$\alpha = \frac{\omega_1 - \omega_2}{\omega} \qquad . \qquad . \qquad . \qquad . \qquad (4.1)$$

and the *coefficient of fluctuation of energy* is the ratio

$$\frac{\text{greatest fluctuation of energy per cycle}}{\text{work done per cycle}}$$

i.e.
$$\beta = \frac{\frac{1}{2}I(\omega_1{}^2 - \omega_2{}^2)}{E} \qquad . \qquad . \qquad . \qquad . \qquad (4.2)$$

If the speed variation is small,

$$\omega_1 + \omega_2 \simeq 2\omega$$

Also
$$\omega_1 - \omega_2 = \alpha\omega$$

$$\therefore \beta = \frac{\frac{1}{2}I.2\omega.\alpha\omega}{E}$$

or
$$I = \frac{\beta E}{\alpha\omega^2} \qquad . \qquad . \qquad . \qquad . \qquad (4.3)$$

1. *A machine running at an average speed of 300 rev/min is driven through a single reduction gear from an engine running at an average speed of 600 rev/min. The moment of inertia of the rotating parts on the machine shaft is equivalent to 110 kg at a radius of 0·3 m and that of the rotating parts on the engine shaft 18 kg at a radius of 0·3 m.*

The torque transmitted to the machine from the engine is $2500 + 675 \sin 2\theta$ N m, where θ is the angle of rotation of the machine from some datum. The torque required to drive the machine is $2500 + 270 \sin \theta$ N m.

Find the coefficient of fluctuation of speed. (U. Lond.)

The torque/crank angle curves for the engine and machine are shown in Fig. 4.4. The engine and resisting torques are equal when

$$2500 + 675 \sin 2\theta = 2500 + 270 \sin \theta$$

i.e. when $5 \sin \theta \cos \theta = \sin \theta$

$\therefore$ either $\sin \theta = 0$ or $\cos \theta = 0\cdot2$

i.e. $\theta = 0, \pi, 2\pi$, etc. and $78° 28', 281° 23'$ etc.

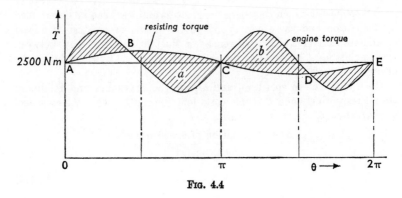

Fig. 4.4

The greatest fluctuation of energy occurs between points B and C, or between C and D, and is represented by the areas a or b,

i.e.   fluctuation of energy $= \displaystyle\int_{78°\ 21'}^{180°} (270 \sin \theta - 675 \sin 2\theta)\, d\theta$

$$= 972 \text{ N m}$$

Equivalent moment of inertia of machine shaft

$$= 110 \times 0{\cdot}3^2 + 18 \times 0{\cdot}3^2 \times \left(\frac{600}{300}\right)^2 = 16{\cdot}38 \text{ kg m}^2$$

$$\therefore 972 = \tfrac{1}{2}I(\omega_1{}^2 - \omega_2{}^2) = \tfrac{1}{2}I \times 2\omega \times (\omega_1 - \omega_2)$$

$$= I\omega^2\!\left(\frac{\omega_1 - \omega_2}{\omega}\right)$$

$$= 16{\cdot}38 \times \left(\frac{2\pi}{60} \times 300\right)^2 \times \alpha$$

$$\therefore \alpha = 0{\cdot}06 \quad \text{or} \quad \underline{6\%}$$

2. *The turning moment diagram for an engine is given by: Torque (N m) = 2100 sin θ + 900 sin 2θ for values of θ, the crank angles, between 0 and π, and by: Torque (N m) = 375 sin θ for values of θ between π and 2π. This is repeated for every revolution of the engine.*

The resisting torque is constant and the speed is 850 rev/min. The total moment of inertia of the rotating parts of the engine and the driven member is 270 kg m². Determine: (i) the power; (ii) the fluctuation in speed; (iii) the maximum instantaneous angular acceleration of the engine, and the value of θ at which it occurs.   (U. Lond.)

Work done per revolution

$$= \int_0^\pi (2100 \sin \theta + 900 \sin 2\theta) \, d\theta + \int_\pi^{2\pi} 375 \sin \theta \, d\theta$$

$$= 3450 \text{ N m}$$

$$\therefore \text{ power} = 3450 \times \left(\frac{2\pi}{60} \times 850 \right) = 307\ 000 \text{ W} \quad \text{or} \quad \underline{307 \text{ kW}}$$

Resisting torque = mean engine torque = $\dfrac{3450}{2\pi} = 549{\cdot}3$ N m

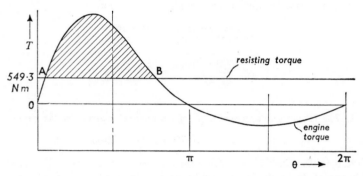

FIG. 4.5

The engine torque/crank angle diagram is shown in Fig. 4.5. The engine and resisting torques are equal when

$$2100 \sin \theta + 900 \sin 2\theta = 549{\cdot}3$$

By trial or plotting, $\theta = 8° 10'$ and $136° 30'$

The greatest fluctuation of energy occurs between points A and B,

i.e. fluctuation of energy $= \displaystyle\int_{8° 10'}^{136° 30'} (2100 \sin \theta + 900 \sin 2\theta) \, d\theta$

$$- 549{\cdot}3(136° 30' - 8° 10') \times \frac{\pi}{180}$$

$$= 2778 \text{ N m}$$

$$\therefore 2778 = \tfrac{1}{2}I({\omega_1}^2 - {\omega_2}^2) = \tfrac{1}{2}I \cdot 2\omega \cdot (\omega_1 - \omega_2)$$

$$= \tfrac{1}{2} \times 270 \times 2 \times 850 \times (N_1 - N_2) \times \left(\frac{2\pi}{60}\right)^2$$

from which $N_1 - N_2 = \underline{1{\cdot}103 \text{ rev/min}}$

It is evident from Fig. 4.5 that the maximum torque occurs within the range 0 to π.

E

For maximum torque, $$\frac{dT}{d\theta} = 0$$

i.e. $$2100 \cos \theta + 1800 \cos 2\theta = 0$$

or $$12 \cos^2 \theta + 7 \cos \theta - 6 = 0$$

from which $$\theta = 61° 45'$$

$$\therefore T_{max} = 2100 \sin 61° 45' + 900 \sin 123° 30'$$

$$= 2600 \text{ N m}$$

$\therefore$ maximum instantaneous angular acceleration

$$= \frac{\text{net accelerating torque}}{I}$$

$$= \frac{2600 - 549 \cdot 3}{270} = \underline{7 \cdot 6 \text{ rad/s}^2}$$

3. *The torque on the crankshaft of an engine is given by the equation:*

$$T(N \text{ m}) = 25\,320 + 12\,600 \sin 2\theta - 15\,650 \cos 2\theta$$

where θ is the crank angle. The resisting torque is uniform. The moment of inertia of the flywheel is 16 000 kg m² and the mean speed of the engine is 150 rev/min.

Calculate :   *(i) the total variation of the energy stored by the flywheel,*

            (ii) the variation of angular velocity of the flywheel,

            (iii) the maximum angular displacement of the flywheel relatively to a disc rotating at a uniform speed of 150 rev/min.   (U. Lond.)

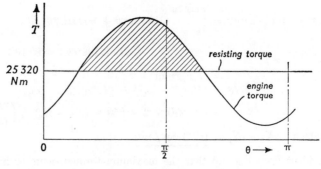

Fɪɢ. 4.6

The torque/crank angle diagram is shown in Fig. 4.6.   The mean torque is 25 320 N m since the areas under the sine and cosine terms between 0 and π are zero.   The terms $12\,600 \sin 2\theta - 15\,650 \cos 2\theta$ may be written

$$20\,090 \sin (2\theta - \phi) \quad \text{where } \phi = \tan^{-1} \frac{15\,650}{12\,600}$$

The engine and resisting torques are equal when $\sin (2\theta - \phi) = 0$

i.e.   when $\theta = \dfrac{\phi}{2}$ and $90° + \dfrac{\phi}{2}$

$\therefore$ maximum fluctuation of energy $= \displaystyle\int_{\frac{\phi}{2}}^{90° + \frac{\phi}{2}} 20\,090 \sin (2\theta - \phi)\, d\theta$

$$= \underline{20\,090 \text{ N m}}$$

$$20\,090 = \tfrac{1}{2} I({\omega_1}^2 - {\omega_2}^2) = \tfrac{1}{2} I . 2\omega . (\omega_1 - \omega_2)$$

$$= \tfrac{1}{2} \times 16\,000 \times 2 \times 150 \times (N_1 - N_2) \times \left(\frac{2\pi}{60}\right)^2$$

from which $N_1 - N_2 = \underline{0.764 \text{ rev/min}}$

Writing $\theta = \omega t$ and neglecting the variation in ω when integrating angular acceleration of flywheel

$$= \frac{T}{I} = \frac{1}{16\,000} \times 20\,090 \sin (2\omega t - \phi)$$

$$= 1.255 \sin (2\omega t - \phi)$$

$\therefore$ angular velocity of flywheel

$$= -\frac{1.255}{2\omega} \cos (2\omega t - \phi) + A$$

$\therefore$ angular displacement of flywheel

$$= -\frac{1.255}{4\omega^2} \sin (2\omega t - \phi) + At + B$$

Whatever initial conditions are assumed for the flywheel will apply also to the disc so that the constants of integration A and B will be the same for each.

Thus the maximum displacement of the flywheel relative to the disc

$$= \frac{1.255}{4 \times \left(150 \times \dfrac{2\pi}{60}\right)^2}$$

$$= 0.001\,274 \text{ rad} = \underline{0.0729°}$$

4. *Fig. 4.7 shows a crank, connecting rod and crosshead. The throw of the crank is 100 mm. The connecting rod has a mass of 18 kg, its length is 400 mm, its centre of gravity is at G, 250 mm from C, and its radius of gyration about an axis through G parallel to the crankshaft is 125 mm. The crosshead has a mass of 9 kg. A spring, of stiffness 2·7 kN/m compression, resists the downward movement of the crosshead. When the crosshead is at the top of its stroke, the force exerted by the spring is 200 N. A flywheel of mass 45 kg, with radius of gyration 150 mm, is keyed to the crankshaft. It is driven at a mean speed of 300 rev/min by an electric motor which provides at all times just enough power to overcome friction.*

Find the coefficient of fluctuation of speed.

Neglect the weight of the crank and that of the spring.

(U. Lond.)

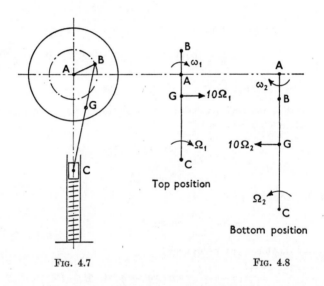

Fig. 4.7   Fig. 4.8

The energy of the system remains constant since the friction losses are made up by the motor. If the angular velocities of the flywheel and connecting rod are ω and Ω respectively, Fig. 4.8, then, at top and bottom positions,

$$\omega \times 100 = \Omega \times 400$$

$$\therefore \Omega = \frac{\omega}{4}$$

Kinetic energy of flywheel

$$= \tfrac{1}{2} \times 45 \times 0 \cdot 15^2 \times \omega^2 = 0 \cdot 506 \omega^2 \text{ J}$$

Kinetic energy of connecting rod

$$= \tfrac{1}{2} \times 18 \times 0.125^2 \times \Omega^2 + \tfrac{1}{2} \times 18 \times (0.25\Omega)^2$$
$$= 0.703\Omega^2 = 0.044\omega^2 \text{ J}$$

Strain energy stored in spring

$$= \tfrac{1}{2}Sx^2 = 1350x^2 \text{ J}$$

$\therefore$ total energy of system

$$= 0.55\omega^2 + 1350x^2 + \text{potential energy of}$$
$$\text{connecting rod and crosshead}$$

At top position,

$$\text{spring compression} = \frac{200}{2700} = 0.074 \text{ m}$$

$$\therefore \text{ total energy} = 0.55\omega_1^2 + 1350 \times 0.074^2 + (9 + 18) \times 9.81 \times 0.2$$
$$= 0.55\omega_1^2 + 60.3 \text{ J}$$

At bottom position,

$$\text{spring compression} = 0.074 + 0.2 = 0.274 \text{ m}$$

$$\therefore \text{ total energy} = 0.55\omega_2^2 + 1350 \times 0.274^2 + 0$$
$$= 0.55\omega_2^2 + 101.4 \text{ J}$$

Equating energies at the extreme positions :

$$0.55\omega_1^2 + 60.3 = 0.55\omega_2^2 + 101.4$$

i.e.
$$0.55(\omega_1^2 - \omega_2^2) = 41.1$$

$$\therefore (\omega_1 + \omega_2)(\omega_1 - \omega_2) = \frac{41.1}{0.55} = 74.6$$

But $\qquad (\omega_1 + \omega_2) = 2(\text{mean angular velocity}) = 2\omega_0$

$\therefore$ coefficient of fluctuation of speed

$$= \frac{\omega_1 - \omega_2}{\omega_0} = \frac{74.6}{2\omega_0^2}$$

$$= \frac{74.6}{2\left(300 \times \dfrac{2\pi}{60}\right)^2} = \underline{0.0378}$$

5. In a six-cylinder petrol engine the accelerating torque in kN m on the crank is given by (a) $Ar(88 \sin 3\theta - 46 \cos 3\theta)$ due to explosion, and (b) $13.5A \sin 3\theta$ due to inertia, where A is the total piston area in m² and r is the crank radius in metres.

If the total piston area is 0.04 m² and the crank radius is 50 mm, find the fluctuation in speed of the flywheel when its mean speed is 1200 rev/min and its moment of inertia is 1.35 kg m². Find also the maximum angle which the flywheel is in advance of an imaginary wheel assumed rotating at constant speed. (*U. Lond.*)   (*Ans.:* 2.27% ; 0.216°)

6. The torque exerted on the crankshaft of an engine is given by the equation
$$T \text{ (N m)} = 10\,500 + 1620 \sin 2\theta - 1340 \cos 2\theta$$
where θ is the crank-angle displacement from the inner dead centre.

Assuming the resisting torque to be constant, determine:

(a) the power of the engine when the speed is 150 rev/min,

(b) the moment of inertia of the flywheel if the speed variation is not to exceed $\pm 0.5\%$ of the mean speed, and

(c) the angular acceleration of the flywheel when the crank has turned through 30° from the inner dead centre. (*U. Lond.*)

(*Ans.:* 165 kW; 851 kg m²; 0·86 rad/s²)

7. The value of the turning moment exerted by a multi-cylinder engine, where θ is a crank angle, is given by $12\,250 - 6000 \sin \theta + 1500 \sin 3\theta$ N m.

Find the radius of gyration of the flywheel, whose mass is $1\frac{1}{4}$ t, if the variation in speed is not to exceed $\pm 1\frac{1}{4}\%$ of the mean speed of 250 rev/min.

Explain a method by which you may determine the angular position of the flywheel relative to one rotating at a constant speed of 250 rev/min. (*U. Lond.*)

(*Ans.:* 0·716 m)

8. The torque T exerted by an engine at any crank angle θ is given by the expression:
$$T = 7500 + 3000 \sin \theta - 1500 \sin 2\theta + 600 \sin 3\theta \text{ N m}$$

The engine works against a uniform resistance, at a mean speed of 160 rev/min. The flywheel has a mass of 1·2 t and a radius of gyration of 1 m.

Confirm that the speed fluctuates cyclically between a minimum at zero crank angle and a maximum at 180° crank angle, and determine the power of the engine and the coefficient of fluctuation of speed.

Describe concisely a method by which the given information may be used to determine the angular position of the flywheel relative to one rotating at a *constant* speed of 160 rev/min. (*U, Lond.*)   (*Ans.:* 126 kW; 1·9%)

9. The torque exerted by a multi-cylinder engine running at a mean speed of 240 rev/min against a uniform resistance, can be expressed by
$$T \text{ N m} = 2500 + 4000 \sin \theta + 600 \sin 2\theta + 60 \sin 3\theta$$
where θ is the angle turned through by the flywheel in time t.

(a) Find the power of the engine, and the minimum mass of the flywheel if its radius of gyration is 0·9 m and the maximum fluctuation of speed is to be $\pm 1\%$.

(b) Explain how, from successive integrations of the intercepted areas of the torque diagram, other curves may be derived showing the variation in speed, and the variation in angular position of the flywheel relative to one of *constant* speed; show particularly how to obtain the scale of each derived curve. (*U. Lond.*)   (*Ans.:* 62·84 kW; 785 kg)

10. A certain machine requires a torque of $(1400 + 200 \sin \theta)$ N m to drive it, where θ is the angle of rotation of its shaft measured from some datum. The machine is direct-coupled to an engine which produces a torque of $(1400 + 250 \sin 2\theta)$ N m. The flywheel and other rotating parts attached to the shaft have a mass of 320 kg, with radius of gyration 400 mm. The mean speed is 160 rev/min.

Calculate: (a) the percentage of fluctuation of speed; (b) the maximum angular acceleration of the flywheel. (*U. Lond.*) (*Ans.:* 3·42%; 7·83 rad/s²)

11. The crank effort of a reciprocating engine is given by the expression $T = 15 + 2.2 \sin 2\theta - 1.8 \cos 2\theta$ kN m, where θ is the angular position of the crank. If the engine runs at a mean speed of 180 rev/min against a constant resisting torque, calculate the power of the engine, and also the least moment of inertia of a flywheel which will limit the fluctuation of speed to ± 0.75 per cent of the mean speed.

Determine the angular acceleration of the flywheel when θ is $45°$. (*U. Lond.*)

(*Ans.*: 283 kW; 533 kg m^2; 4·12 rad/s^2)

12. The crankshaft torque of a multi-cylinder engine is given by Q_E (N m) $= 60 + 8 \sin 3\theta$, where θ is the crankshaft angle measured from a convenient datum. The engine is coupled directly to a machine which requires a torque given by Q_M (N m) $= 60 + 32 \sin \theta$. All the components rotating with the crankshaft together have a mass of 20 kg and a radius of gyration of 200 mm.

If when $\theta = 0$ the crankshaft speed is 300 rev/min, find the speed when $\theta = 60°$. Find also the maximum fluctuation of energy and the maximum angular acceleration of the crankshaft during any one complete cycle of operations. (*U. Lond.*) (*Ans.*: 295·9 rev/min; 58·7 J; 49·9 rad/s^2)

13. A certain engine develops an output torque of $(1600 + 300 \sin 2\theta)$ N m, where θ is the crank angle measured from some datum. This engine drives a machine which requires a driving torque of $(1600 + 170 \sin \theta)$ N m. The rotating parts have a mass of 240 kg, with radius of gyration 0·5 m. The maximum speed of the rotating parts is observed to be 200 rev/min.

Find (*a*) the minimum speed of the rotating parts, (*b*) the coefficient of fluctuation of speed, (*c*) the mean power developed by the engine. (*U. Lond.*)

(*Ans.*: 196·2 rev/min; 0·0192; 33·2 kW)

14. An engine is fitted with a flywheel which has a mass of 22·5 kg and has a radius of gyration of 0·2 m. The load torque is constant and the crankshaft torque can be represented by the expression

$$\text{Torque} = 4a + a \sin \theta + a \sin 2\theta$$

where θ is the crank angle.

When the engine develops 7·5 kW at 600 rev/min, find:

(*a*) the crank angle at which the engine torque is equal to the load torque,

(*b*) the coefficient of cyclical speed fluctuation,

(*c*) the maximum angular acceleration of the flywheel. (*U. Lond.*)

(*Ans.*: 0, 120°, 180°, 240°, 360°; 1·89%; 58·3 rad/s^2)

15. The output torque of a multi-cylinder engine in N m is given by $400 + 240 \sin 3\theta$, where θ is the angle turned by the crank. Sketch the corresponding curve and determine the variation in the kinetic energy of the flywheel. If the mean speed is 1200 rev/min, the total speed variation 20 rev/min and the radius of gyration of the flywheel 200 mm, find its mass. Calculate also the power developed by the engine. (*I. Mech. E.*)

(*Ans.*: 160 J; 15·2 kg; 50·27 kW)

16. The shafts of an electric generator and an electric motor are coupled together and a flywheel is mounted on the common shaft. The generator, which supplies electric power to a rolling mill, absorbs 750 kW from the flywheel shaft during a period of 10 s and 60 kW during the next 15 s, after which the cycle is repeated. The motor drives the flywheel and is adjusted so that the power developed on the shaft is constant. The greatest and least speeds of the shaft are 500 and 400 rev/min. Calculate the mass of the flywheel, which has radius of gyration 1·2 m.

Give expressions for the speed of the flywheel in terms of time during the periods of increasing and decreasing speed, and calculate the greatest angular retardation of the flywheel. (*U. Lond.*)

(*Ans.*: 5830 kg; $400\sqrt{1 + 0.0375t}$ rev/min; $500\sqrt{1 - 0.036t}$ rev/min; 1·18 rad/s^2)

17. The variation in turning moment with crank angle for a six-cylinder engine is shown in Fig. 4.9. The flywheel is equivalent to a mass of 15 kg concentrated at a radius of 0·3 m, and the mean speed of rotation is 1200 rev/min. Sketch qualitatively the speed/crank-angle curve over a revolution and estimate the difference between maximum and minimum speeds.

If the crank-angle base can now be assumed to be a time base and also the speed changes to be linear with time, estimate the angle through which the crank advances and falls back relative to an imaginary shaft rotating uniformly at the mean speed. (*I. Mech.*)   (*Ans.*: 29·45 rev/min ; 0·365°)

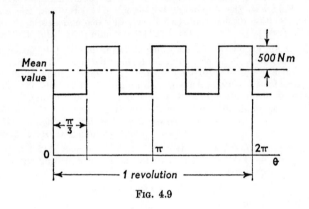

Fig. 4.9

CHAPTER 5

HOOKE'S JOINT

5.1 Velocity and acceleration. A Hooke's joint, or universal coupling, is used to connect two co-planar, non-parallel shafts, as shown in Fig. 5.1, where the axes of the driver A and the follower B are inclined to each other at an angle δ. Each shaft has a semicircular forked end which is connected by pin-joints to a central cross member, as shown.

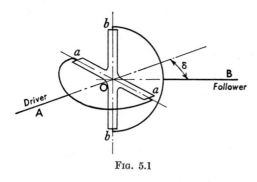

FIG. 5.1

Fig. 5.2 shows the plan and end elevation of the joint, looking along the axis of A, the lines aa and bb representing the forks on A and B respectively. When the shafts are rotated, aa will trace out the circle of that diameter while bb will trace out the ellipse shown dotted.

If shaft A now turns through an angle α from aa to a_1a_1, then the projection of bb will also turn through an angle α to b_1b_1. The angle β turned through by shaft B is found by obtaining the true position of b_1b_1 when looking along the axis of B. Point b_1 is projected into the plan view to give b_1'. This projection is then rabbated into the plane of aa and projected back into the elevation to give point b_2.

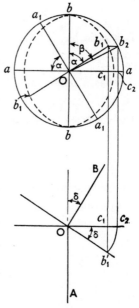

FIG. 5.2

129

If the projections of b_1 and b_2 on aa are c_1 and c_2 respectively, then

$$\tan \alpha = \frac{oc_1}{b_1 c_1} \quad \text{and} \quad \tan \beta = \frac{oc_2}{b_2 c_2} = \frac{oc_2}{b_1 c_1}$$

$$\therefore \frac{\tan \alpha}{\tan \beta} = \frac{oc_1}{oc_2} = \frac{oc_1}{ob_1'} = \cos \delta, \text{ from the plan view}$$

$$\therefore \tan \alpha = \tan \beta \cos \delta \quad . \quad . \quad . \quad . \quad . \quad . \quad (5.1)$$

Differentiating both sides with respect to time,

$$\sec^2 \alpha \frac{d\alpha}{dt} = \sec^2 \beta \frac{d\beta}{dt} \cos \delta$$

i.e.
$$\sec^2 \alpha \, \omega_a = \sec^2 \beta \, \omega_b \cos \delta$$

$$= (1 + \tan^2 \beta) \omega_b \cos \delta$$

$$= \left(1 + \frac{\tan^2 \alpha}{\cos^2 \delta}\right) \omega_b \cos \delta \quad . \quad \text{from equation (5.1)}$$

$$\therefore \omega_a = \frac{(\cos^2 \delta + \tan^2 \alpha)}{\cos \delta} \cos^2 \alpha \, \omega_b$$

$$= \frac{(1 - \sin^2 \delta) \cos^2 \alpha + \sin^2 \alpha}{\cos \delta} \omega_b$$

or
$$\frac{\omega_b}{\omega_a} = \frac{\cos \delta}{1 - \sin^2 \delta \cos^2 \alpha} \quad . \quad . \quad . \quad . \quad (5.2)$$

This ratio has a maximum value of $\dfrac{1}{\cos \delta}$ when $\cos \alpha = \pm 1$,

i.e.   when $\alpha = 0$, $180°$, etc.

and has a minimum value of $\cos \delta$ when $\cos \alpha = 0$,

i.e.   when $\alpha = 90°$, $270°$, etc.

The coefficient of fluctuation of speed

$$= \frac{\dfrac{1}{\cos \delta} - \cos \delta}{1} = \sin \delta \tan \delta \quad . \quad (5.3)$$

The driving and driven shafts have the same speed when

$$\frac{\cos \delta}{1 - \sin^2 \delta \cos^2 \alpha} = 1$$

$$\therefore \cos^2 \alpha = \frac{1 - \cos \delta}{\sin^2 \delta} = \frac{1}{1 + \cos \delta}$$

$$\therefore \tan^2 \alpha = \sec^2 \alpha - 1 = (1 + \cos \delta) - 1$$

$$= \cos \delta$$

i.e.   when $\tan \alpha = \pm \sqrt{\cos \delta} \quad . \quad . \quad . \quad . \quad (5.4)$

There are four values of α during each revolution when the speeds of the driving and driven shafts are equal.

From equation (5.2),

$$\omega_b = \frac{\cos \delta}{1 - \sin^2 \delta \cos^2 \alpha} \omega_a$$

Assuming ω_a and δ to be constants, acceleration of B

$$= \frac{d\omega_b}{dt} = - \frac{\cos \delta \left(\sin^2 \delta \times 2 \cos \alpha \sin \alpha \times \frac{d\alpha}{dt} \right) \omega_a}{(1 - \sin^2 \delta \cos^2 \alpha)^2} \, .$$

$$= -\omega_a^2 \frac{\cos \delta \sin^2 \delta \sin 2\alpha}{(1 - \sin^2 \delta \cos^2 \alpha)^2} \qquad . \qquad . \qquad . \qquad (5.5)$$

For maximum acceleration,

$$\frac{d}{d\alpha} \left\{ \frac{\sin 2\alpha}{(1 - \sin^2 \delta \cos^2 \alpha)^2} \right\} = 0$$

from which $\quad (1 - \sin^2 \delta \cos^2 \alpha) \cos 2\alpha = \sin^2 2\alpha \sin^2 \delta$

i.e. $\quad \{1 - \tfrac{1}{2} \sin^2 \delta (1 + \cos 2\alpha)\} \cos 2\alpha = (1 - \cos^2 2\alpha) \sin^2 \delta$

i.e.

$$2 \cos 2\alpha - \sin^2 \delta \cos 2\alpha - \sin^2 \delta \cos^2 2\alpha = 2 \sin^2 \delta - 2 \sin^2 \delta \cos^2 2\alpha$$

from which $\qquad\qquad\qquad \cos 2\alpha = \frac{\sin^2 \delta (2 - \cos^2 2\alpha)}{2 - \sin^2 \delta}$

For practical values of δ (i.e. up to about 30°), the solution of this equation gives $\alpha \simeq 45°$, for which $\cos^2 2\alpha$ is very small in comparison with 2.

Thus $\qquad\qquad\qquad \cos 2\alpha \simeq \frac{2 \sin^2 \delta}{2 - \sin^2 \delta} \qquad . \qquad . \qquad . \qquad (5.6)$

5.2 Double Hooke's joint. The driven shaft can be made to revolve at the same speed as the driver at all instants by the use of an intermediate shaft and two Hooke's joints, as shown in Fig. 5.3. The two forks at the ends of the intermediate shaft C must lie in the same plane and shafts A and B must be equally inclined to shaft C. There are thus two possible positions for shaft B.

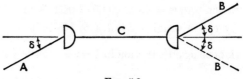

Fɪɢ. 5.3

The relation between ω_a and ω_b may be obtained as follows:

From equation (5.1), $\tan \alpha = \tan \gamma \cos \delta$

where γ is the angle turned through by C.

Also $\tan \beta = \tan \gamma \cos \delta$

$$\therefore \alpha = \beta$$

i.e. A and B both turn through the same angle in the same time, so that

$$\omega_a = \omega_b$$

If the forks on C are set at right angles, then ω_b will fluctuate between

$\omega_a \cos^2 \delta$ and $\dfrac{\omega_a}{\cos^2 \delta}$.

1. *Two shafts are connected by a Hooke's joint of the ordinary form. The angle between the centre lines of the shafts is 20°. One shaft is driven uniformly at 100 rev/min. Calculate the highest and lowest speeds of the second shaft.*

The four pins which connect the two shafts to the centre member are each 10 mm diameter. The driving force on each pin can be considered to be 600 N acting at a distance of 40 mm from the shaft axis. Taking the coefficient of friction on the pins as 0·08, estimate the average efficiency of the joint.

(U. Lond.)

From equation (5.2), maximum value of speed ratio

$$\frac{N_b}{N_a} = \frac{1}{\cos 20°}$$

i.e. $N_b = \dfrac{100}{0·9397} = \underline{106·42 \text{ rev/min}}$

Minimum value of speed ratio

$$\frac{N_b}{N_a} = \cos 20°$$

i.e. $N_b = 100 \times 0·9397 = \underline{93·97 \text{ rev/min}}$

At each pin,   normal force = 600 N

$$\therefore \text{friction force} = 600 \times 0·08 = 48 \text{ N}$$

$$\therefore \text{friction torque} = 48 \times 0·005 = 0·24 \text{ N m}$$

Angle moved by centre member relative to pin in $\frac{1}{2}$ rev = 40°

$$\therefore \text{work done against friction in } \tfrac{1}{2} \text{ rev} = 0·24 \times \left(40 \times \frac{\pi}{180}\right)$$

$$= 0·1675 \text{ N m}$$

$\therefore$ total work done at 4 pins in $\frac{1}{2}$ rev $= 0 \cdot 1675 \times 4$

$$= 0 \cdot 67 \text{ J}$$

Input work done at each pin in $\frac{1}{2}$ rev $= 600 \times \pi \times 0 \cdot 04$

$$= 75 \cdot 4 \text{ J}$$

$\therefore$ total work done by input shaft in $\frac{1}{2}$ rev $= 2 \times 75 \cdot 4$

$$= 150 \cdot 8 \text{ J}$$

$$\therefore \text{ efficiency} = \frac{150 \cdot 8 - 0 \cdot 67}{150 \cdot 8} \times 100$$

$$= \underline{99 \cdot 56 \%}$$

2. *Derive from first principles an expression for the velocity ratio of two shafts connected by a Hooke's joint, giving the expression in terms of the angle δ between the axes of the shafts and the angle of rotation α of one shaft from a given datum.*

Three shafts A, B and C are supported in bearings and are connected end to end, by two Hooke's joints. The axes of A and C are parallel but not in the same line, and the axis of B makes angles of 20° with the axes of A and C. The two joints are arranged in the appropriate relative position so that the rotational speeds of A and C are equal. The moments of inertia of the three shafts, with the rotating parts mounted upon them, are : for A, 0·9 g m², for B, 0·15 g m², for C, 1·2 g m². If the shafts are set in motion and allowed to rotate freely without friction, what is the percentage fluctuation of speed of the shafts A and C ? (U. Lond.)

From equation (5.2),

$$\frac{\omega_b}{\omega_a} = \frac{\cos \delta}{1 - \sin^2 \delta \cos^2 \alpha}$$

Total K.E. of system

$$= \left\{ \frac{0 \cdot 9}{2} \omega_a^2 + \frac{0 \cdot 15}{2} \omega_b^2 + \frac{1 \cdot 2}{2} \omega_c^2 \right\} \times 10^{-3} \text{ J}$$

But $\qquad \omega_b = \omega_a \left(\dfrac{\cos \delta}{1 - \sin^2 \delta \cos^2 \alpha} \right)$

and $\qquad \omega_c = \omega_a$

$\therefore$ total K.E. $= \dfrac{\omega_a^2}{2} \left\{ 0 \cdot 9 + 0 \cdot 15 \left(\dfrac{\cos 20°}{1 - \sin^2 20° \cos^2 \alpha} \right)^2 + 1 \cdot 2 \right\} \times 10^{-3}$

$$= \frac{\omega_a^2}{2} \left\{ 2 \cdot 1 + \frac{0 \cdot 1325}{(1 - 0 \cdot 1169 \cos^2 \alpha)^2} \right\} \times 10^{-3} \text{ J}$$

which remains constant.

$$\therefore \; \omega_a = \frac{c}{\sqrt{\left\{2 \cdot 1 + \dfrac{0 \cdot 1325}{(1 - 0 \cdot 1169 \cos^2 \alpha)^2}\right\}}}$$

where c is a constant.

The maximum value of ω_a occurs when $\alpha = 90°$,

i.e.   maximum $\omega_a = \dfrac{c}{\sqrt{\left\{2 \cdot 1 + \dfrac{0 \cdot 1325}{(1 - 0)^2}\right\}}} = 0 \cdot 6694c$

The minimum value of ω_a occurs when $\alpha = 0$,

i.e.   minimum $\omega_a = \dfrac{c}{\sqrt{\left\{2 \cdot 1 + \dfrac{0 \cdot 1325}{(1 - 0 \cdot 1169)^2}\right\}}} = 0 \cdot 6637c$

Over this small range of speed, the mean speed may be taken as

$$\tfrac{1}{2}(0 \cdot 6694c + 0 \cdot 6637c) = 0 \cdot 666\,55c$$

$$\therefore \text{ percentage fluctuation of speed} = \frac{0 \cdot 6694c - 0 \cdot 6637c}{0 \cdot 666\,55c} \times 100$$

$$= \underline{0 \cdot 855}$$

The small fluctuation is due to the small inertia of shaft B in relation to that of shafts A and C.

3. *A Hooke's joint is used to connect two non-parallel intersecting shafts, the axes of which are inclined at 30°. The driving shaft runs at a uniform speed of 250 rev/min and the driven shaft carries a rotor of moment of inertia 1·25 kg m².*

Working from first principles, find the torque on the driving shaft due to the acceleration of the driven shaft at the instant when the acceleration is a maximum. (U. Lond.)

For maximum acceleration,

$$\cos 2\alpha = \frac{2 \sin^2 \delta}{2 - \sin^2 \delta} \qquad . \qquad \text{from equation (5.6)}$$

$$= \frac{2 \sin^2 30°}{2 - \sin^2 30°} = \frac{2}{7}$$

$$\therefore \; 2\alpha = 73° \; 24'$$

$$\therefore \; \alpha = 36° \; 42'$$

$$\frac{d\omega_b}{dt} = \omega_a^2 \frac{\cos \delta \sin^2 \delta \sin 2\alpha}{(1 - \sin^2 \delta \cos^2 \alpha)^2} \qquad . \qquad \text{from equation (5.5)}$$

∴ maximum acceleration of B

$$= \left(250 \times \frac{2\pi}{60}\right)^2 \frac{\cos 30° \sin^2 30° \sin 73° 24'}{(1 - \sin^2 30° \cos^2 36° 42')^2}$$

$$= 202 \text{ rad/s}^2$$

$$\therefore T_b = 1{\cdot}25 \times 202 = 252{\cdot}5 \text{ N m}$$

$$T_a \omega_a = T_b \omega_b$$

$$\therefore T_a = T_b \times \frac{\cos \delta}{1 - \sin^2 \delta \cos^2 \alpha} \qquad . \qquad \text{from equation (5.2)}$$

$$= 252{\cdot}5 \times \frac{\cos 30°}{1 - \sin^2 30° \cos^2 36° 42'}$$

$$= \underline{261 \text{ N m}}$$

4. *Fig. 5.4 shows an arrangement to test a torsional damper, one side of which is connected to a shaft which rotates uniformly at 1000 rev/min. The other side of the damper is connected by means of a Hooke's joint to another*

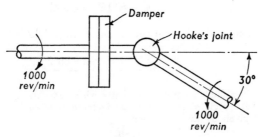

Fig. 5.4

shaft also rotating uniformly at 1000 rev/min. Both shafts rotate in the same direction and the angle between them is 30°.

The damper is of the friction type and produces a constant torque of 15 N m. Determine the maximum angular velocity of one half of the damper relative to the other. Find also the maximum relative angular displacement and hence the total work done on the damper during one revolution of the driving shaft.

Derive any formulae which you use. (U. Lond.)

The maximum relative velocity of one half of the damper to the other

$$= 1000\left(\frac{1}{\cos 30°} - 1\right) = \underline{154{\cdot}7 \text{ rev/min}}$$

The maximum displacement of one half of the damper relative to the other will occur when the two sides have the same speed, i.e.   when $\tan \alpha = \sqrt{\cos \delta}$.   from equation (5.4)

$$= \sqrt{\cos 30°} = 0.9306$$

i.e.   when $\alpha = 42° 56'$

$$\tan \beta = \frac{\tan \alpha}{\cos \delta} \quad . \quad . \quad \text{from equation (5.1)}$$

$$= \frac{0.9306}{0.866} = 1.075$$

$$\therefore \beta = 47° 4'$$

$\therefore$ relative angular displacement $= 47° 4' - 42° 56'$

$$= \underline{4° 8'}$$

There are four points in the revolution at which the speeds are equal, i.e. $\alpha = \pm 42° 56'$ and $180° \pm 42° 56'$, and in moving between two consecutive positions, the relative movement between the two sides is $2 \times 4° 8' = 8° 16'$. The total relative movement between the two sides during a complete revolution is therefore

$$4 \times 8° 16' = 33° 4'$$

$\therefore$ work done in damper per revolution $= 15 \times \left(33.067 \times \dfrac{\pi}{180} \right)$

$$= \underline{8.65 \text{ J}}$$

5. If a Hooke's joint connects two shafts whose axes are inclined to each other at an acute angle α, show that the instantaneous speed ratio between these shafts is given by the expression

$$\frac{\sec^2 \theta}{\sec^2 \phi \cos \alpha}$$

where θ and ϕ are the angles through which the two halves of the joint have turned respectively from some datum. State what this datum is.

Determine the maximum and minimum values of this ratio when $\alpha = 10°$. (*U. Lond.*)   (*Ans.:* 1.0154 ; 0.9848)

6. A Hooke's joint is used to couple two shafts together. The driving shaft rotates at a uniform speed of 1000 rev/min. Working from first principles, determine the greatest permissible angle between the shaft axes so that the total fluctuation of speed of the driven shaft does not exceed 150 rev/min. What will then be the maximum speed of the driven shaft ? (*U. Lond.*)

(*Ans.:* 21.9° ; 1078 rev/min)

7. Two shafts are to be connected by a Hooke's joint ; the driving shaft rotates at a uniform speed of 500 rev/min, and the speed of the driven shaft must lie between 475 and 525 rev/min.

Determine, from first principles, the maximum permissible angle between the shafts, and prove that by a suitable arrangement of two such joints the driven and driving shafts can have the same rotational speeds. (*U. Lond.*)

(*Ans.:* 18°)

8. A driving shaft having a uniform speed of 300 rev/min is coupled to a driven shaft by a Hooke's joint. If the speed of the driven shaft must always be between 315 and 285 rev/min, find the greatest permissible angle between the shafts. What will then be the actual maximum and minimum speeds of the driven shaft ? Any formula used must be proved. (*U. Lond.*)

(*Ans.:* 17° 46′ ; · 315 rev/min ; 285·7 rev/min)

9. Two horizontal shafts A and B, whose axes intersect and are inclined at an angle α, are connected by a Hooke's joint. If A is the driving shaft and rotates at a uniform angular speed of ω rad/s, find an expression for the angular speed of the shaft B when one of the cross arms on A has turned through an angle θ from its vertical position.

Hence determine for the case when $\alpha = 30°$ the maximum and minimum velocity ratios and the ratio of the fluctuation of the speed of B to the mean speed. (*U. Lond.*) (*Ans.:* 1·1547 ; 0·866 ; 0·2887)

10. In a single Hooke's joint where the angle between the shafts is θ, show that the ratio of the fluctuation of speed to the mean speed is $\sin \theta \tan \theta$ if the angular velocity of the driving shaft is constant and the axes of the pins in the joint intersect.

When a double coupling is used, explain what conditions are necessary for the driven shaft to have uniform angular velocity if that of the driving shaft is constant.

In such a double coupling, the driving and driven shafts are parallel and the angle between each and the intermediate shaft is 20°. Find the maximum and minimum velocities of the driven shaft if the axis of the driving pin carried by the intermediate shaft has inadvertently been placed 90° in advance of the correct position. The driving shaft rotates uniformly at 200 rev/min. (*U. Lond.*)

(*Ans.:* 226·3 rev/min ; 176·7 rev/min)

11. Show that, for two shafts connected by a Hooke's joint, the ratio of the angular velocities is given by

$$\frac{\dot{\omega}_2}{\omega_1} = \frac{\cos \alpha}{1 - \sin^2 \theta \sin^2 \alpha}$$

where θ is the angle of rotation of shaft 1 from the position where the forked end is perpendicular to the plane containing the shaft axes, and α is the angle of deviation of the drive.

Two shafts A and B are connected by a Hooke's joint, the angle of deviation being 20°. Masses on shafts A and B have moments of inertia of 3 and 6 g m² respectively. When the masses are set in motion and allowed to rotate freely without friction, find the fluctuation of speed of shaft A, as a percentage of its mean speed. (*U. Lond.*) (*Ans.:* 8·3%)

12. Two shafts, the axes of which intersect but are inclined at 20° to each other, are connected by a Hooke's joint. If the driving shaft has a uniform speed of 1000 rev/min, find from first principles the variation in speed of the driven shaft.

The driven shaft carries a rotating mass which has a mass of 15 kg and a radius of gyration of 250 mm. Find the accelerating torque on the driven shaft for the position when the driven shaft has turned 45° from the position in which its fork end is in the plane containing the two shafts. (*U. Lond.*)

(*Ans.:* 126·5 rev/min ; 1277 N m)

13. For two shafts connected by a Hooke's joint, show that, if shaft 1 has uniform angular velocity ω_1, the angular acceleration of shaft 2 is given by

$$\dot{\omega}_2 = -\frac{\omega_1{}^2 \cos \alpha \sin^2 \alpha \sin 2\theta_1}{(1 - \sin^2 \alpha \cos^2 \theta_1)^2}$$

where θ_1 is the angle of rotation of shaft 1 from the position where its forked end is in the plane containing the shaft axes and α is the angle of deviation of the drive.

In a particular case, shaft 1 is driven at a constant speed of 500 rev/min. Shaft 2 carries a rotor, of moment of inertia 1 kg m², and is subjected to a constant resisting torque of 275 N m. Find the torque input to shaft 1 at the instant when $\theta_1 = 30°$, $\alpha = 25°$. (*U. Lond.*)   (*Ans.:* -248 N m)

14. Two parallel shafts are connected by an intermediate shaft with a Hooke's joint at each end. Show how the joints should be arranged to obtain a constant velocity ratio between the driving and driven shafts.

The intermediate shaft has a moment of inertia of 3 g m² and is inclined at 30° to the axes of the driving and driven shafts. If the driving shaft rotates uniformly at 2400 rev/min with a steady input torque of 300 N m, determine the maximum fluctuation of the output torque.

(The approximate expression $\cos 2\theta = \dfrac{2 \sin^2 \alpha}{2 - \sin^2 \alpha}$ may be used to find the angle at which the maximum acceleration occurs, but all other formulae should be derived.) (*U. Lond.*)   (*Ans.:* $\pm57\cdot5$ N m)

15. Prove the relation for a Hooke's universal coupling:

$$\omega_2 = \frac{\omega_1 \cos \delta}{1 - \cos^2 \theta \sin^2 \delta}$$

and deduce the acceleration of the driven shaft; ω_1 and ω_2 are the speeds of the driving shaft (assumed constant) and driven shaft respectively, θ is the angular displacement of the driving shaft (from the position where the plane containing the arms is perpendicular to that containing the axes of the two shafts), and δ is the divergence of the driven shaft.

If the angle of divergence is $22\frac{1}{2}°$ and a steady torque of 250 N m is applied to the driving shaft while it is rotating at 120 rev/min, what must be the mass of the flywheel (radius of gyration 250 mm) attached to the driven shaft, if the output torque does not vary by more than $\pm25\%$?

The value of θ for maximum acceleration of the driven shaft may be taken from

$$\cos 2\theta = \frac{2 \sin^2 \delta}{2 - \sin^2 \delta}.$$

(*U. Lond.*)   (*Ans.:* Maximum mass = 7·4 kg)

CHAPTER 6

GOVERNORS

6.1 Function of a governor. The function of an engine governor is to control the mean speed, as distinct from that of a flywheel which controls only cyclic fluctuations in speed.   If the mean speed varies due to a variation in the load, the governor adjusts the fuel supply to the engine and restores the speed to its former value.

6.2 Centrifugal governors. In a centrifugal governor, the effect of centrifugal force on the rotating balls causes the sleeve to rise until equilibrium is obtained.   The sleeve controls the fuel supply by means of a linkage and any change in speed produces a change in the sleeve position, which adjusts the fuel supply accordingly.

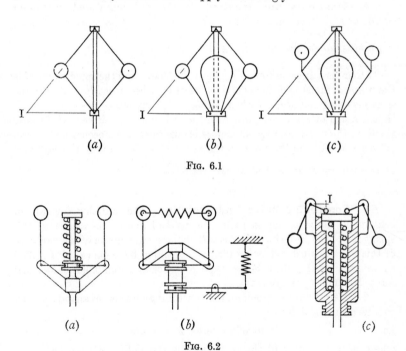

(a) (b) (c)

Fig. 6.1

(a) (b) (c)

Fig. 6.2

139

Centrifugal governors may be divided into (a) dead weight governors, such as the Watt, Porter and Proell types shown in Fig. 6.1(a), (b) and (c) respectively, and (b) spring-controlled governors as shown in Fig. 6.2(a), (b) and (c), the former being known as the Hartnell governor.

In the dead weight types, an equation of equilibrium is obtained by taking moments of forces about the instantaneous centre, I, of the lower link, thus eliminating the tension in the upper link and the side thrust at the sleeve. In the spring-controlled types shown in Fig. 6.2(a) and (b), moments are taken about the fulcrum of the bell-crank levers, eliminating the reaction at that point, and in the type shown in Fig. 6.2(c), moments are taken about the instantaneous centre, I, for the upper bell-crank arm to eliminate the reaction between the roller and the top of the spindle.

6.3 Effort and power.

The *effort* of a governor is the force exerted at the sleeve for a given fractional change in speed.

The *power* of a governor is the work done at the sleeve for a given fractional change in speed,

i.e.   power = mean effort × sleeve movement

6.4 Sensitivity and friction.

If the maximum and minimum speeds of a governor are ω_1 and ω_2 respectively and its mean speed is ω, the *sensitivity* of the governor is defined as $\dfrac{\omega}{\omega_1 - \omega_2}$.

If there is a friction force f between the sleeve and the spindle and M is the mass at the sleeve, the effective sleeve load becomes $Mg + f$ when the sleeve is rising and $Mg - f$ when falling. For any sleeve position, there is thus a range of speed over which the governor is insensitive. If, for any position, the maximum speeds before sleeve movement occurs are ω' and ω'' respectively and the speed in the absence of friction is ω, the *coefficient of insensitiveness* is defined as the ratio $\dfrac{\omega' - \omega''}{\omega}$.

6.5 Controlling force and stability.

The radially inward, or centripetal, force acting on each rotating ball due to the sleeve weight, spring force, etc., is termed the *controlling force* and a graph of the variation of this force against radius is called a controlling force curve. Fig. 6.3(a) shows such a curve for a Porter type governor and Fig. 6.3(b) shows that for a Hartnell type governor.

At any equilibrium speed, ω, the controlling force is equal and opposite to the centrifugal force,

i.e.   controlling force, $F = m\omega^2 r$

where m is the mass of the ball and r is the radius of rotation.

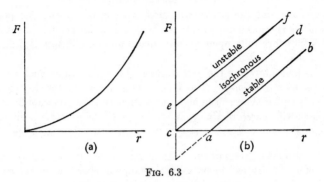

FIG. 6.3

A governor is *stable* if r increases as ω increases. But

$$\omega = \sqrt{\frac{F}{mr}}$$

i.e. for stability, $\dfrac{F}{r}$ must increase as r increases, i.e. the slope of the controlling force curve, $\dfrac{dF}{dr}$, must be greater than the mean slope, $\dfrac{F}{r}$, Fig. 6.4,

i.e.
$$\frac{dF}{dr} > \frac{F}{r} \; . \qquad . \qquad . \qquad . \qquad . \qquad (6.1)$$

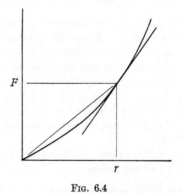

FIG. 6.4

This condition is satisfied in the case of the Porter governor curve, but for the Hartnell governor, it is only satisfied if the straight line intercepts the vertical axis below the origin, such as ab.

If the line passes through the origin, such as cd, F/r is a constant and the equilibrium speed is the same for all radii of rotation. The governor is *isochronous* and the sleeve will move to one of its extreme positions immediately the speed deviates from the isochronous speed. The fuel inlet valve is therefore fully open or fully closed and equilibrium is unlikely to be achieved.

If the line intercepts the vertical axis above the origin, such as *ef*, *F/r* decreases as *r* increases. The equilibrium speed at the lowest sleeve position is higher than at the highest position and the governor is *unstable*. The sleeve will therefore always move to one of the extreme positions and the engine speed will not settle between the two limiting speeds.

6.6 Inertia governors. An inertia governor is operated by *rate of change* of speed, instead of by a finite change of speed, as in the case of a centrifugal governor and is therefore more sensitive than the centrifugal type. Fig. 6.5 shows a very simple form of inertia governor in which an arm, carrying masses at each end, is pivoted on the engine axis; when running at a uniform speed, the arm rotates with the shaft, but immediately there is a change in speed, inertia forces, *P*, act on the masses in a tangential direction and cause the arm to rotate relative to the shaft against the action of a torsion spring. This action operates a valve which adjusts the fuel supply to the engine.

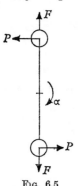
FIG. 6.5

In more advanced types of governor, the operating torque depends on the direction of rotation and on whether the shaft is accelerating or de-celerating.

1. *A spring-loaded governor of the Hartnell type has arms of equal length. The masses rotate in a circle of 160 mm diameter when the sleeve is in its mid-position and the mass arms are vertical. The equilibrium speed for this position is 450 rev/min, neglecting friction. The maximum sleeve movement is to be 40 mm, and the maximum variation of speed (allowing for friction) is to be ±5 per cent of the mid-position speed. The mass of the sleeve is 3·4 kg and friction may be considered equivalent to 27 N at the sleeve. The power of the governor must be sufficient to overcome the friction by a 1 per cent change of speed at mid-position.*

Determine the magnitude of each rotating mass, the spring stiffness, and the initial compression of the spring. (U. Lond.)

When the sleeve is rising, effective sleeve load $= 3\cdot4 \times 9\cdot81 + 27$ N
When the sleeve is falling, effective sleeve load $= 3\cdot4 \times 9\cdot81 - 27$ N

If P is the spring force and F the centrifugal force, then, taking moments about the fulcrum of the bell-crank lever, Fig. 6.2(*a*):

$$F = \tfrac{1}{2}(P + 33\cdot4 \pm 27)$$

since the arms are of equal length.

i.e.
$$m\left(N \times \frac{2\pi}{60}\right)^2 r = \tfrac{1}{2}(P + 33\cdot4 \pm 27)$$

$$\therefore\ mN^2r = 45\cdot6(P + 33\cdot4 \pm 27) \quad . \qquad . \qquad (1)$$

In mid-position, neglecting friction,

$$m \times 450^2 \times 0\cdot08 = 45\cdot6(P + 33\cdot4)$$

$$\therefore\ m = 0\cdot002\ 81(P + 33\cdot4) \qquad . \qquad . \qquad (2)$$

The sleeve is about to move up from the mid-position when the speed rises by 1%.

Therefore, from equation (1),

$$m \times (450 \times 1\cdot01)^2 \times 0\cdot08 = 45\cdot6(P + 60\cdot4)$$

$$\therefore\ m = 0\cdot002\ 76(P + 60\cdot4) \qquad . \qquad . \qquad (3)$$

Therefore, from equations (2) and (3),

$$m = \underline{4\cdot18\ \text{kg}}$$

At top of movement, $N = 472\cdot5$ rev/min and $r = 0\cdot1$ m

$$\therefore\ 4\cdot18 \times 472\cdot5^2 \times 0\cdot1 = 45\cdot6(P_1 + 60\cdot4)$$

$$\therefore\ P_1 = 1990\ \text{N}$$

At bottom of movement, $N = 427\cdot5$ rev/min and $r = 0\cdot06$ m

$$\therefore\ 4\cdot18 \times 427\cdot5^2 \times 0\cdot06 = 45\cdot6(P_2 + 6\cdot4)$$

$$\therefore\ P_2 = 999\ \text{N}$$

$$\therefore\ \text{spring stiffness} = \frac{1990 - 999}{0\cdot04}$$

$$= 24\ 750\ \text{N/m} \quad \text{or} \quad \underline{24\cdot75\ \text{kN/m}}$$

$$\text{Initial compression} = \frac{999}{24\ 750} = 0\cdot040\ 35\ \text{m} \quad \text{or} \quad \underline{40\cdot35\ \text{mm}}$$

2. *In a spring-controlled governor of the Hartnell type, the ball and sleeve arms of the bell-crank levers are at right angles, and are of length a and b respectively. When the ball arms are parallel to the spindle, the radius of rotation of the balls is r_0 and the central spring is compressed through a distance x from its unstrained position. Derive an expression for the controlling force at any other radius of rotation r, if the spring stiffness is S. Hence show that the governor is stable if*

$$x < \frac{b}{a}.r_0$$

In a given Hartnell governor, the mass of each rotating ball is 2 kg, a = 125 mm, b = 75 mm, $r_0 = 100$ mm. If the equilibrium speed at this radius is 250 rev/min and an increase in speed of 8 per cent is to correspond to a sleeve movement of 12 mm, determine the stiffness of the spring and the initial spring load. (U. Lond.)

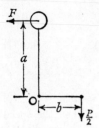

FIG. 6.6

If the ball radius increases from r_0 to r, Fig. 6.6,

$$\text{compression of spring} = \frac{b}{a}(r - r_0)$$

$$\therefore \text{spring force, } P = \left\{ x + \frac{b}{a}(r - r_0) \right\} S$$

Taking moments about the fulcrum O,

$$F \times a = \frac{P}{2} \times b \quad . \quad . \quad . \quad (1)$$

$$\therefore \text{controlling force on ball} = F = \left\{ x + \frac{b}{a}(r - r_0) \right\} \frac{S}{2} \cdot \frac{b}{a}$$

$$\therefore \frac{F}{r} = \left\{ \frac{x}{r} + \frac{b}{a}\left(1 - \frac{r_0}{r} \right) \right\} \frac{S}{2} \cdot \frac{b}{a}$$

and

$$\frac{dF}{dr} = \frac{S}{2} \cdot \frac{b^2}{a^2}$$

For stability,

$$\frac{dF}{dr} > \frac{F}{r} \quad . \qquad \text{from equation (6.1)}$$

i.e.

$$\frac{S}{2} \cdot \frac{b^2}{a^2} > \left\{ \frac{x}{r} + \frac{b}{a}\left(1 - \frac{r_0}{r} \right) \right\} \frac{S}{2} \cdot \frac{b}{a}$$

from which

$$\underline{x < \frac{b}{a} r_0}$$

At 250 rev/min,

$$2\left(250 \times \frac{2\pi}{60} \right)^2 \times 0 \cdot 1 \times 0 \cdot 125 = \frac{P_1}{2} \times 0 \cdot 075 \quad \text{from equation} \quad (1)$$

$$\therefore \text{initial load} = P_1 = \underline{457 \text{ N}}$$

After an 8% increase in speed, $N = 270$ rev/min

and

$$r = 0 \cdot 1 + 0 \cdot 012 \times \frac{0 \cdot 125}{0 \cdot 075} = 0 \cdot 12 \text{ m}$$

$$\therefore 2\left(270 \times \frac{2\pi}{60} \right)^2 \times 0 \cdot 12 \times 0 \cdot 125 = \frac{P_2}{2} \times 0 \cdot 075$$

$$\therefore P_2 = 640 \text{ N}$$

$$\therefore \text{stiffness} = \frac{640 - 457}{0 \cdot 012} = 15\,250 \text{ N/m} \quad \text{or} \quad \underline{15 \cdot 25 \text{ kN/m}}$$

3. *Fig. 6.7 shows diagrammatically a governor in which the two control springs are directly connected between the balls: further particulars are as follows: tension in each spring = 300 N for position shown; stiffness of each spring = 3·5 kN/m; mass of each spring = 1·4 kg; mass of each ball = 3·6 kg; mass of each bell-crank = 0·9 kg; radius of gyration of each bell-crank about fulcrum = 40 mm; mass of sleeve = 2·4 kg; mass of lever = 3·2 kg; radius of gyration of lever about fulcrum = 135 mm; mass of operating link = 4·5 kg.*

Find the speed of the governor under these conditions and estimate the frequency of the vibrations resulting if the governor is slightly disturbed when running at this speed. (U. Lond.)

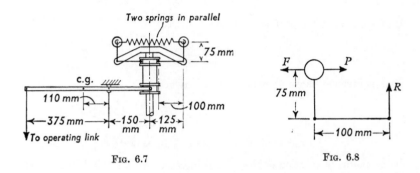

FIG. 6.7   FIG. 6.8

Let F be the centrifugal force on the ball, P the spring force and R the resultant force on the end of the horizontal arm of the bell-crank lever, Fig. 6.8.

Then $P = 2 \times 300 = 600$ N

and $R = \frac{1}{2}\left\{3\cdot2 \times 9\cdot81 \times \dfrac{0\cdot110}{0\cdot150} + 4\cdot5 \times 9\cdot81 \times \dfrac{0\cdot375}{0\cdot150} - 2\cdot4 \times 9\cdot81\right\}$

$= 54\cdot9$ N (upward)

Taking moments about the fulcrum:

$$R \times 0\cdot1 = (P - F) \times 0\cdot075$$

i.e. $\qquad 54\cdot9 \times 0\cdot1 = (600 - F) \times 0\cdot075$

$$\therefore F = 526\cdot8 \text{ N} = 3\cdot6\left(N \times \frac{2\pi}{60}\right)^2 \times 0\cdot125$$

from which $\qquad N = \underline{\underline{327 \text{ rev/min}}}$

Total mass of springs $= 2 \cdot 8$ kg

$\therefore$ equivalent mass of springs attached to each ball

$$= \frac{1}{3} \times \frac{2 \cdot 8}{2} = 0 \cdot 467 \text{ kg}$$

Equivalent mass of bell-crank referred to ball

$$= 0 \cdot 9 \times \left(\frac{0 \cdot 04}{0 \cdot 075} \right)^2 = 0 \cdot 256 \text{ kg}$$

Half equivalent mass of sleeve referred to ball

$$= \frac{2 \cdot 4}{2} \times \left(\frac{0 \cdot 1}{0 \cdot 075} \right)^2 = 2 \cdot 133 \text{ kg}$$

Half equivalent mass of lever referred to ball

$$= \frac{3 \cdot 2}{2} \times \left(\frac{0 \cdot 135}{0 \cdot 15} \times \frac{0 \cdot 1}{0 \cdot 075} \right)^2 = 2 \cdot 305 \text{ kg}$$

Half equivalent mass of operating link referred to ball

$$= \frac{4 \cdot 5}{2} \times \left(\frac{0 \cdot 375}{0 \cdot 15} \times \frac{0 \cdot 1}{0 \cdot 075} \right)^2 = 25 \cdot 0 \text{ kg}$$

$\therefore$ total equivalent mass of ball $= 30 \cdot 16$ kg

If the radius of rotation of the balls is increased by a small distance x m,

$$\text{increase in centrifugal force} = 3 \cdot 6 \left(316 \times \frac{2\pi}{60} \right)^2 x = 3945x \text{ N}$$

and $\qquad$ increase in spring force $= 2x \times 3 \cdot 5 \times 10^3 = 7000x$ N

$\therefore$ net restoring force $= 7000x - 3945x = 3055x$ N

$$\therefore 3055x = 30 \cdot 16f$$

$$\therefore \frac{f}{x} = 100 \cdot 1$$

$$\therefore \text{frequency} = \frac{1}{2\pi} \sqrt{100 \cdot 1} = \underline{1 \cdot 59 \text{ Hz}}$$

NOTE. For equivalence, a mass referred to the ball must have the same K.E. at any instant of the vibration as it possesses in its actual position and must therefore be multiplied by the ratio $\left(\dfrac{\text{mass velocity}}{\text{ball velocity}} \right)^2$. In the case of a body having angular motion, the mass may be regarded as concentrated at the radius of gyration.

4. *In the governor shown in Fig. 6.9 the pivots for the bell-crank levers are carried by the sleeve, which is capable of an axial movement relatively to the governor spindle A which has a cap fixed to its upper end. The spring is compressed between the sleeve and the cap, and the outward movement of the balls with increase of speed causes the sleeve to be raised by the compression of the spring due to pressure of the roller on the ends of the short arms of the bell-crank lever.*

If the mass of the sleeve is 13·5 kg and of each ball is 2·7 kg, and the minimum radius of rotation of the balls is 100 mm, find the initial compression in the spring and also the stiffness of the spring in order that the sleeve shall begin to rise at 240 rev/min and rise 6 mm when the speed increases to 264 rev/min. (U. Lond.)

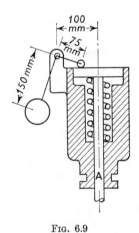

FIG. 6.9

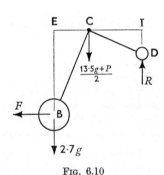

FIG. 6.10

The bell-crank lever BCD, Fig. 6.10, is in equilibrium under the centrifugal force, F, the weight of the ball, half the spring force, P, half the sleeve weight and the reaction, R, between the top of the spindle and the roller.

Taking moments about I to eliminate the reaction R :

$$F \times BE = 2·7 \times 9·81 \times EI + \frac{13·5 \times 9·81 + P}{2} \times CI$$

i.e. $2·7\left(N \times \frac{2\pi}{60}\right)^2 r \times BE = 26·5 \times EI + \frac{132·5 + P}{2} \times CI$

i.e. $0·0294N^2r \times BE = 26·5 \times EI + \frac{132·5 + P}{2} \times CI$

At 240 rev/min, $r = 0.1$ m, BE $= 0.15$ m, CI $=$ EI $= 0.075$ m

$\therefore\ 0.0294 \times 240^2 \times 0.1 \times 0.15$

$$= 26.5 \times 0.075 + \frac{132.5 + P_1}{2} \times 0.075$$

from which $P_1 = 491.5$ N

At 264 rev/min,   DI $= 0.006$ m, CI $= 0.0747$ m, EI $= 0.0867$ m,

BE $= 0.1493$ m, and $r = 0.112$ m

$\therefore\ 0.0294 \times 264^2 \times 0.112 \times 0.1493$

$$= 26.5 \times 0.0867 + \frac{132.5 + P_2}{2} \times 0.0747$$

from which $P_2 = 724$ N

$$\therefore\ \text{spring stiffness, } S = \frac{724 - 491.5}{0.006} = 38\ 750 \text{ N/m}$$

$$= 38.75 \text{ kN/m}$$

$$\text{Initial compression} = \frac{P_1}{S} = \frac{491.5}{38\ 750} = 0.012\,66 \text{ m} = \underline{12.66 \text{ mm}}$$

5. *Fig. 6.11 shows an arrangement for a speed indicator. AB is a uniform rod 75 mm long, having a mass of 0·1 kg, which is pivoted at the middle of its length C to a vertical spindle DE, which is rotated about its axis by the machine whose speed is to be measured.   The rod AB is controlled by torsion springs F and F'.   The angle α is to vary from 30° to 60° whilst the speed of DE varies from 200 to 600 rev/min.*

Determine the controlling couple which must be exerted by the springs for the extreme positions of AB and find the value of α corresponding to a speed of 400 rev/min.   (U. Lond.)

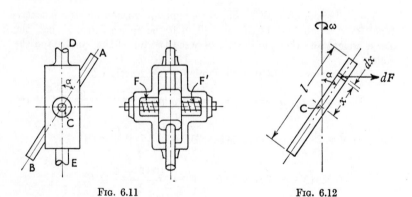

FIG. 6.11   FIG. 6.12

If m is the mass of the rod per unit length, then the centrifugal force acting on an element of length dx, Fig. 6.12,

$$= m \, dx \, \omega^2(x \sin \alpha)$$

Moment of this force about $C = m \, dx \, \omega^2(x \sin \alpha) \times (x \cos \alpha)$

$$\therefore \text{ total moment about } C = 2m\omega^2 \sin \alpha \cos \alpha \int_0^{\frac{l}{2}} x^2 \, dx$$

$$= m\omega^2 \sin 2\alpha \frac{l^3}{24}$$

$$= \frac{\left(\dfrac{0 \cdot 1}{0 \cdot 075}\right)\left(N \times \dfrac{2\pi}{60}\right)^2 \sin 2\alpha \times 0 \cdot 075^3}{24}$$

i.e. couple exerted by springs, $C = 0 \cdot 257 \times 10^{-6} N^2 \sin 2\alpha$ N m

At 200 rev/min $\qquad C = 0 \cdot 257 \times 10^{-6} \times 200^2 \times \sin 60°$

$$= 0 \cdot 0089 \text{ N m}$$

At 600 rev/min, $\qquad C = 0 \cdot 257 \times 10^{-6} \times 600^2 \times \sin 120°$

$$= 0 \cdot 0801 \text{ N m}$$

The couple exerted by the springs is proportional to the angle of twist of the springs. Hence the graph of C against α, Fig. 6.13, is a straight line, not passing through the origin owing to an initial twist when α is zero. Inserting the values of the couples corresponding to angles of 30° and 60°, the equation of this line is found to be $C = 0 \cdot 136\alpha - 0 \cdot 0623$

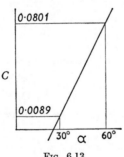

FIG. 6.13

Hence, at 400 rev/min,

$$0 \cdot 136\alpha - 0 \cdot 0623 = 0 \cdot 257 \times 10^{-6} \times 400^2 \sin 2\alpha$$

or $\qquad \alpha - 0 \cdot 3022 \sin 2\alpha - 0 \cdot 4581 = 0$

Solving this equation by Newton's approximation, or otherwise,

$$\alpha = 0 \cdot 760 \text{ rad} \quad \text{or} \quad \underline{43° \ 33'}$$

6. *Fig. 6.14 shows the arrangement of a governor. Two weighted arms, A, A', are pivoted on pins B, B', 60 mm apart attached to a plate C which rotates about its centre. Each weighted arm has a mass of 225 g and the centres of gravity G, G', are 48 mm from the centres of the pivots. Points S, S', on the arms, at 24 mm distance from the pivots are connected by a spring. A linkage (not shown) ensures that the two angles θ and θ' are equal. The stiffness of the spring is 700 N/m.*

(a) Find the tension required in the spring so that the angles θ, θ' shall be 30° when the governor speed is 300 rev/min.

(b) If the governor, rotating in the anticlockwise direction, accelerates at the rate of 50 rad/s², at what speed of rotation will the angles be 45°?

(U. Lond.)

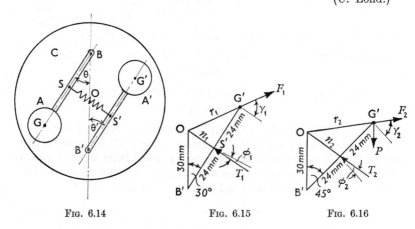

FIG. 6.14   FIG. 6.15   FIG. 6.16

(a) From Fig. 6.15,

$$r_1 = 26 \cdot 64 \text{ mm}, \quad n_1 = 15 \cdot 12 \text{ mm}, \quad \gamma_1 = 55° 43', \quad \phi_1 = 6° 30',$$

Centrifugal force, $F_1 = 0 \cdot 225 \left(300 \times \dfrac{2\pi}{60} \right)^2 \times 0 \cdot 026 \, 64 = 5 \cdot 92$ N

If the spring tension is T_1 then, taking moments about B' of the forces on the arm B'G',

$$5 \cdot 92 \times \cos 55° 43' \times 0 \cdot 048 = T_1 \times \cos 6° 30' \times 0 \cdot 024$$

$$\therefore T_1 = \underline{6 \cdot 7 \text{ N}}$$

(b) From Fig. 6.16,

$$r_2 = 34 \cdot 18 \text{ mm}, \quad n_2 = 21 \cdot 41 \text{ mm}, \quad \gamma_2 = 52° 42', \quad \phi_2 = 7° 30',$$

Centrifugal force, $F_2 = 0 \cdot 225 \left(N \times \dfrac{2\pi}{60} \right)^2 \times 0 \cdot 034 \, 18$

$$= 0 \cdot 000 \, 084 \, 3N^2 \text{ N}$$

Spring tension, $T_2 = 6 \cdot 7 + 2(0 \cdot 021\ 41 - 0 \cdot 015\ 12) \times 700$

$\qquad = 15 \cdot 51$ N

Linear acceleration of $G' = 50 \times 0 \cdot 034\ 18 = 1 \cdot 71$ m/s^2

$\therefore$ inertia force, $P = 0 \cdot 225 \times 1 \cdot 71$

$\qquad\qquad = 0 \cdot 385$ N, acting in the opposite direction to the acceleration

Taking moments about B',

$\{0 \cdot 000\ 084\ 3 N^2 \cos 52° 42' + 0 \cdot 385 \sin 52° 42'\} \times 0 \cdot 048$

$\qquad\qquad = 15 \cdot 51 \cos 7° 30' \times 0 \cdot 024$

from which $\qquad\qquad N = \underline{380 \text{ rev/min}}$

7. A conical pendulum consists of a uniform straight bar 350 mm long and of mass 2 kg, with a bob of mass 4 kg at its lower end. It rotates at 70 rev/min about a vertical axis through the upper end of the bar. Find the 'height' of the pendulum and the bending moment at the mid-point of the bar. (*U. Lond.*)

$\qquad\qquad$ (*Ans.:* 195·5 mm ; 0·33 N m)

8. A uniform thin rigid rod AB of mass/length m forms a quarter of a circle of radius r. To one end B is fixed a bob of mass M. The other end A is hinged to a vertical spindle so that the rod and bob can move in a vertical plane. When the assembly rotates with constant angular velocity ω about the axis of the spindle the bob rotates in a circle of radius r, in a plane distant r below the hinge, the tangent to the rod at A being horizontal.

(*a*) Show that

$$\omega^2 = \frac{g}{r} \cdot \frac{M + mr}{M + \dfrac{mr}{2}}$$

(*b*) Write down an expression for the bending moment at the mid-point of the rod. (*U. Lond.*)

$\qquad\qquad$ (*Ans.:* $gr(0 \cdot 2929M + 0 \cdot 1516mr) - \omega^2 r^2 (0 \cdot 7071M + 0 \cdot 25mr)$)

9. In a centrifugal governor, if F is the controlling force corresponding to a radius of rotation r of the balls, show that a necessary condition for the stability of the governor is that

$$\frac{dF}{dr} > \frac{F}{r}$$

Hence show that a Porter governor having arms of equal length pivoted at the axis of rotation is always stable.

If the rotating masses of such a governor are each 2·5 kg, the central mass 15 kg, and the length of the arms is 250 mm, determine the equilibrium speed corresponding to a radius of rotation of 175 mm. (*U. Lond.*)

$\qquad\qquad$ (*Ans.:* 187·4 rev/min)

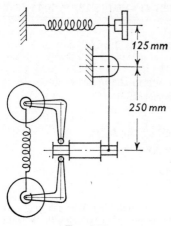

Fig. 6.17

10. Fig. 6.17 shows a throttle valve governor. Each ball has a mass of 6·5 kg and the arms are of equal length. The speed range is 425 to 440 rev/min with a sleeve movement of 6 mm. The minimum ball path radius is 122 mm. If the combined strength of the ball springs is 15 kN/m of extension, find the rate of the auxiliary spring. If, with the auxiliary spring removed, the speed at maximum radius is 418 rev/min, find the unstretched length of the ball springs and the required extension of the auxiliary spring at maximum radius to give the running conditions.

Describe briefly the purpose of the auxiliary spring and its possible effect on stability. (*I. Mech. E.*)

(*Ans.:* 20·7 kN/m; 150 mm; 33·2 mm)

11. A form of spring-controlled governor is shown diagrammatically in Fig. 6.18. The governor has two rotating balls, their outward movement being controlled by two tension springs connected in parallel across the balls. Each spring has a stiffness of 1 kN/m and a free length of 110 mm. The central sleeve has a mass of 6·5 kg and the combined mass of the lever PQ and throttle spindle T is 2·75 kg. The effective centre of gravity of the lever and throttle spindle is at G.

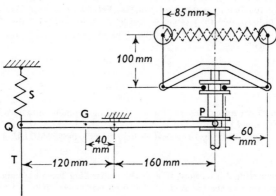

Fig. 6.18

The governor is to be designed to satisfy the following requirements:

(*i*) In the mid-position shown the equilibrium speed of the governor is to be 300 rev/min, ignoring friction.

(*ii*) In this mid-position a 0·5% increase in speed must be sufficient to overcome a friction force at the throttle spindle of 7·5 N.

(*iii*) A 7% change in speed must produce a movement of the throttle spindle of 5 mm, taking account of the friction resistance of 7·5 N but ignoring the effect of gravity on the balls.

Find the mass of each ball, the stiffness of the auxiliary tension spring S and the force to be exerted by this spring when the governor is in the mid-position. (*U. Lond*). (*Ans.:* 2·01 kg ; 140 N ; 6·76 kN/m)

12. In a governor of the Hartnell type the arms of the bell-crank levers are equal in length, and those carrying the operating masses are vertical when the governor is rotating at its mean speed of 775 rev/min, with the masses moving in a circle of 175 mm diameter. The usual central controlling spring is replaced by two parallel tension springs directly connecting the operating masses.

Find :

(*a*) the magnitude of each operating mass if a force of 90 N is required at the sleeve to maintain it in the mean speed position when the speed is increased from 775 to 800 rev/min ;

(*b*) the stiffness, or rate, of each spring if the ratio of sleeve movement to increase of speed is 1 mm to 10 rev/min when in the mean speed position. (*U. Lond.*) (*Ans.:* 1·19 kg ; 6·42 kN/m)

13. A Hartnell type governor with a vertical axis has two rotating masses of 1·4 kg carried on right-angled bell-crank levers, in which the mass arm is 60 mm and the sleeve arm is 50 mm long. The sleeve has a total movement of 25 mm and is in mid-position when the sleeve arm is horizontal ; when the sleeve is in the mid-position the masses revolve in a circle of 80 mm radius.

Owing to the maladjustment of the controlling spring it is found that the equilibrium speed at the top stop is lower than that at the bottom stop, being 420 and 435 rev/min respectively.

Determine (*a*) the stiffness of the spring and the compression at the bottom stop, (*b*) the initial compression which would give a top stop equilibrium speed 12 rev/min greater than the bottom stop speed. (*U. Lond*) (*Ans.:* 6·56 kN/m ; 69·1 mm ; 44·2 mm)

14. A governor with the control spring directly connected between the masses operates a hanging link through a lever as shown in Fig. 6.19. For the position given, the particulars are as follows : spring tension 650 N ; elastic force of spring 11 kN/m ; mass of spring 2·2 kg ; mass of each ball 2·7 kg ; mass of each bell-crank 0·9 kg ; radius of gyration of bell-crank about fulcrum 30 mm ; mass of sleeve 1·6 kg ; mass of lever 2·7 kg ; c.g. of lever as marked ; radius of gyration of lever about fulcrum 70 mm ; mass of hanging link 3·6 kg.

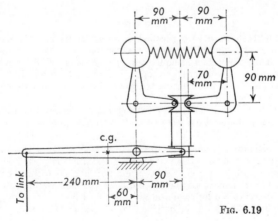

FIG. 6.19

F

Determine (a) the governor speed for the position shown, neglecting the control effects of the bell-cranks, and (b) the frequency of vibration if disturbed when running at this speed. (*U. Lond.*)   (*Ans.:* 512 rev/min ; 5·56 Hz)

15. The main elements of a spring-controlled governor with an external auxiliary spring adjustment are shown in Fig. 6.20. Two springs connect the rotating masses and the stiffness of each spring is 4·5 kN/m. The stiffness of the auxiliary spring is 13·5 kN/m. The magnitude of each rotating mass is 3·6 kg, and when the governor operates, a force of 140 N acts on the 260 mm lever as shown. If the movement of the sleeve is 13 mm and the speeds of the governor when in its mid-position are between 350 and 500 rev/min, determine the initial tensions in the two main springs, and the adjustable range for the auxiliary spring. It may be assumed that the initial tension of the auxiliary spring is 45 N.

With the auxiliary spring set so that the speed of the governor is 500 rev/min in its mid-position, find the range of speed. (*U. Lond.*)

(*Ans.:* 234 N (total) ; 110·8 to 1219 N ; 477 to 519 rev/min)

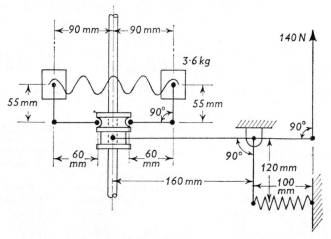

Fɪɢ. 6.20

16. Fig. 6.21 shows a spring-loaded governor in its inner position. Each ball has a mass of 5 kg and the sleeve mass is 2·5 kg. The lever and the gear attached to it may be taken as equivalent to a mass m, of 18 kg, acting in the position indicated. The internal spring S_1 has an elastic force of 22 kN/m and, for a ball radius of 80 mm, exerts a tension of 1·8 kN. The external spring S_2 has an elastic force of 50 kN/m.

Calculate the tension, P, of the external spring for the governor in the position shown if, when the ball radius increases to 100 mm, the speed is to be 750 rev/min. (*U. Lond.*)   (*Ans.:* 603·6 N)

17. The bell-crank levers of a spring-controlled governor are carried by a rotating casing, Fig. 6.22. Each ball arm is 100 mm long, each sleeve arm is 125 mm long and when the sleeve is in its mid-position, the ball arms are vertical, the sleeve arms are horizontal and the radius of the ball path is 150 mm. The balls have a mass of 3·6 kg each and are controlled by helical springs in initial compression in contact with the outside of the balls and the inside of the casing.

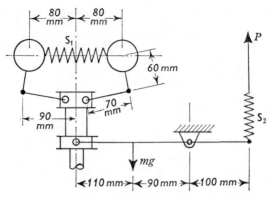

FIG. 6.21

The operating gear—reduced to the sleeve—has a mass of 9 kg. Each of the springs has a stiffness of 10 kN/m and is initially compressed 50 mm for the bottom position of the sleeve.

Determine the range of speed of operation of the governor for a sleeve lift of 25 mm.

When the sleeve is in its mid-position, determine the frequency of the vibrations if disturbed when the governor is running at its equilibrium speed. (*U. Lond.*)   (*Ans.:* 332·7 to 346 rev/min ; 3·66 Hz)

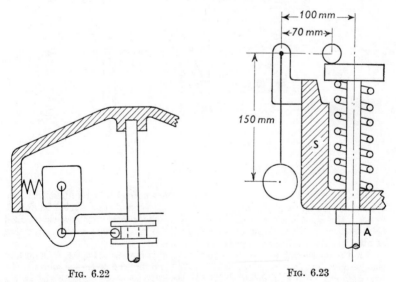

FIG. 6.22   FIG. 6.23

18. Fig. 6.23 shows the arrangement of a governor. The central spindle A does not move axially. The sleeve S has a mass of 18 kg and the frictional resistance to its movement may be taken as 20 N. There are two right-angled bell-crank levers, each having a ball of mass 4 kg attached to its lower end. The masses of the levers themselves and the rollers on their horizontal arms may be neglected but the dead weight of the balls must not be neglected. The spring has a stiffness of 50 kN/m compression. When the sleeve is in its lowest position, the balls rotate in a circle of 100 mm radius, the ball arms are vertical and the spring force is 550 N.

Calculate (a) the speed at which the sleeve will begin to rise from its lowest position, and (b) the range within which the speed must lie when the sleeve is 10 mm above its lowest position. (*U. Lond.*)

(*Ans.*: 208·5 rev/min ; 239·4 to 243·2 rev/min)

19. A speed indicator is constructed as shown in Fig. 6.24. The fly-weights can be taken as thin uniform bars of mass m kg/m, pivoted at the lower ends, and the inertia of the other parts of the arms neglected. The control is by a spring acting on the sleeve. By considering an element of the mass, show that the force required at the sleeve when the speed is ω rad/s and the inclination of the masses to the axis is α, is given by

$$P = \frac{2}{a} m\omega^2 b^2 \left\{ \frac{c}{2} + \frac{b}{3} \sin \alpha \right\}$$

If $m = 4\cdot5$ kg/m, $a = 75$ mm, $b = 125$ mm, $c = 75$ mm, and for $\alpha = 0°$, rev/min = 240, for $\alpha = 45°$, rev/min = 360, find the stiffness of the controlling spring. (*I. Mech. E.*)   (*Ans.*: 2·53 kN/m)

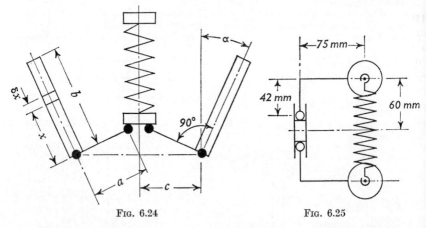

Fig. 6.24   Fig. 6.25

20. The centrifugal governor, shown diagrammatically in Fig. 6.25, has two operating masses, each of mass m, carried at the ends of right-angled bell-crank levers ; the masses are connected directly to one another by two close-coiled helical springs. In the position shown, with the mass arms parallel to the axis of rotation, the equilibrium speed is 850 rev/min.

Find : (a) the value of m if, when the speed is increased by 1% without any change of radius from the given position, an axial force of 24 N is required at the sleeve to maintain equilibrium ; (b) the strength and extension of the spring if the rate of sleeve movement when in mid-position is 25 mm per 600 rev/min change of speed. (*U. Lond.*)

(*Ans.*: 0·706 kg ; 4·04 kN/m per spring ; 41·5 mm)

21. A tachometer consists of a thin bar turning about an axle through the centre O perpendicular to its length, Fig. 6.26. The axle is in the horizontal plane and it rotates about the vertical through O at the speed ω to be measured. The bar has a mass m and length l and a control spring provides the torque T about the axle. If α is the inclination of the bar to the axis of rotation, show that the controlling torque required is $T = ml^2\omega^2 \sin 2\alpha/24$.

If $m = 0.225$ kg, $l = 88$ mm and $T = 0.009\alpha - 0.22$ N m, where α is measured in degrees, draw carefully the calibration curve showing the speed in rev/min for values of α between $30°$ and $60°$. (*I. Mech. E.*)

22. Fig. 6.27 shows a centrifugal governor. Six steel balls, each of mass 3·6 g, are driven round by a spider attached to the governor shaft. At the operating speed, the balls move outward lifting the conical plate C against four compression springs. Each of these four springs exerts an initial force of 40 N

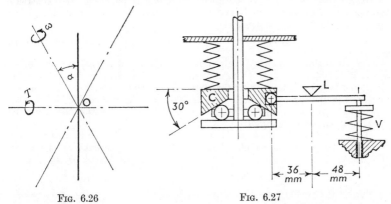

FIG. 6.26   FIG. 6.27

with the balls at their innermost radius of 14 mm and each spring has a stiffness of 6·3 kN/m. As the conical plate C lifts, it operates a valve V through the toggle lever L. The movement of the valve V is restrained by a compression spring.

Find the initial force required in this spring for the valve to begin to move when the speed reaches 5500 rev/min, and the stiffness of the spring for the valve travel to be 6 mm at a speed of 6000 rev/min. (*U. Lond.*)

(*Ans.:* 10·28 N; 4·28 kN/m)

23. (*a*) Shaft governors may be of the centrifugal or of the inertia type. Distinguish between the actions of these two types in controlling the speed of an engine.

(*b*) A shaft inertia governor consists of an arm AB pivoted at C, C being a fixed point on a disc concentric with and rigidly attached to the engine shaft. C is offset from the shaft axis O by 75 mm. The arm AB is 350 mm long, and is symmetrical about C, that is AC = CB = 175 mm. A mass of 7 kg is attached to each end of the arm at A and B, these masses being in the form of circular discs, each 150 mm in diameter, with their axes parallel to that of the shaft. In the normal position the arm ACB is at right angles to the radius OC.

If the speed of the engine increases by 15 rev/min in 2 seconds, this increase being at a uniform rate, determine the torque about C needed to hold the arm stationary relative to the concentric disc. Neglect the mass of the arm, but explain carefully the reasoning behind any equations you may employ. (*U. Lond.*)   (*Ans.:* 0·368 N m)

24. An inertia governor mounted on a flywheel, consists of a concentrated mass of 5 kg attached rigidly to an arm which is pivoted at a distance of 175 mm from the centre of the crankshaft. The effective length of the arm, measured from the centre of the pivot to the centre of the mass, is 200 mm, and when the flywheel is rotating at a uniform speed of 90 rev/min, the line joining the pivot centre to the crankshaft centre makes an angle of 135° with the arm. If the flywheel suddenly starts to change its speed at a rate of 2·5 rad/s², what will be the torque immediately available for operating the governor system ? Discuss the effect for both directions of rotation, also for positive and negative changes of speed.

Give a description of a more developed type of inertia governor as adopted in practice, explaining its performance as compared with a centrifugal governor. (*U. Lond.*)

(*Ans.:* 10·2 N m for anticlockwise rotation, accelerating, and clockwise rotation, decelerating ; 11·8 N m for anticlockwise rotation, decelerating, and clockwise rotation, accelerating.)

25. Distinguish between the actions of a centrifugal and of an inertia type governor.

The masses of an inertia type governor are attached to the two ends of an arm POQ, pivoted at O. O is a fixed point on a disc, rigidly attached to a turbine shaft, and is 100 mm from that shaft axis. PO = OQ = 150 mm. The masses are made in the form of circular discs of 100 mm diameter each of mass 5 kg, having axes parallel to the shaft axis R. In the normal position the arm POQ i ι at right angles to a radial line RO.

If the shaft speed increases at a uniform rate of 30 rev/min per second determine the torque available about O. Neglect the masses of the arms. (*U. Lond.*)   (*Ans.:* 0·746 N m)

CHAPTER 7

BALANCING

7.1 Inertia force on a reciprocating mass. If R is the mass of the reciprocating parts of a single-cylinder engine, r the crank radius, l the length of the connecting rod, ω the crank speed and θ the angle made by the crank to the i.d.c., the inertia force on the reciprocating parts is given approximately by

$$P = R\omega^2 r \left(\cos\theta + \frac{\cos 2\theta}{n} \right), \quad \text{where } n = \frac{l}{r}$$

The first term $R\omega^2 r \cos\theta$ is called the *primary* inertia force and reaches a maximum value of $R\omega^2 r$ twice per revolution, when $\theta = 0$ and 180°. The second term $R\omega^2 r \dfrac{\cos 2\theta}{n}$ is called the *secondary* inertia force and reaches a maximum value of $R\dfrac{\omega^2 r}{n}$ four times per revolution, when $\theta = 0$, 90°, 180°, and 270°.

Each of these terms represents a force which is constant in direction but varying in magnitude and therefore cannot be completely balanced by a rotating mass, on which the inertia force is varying in direction but is constant in magnitude.

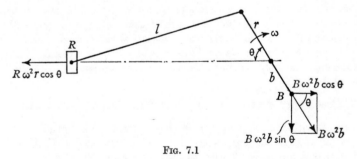

Fig. 7.1

In the case of the primary inertia force, this can be balanced by a balance mass B, Fig. 7.1, rotating at a radius b such that $Bb = Rr$, the component of the inertia force on B in the line of stroke being

$$B\omega^2 b \cos\theta,$$

159

but a force $B\omega^2 b \sin\theta$ is then introduced in a direction perpendicular to
the line of stroke. In many cases, the reciprocating mass is partially
balanced by a rotating mass, reducing the inertia force in the line of stroke
and introducing a force smaller than $R\omega^2 r$ in the perpendicular direction.

A similar partial balance of the secondary inertia force could only be
achieved by a balance mass which rotates such that the radius makes an
angle 2θ with the i.d.c., i.e. rotates at twice the speed of the crankshaft.
Due to the practical difficulties of such an arrangement and the relatively
small magnitude of the secondary inertia force, such balance is not normally attempted.

**7.2 Primary and secondary balance of multi-cylinder in-line
engines.** For each cylinder, the primary inertia force is identical with
the component in the line of stroke of an imaginary mass R rotating at the
crankpin, Fig. 7.2. A multi-cylinder engine can therefore be treated in the

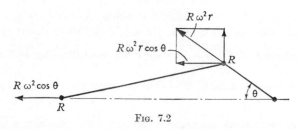

Fig. 7.2

same way as a shaft carrying a number of rotating masses* ; if the force
or couple polygons close, the forces or couples are balanced but otherwise
the closing side gives the maximum value of the unbalanced force or
couple, the actual unbalanced force or couple at any crank position being
the component along the line of stroke of the closing side. Since these
polygons rotate with the cranks, the positions of the cranks at which the
unbalanced forces or couples reach their maximum values can be obtained
by rotating these polygons until the closing sides are parallel with the line
of stroke.

The secondary inertia force may be written $R(2\omega)^2 \dfrac{r}{4n} \cos 2\theta$ and this
is identical with the component in the line of stroke of an imaginary mass
R rotating at *twice* the crank speed at a radius $r/4n$, such that the radius
always makes an angle 2θ with the line of stroke, Fig. 7.3.

For multi-cylinder engines, out-of-balance secondary forces and couples
are obtained from force and couple polygons in the same way as for the
primary forces and couples, except that the sides of these polygons are
drawn parallel with the imaginary 'secondary cranks', which always

* See the authors' *Mechanics of Machines, Elementary Theory and Examples.*

make twice the angle made with the line of stroke by the actual cranks. To determine the actual crank positions at which the maximum unbalanced secondary forces or couples occur, the cranks must be rotated

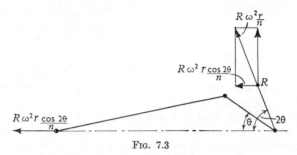

FIG. 7.3

through *half* the angle necessary to bring the closing side of the force or couple polygon into the line of stroke.

When tabulating values of the forces or couples, it is customary to omit the constant factor ω^2 and provided n is the same for all cylinders this can also be omitted from the table, enabling it to be used for both primary and secondary polygons.

If there is also a rotating mass M at the crankpin, this can be included with R for the purpose of drawing the primary force and couple polygons but has no effect on the secondary forces or couples.

7.3 Firing order. In multi-cylinder engines, there are frequently several possible firing orders for the cylinders, each having a different effect on the balance of the engine. The firing order adopted may be influenced by other factors, however, such as torsional vibrations, fuel and exhaust distributions, etc. In a two-stroke engine, the cycle of operations is complete in one revolution of the crankshaft and so the interval between the cranks is $360°/N$, where N is the number of cylinders; the order of the cranks corresponds to the given firing order. In a four-stroke engine, the cycle of operations requires two revolutions of the crankshaft and therefore the interval between the cranks is $720°/N$.

7.4 Exact expression for inertia force. The inertia, or *shaking*, force on the reciprocating parts may be expressed as a Fourier series of the form*

$$R\omega^2 r(\cos \theta + A_2 \cos 2\theta + A_4 \cos 4\theta + A_6 \cos 6\theta + \ldots)$$

where the *harmonic coefficients* A_2, A_4, A_6, etc., are functions of n. Usually only the first two terms are of importance but higher harmonics may become important when in phase with the resonant frequency of the engine mounting. The higher harmonics are treated in exactly the same way as

* See Art. 2.6.

the secondary inertia force, e.g., the fourth harmonic $R\omega^2 r A_4 \cos 4\theta$ may be written $R(4\omega)^2 \dfrac{r A_4}{16} \cos 4\theta$ and is thus identical with the component in the line of stroke of an imaginary mass R rotating at four times the crank speed at a radius $r A_4/16$. The sides of the force and couple polygons are then parallel with the fourth order cranks, which each make an angle 4θ with the line of stroke.

7.5 Direct and reverse cranks. For engines in which the cylinders are not in line, such as V-engines and radial engines, the reciprocating mass must be replaced by an exactly equivalent system of rotating masses, not merely by one in which only a component of the inertia force on the rotating mass is equivalent to the actual disturbing force. This is obtained by replacing the reciprocating mass R by $R/2$ at the crankpin and $R/2$ at the end of an imaginary crank of the same radius and always making an angle θ on the opposite side of the centre line, Fig. 7.4 ; the components in the line of stroke of the inertia forces on the two masses are each $\dfrac{R}{2}\omega^2 r \cos \theta$ and the components perpendicular to the line of stroke

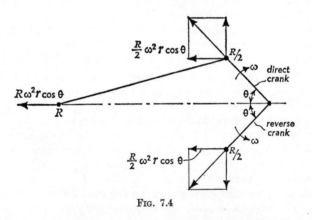

Fig. 7.4

balance each other. The actual and imaginary cranks are referred to as the direct and reverse cranks respectively.* The secondary inertia forces can similarly be simulated by masses $R/2$ at the ends of direct and reverse cranks, each of radius $r/4n$, rotating at twice the engine speed ; the fourth order, sixth order, etc., forces can be treated in a like manner.

* The reverse crank is a mirror image of the actual crank across the line of stroke.

1. *An engine having five cylinders in line has successive cranks 144°
apart, the distance between cylinder centre lines being 450 mm. The reciprocating mass for each cylinder is 16 kg, the crank radius is 135 mm and
the connecting rod length is 540 mm. The engine runs at 600 rev/min.*

*Examine the engine for balance of primary and secondary forces and
couples. Determine the maximum values of these and the position of the
central crank at which these maximum values occur.* (U. Lond.)

Crank No. 1 has been taken as coinciding with the line of stroke and
the other cranks have then been placed in the order 1—2—3—4—5 at
144° intervals (which corresponds to a four-stroke cycle of operation).
The positions of the secondary cranks are obtained by doubling the angle
which each crank makes with the line of stroke.

Using the given data, the following table is compiled, taking the central
plane of the engine as the reference plane for couples.

Plane	R (kg)	r (m)	Rr (kg m)	x (m)	Rrx (kg m²)
1	16	0·135	2·16	−0·9	−1·944
2	16	0·135	2·16	−0·45	−0·972
3	16	0·135	2·16	0	0
4	16	0·135	2·16	+0·45	+0·972
5	16	0·135	2·16	+0·9	+1·944

From the symmetry of the primary and secondary cranks, Fig. 7.5(*a*)
and (*b*), the primary and secondary forces are in balance.

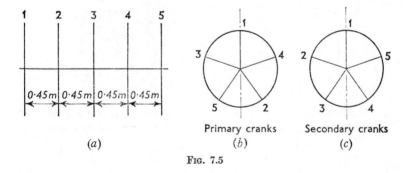

(*a*) Primary cranks (*b*) Secondary cranks (*c*)

Fɪɢ. 7.5

From the primary couple polygon, Fig. 7.6, the closing side, 05, represents 2·57 kg m², which must be multiplied by ω^2 to give the actual
unbalanced couple, i.e.

$$\text{maximum unbalanced primary couple} = 2\cdot57 \times \left(600 \times \frac{2\pi}{60}\right)^2$$

$$= \underline{10\ 150\ \text{N m}}$$

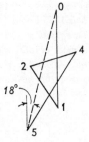

Primary couple polygon

FIG. 7.6

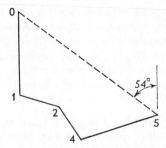

Secondary couple polygon

FIG. 7.7

The vector 05 is parallel with the line of stroke when the couple polygon has rotated through 18° and 198° anticlockwise from the position shown, i.e. when crank No. 3 is at 90° on either side of the line of stroke, Fig. 7.8(a).

From the secondary couple polygon, Fig. 7.7, the closing side, 05, represents 4·15 kg m², which must be multiplied by ω^2/n to give the actual unbalanced couple, i.e.

$$\text{maximum unbalanced secondary couple} = 4\cdot15 \times \left(600 \times \frac{2\pi}{60}\right)^2 \times \frac{0\cdot135}{0\cdot54}$$

$$= \underline{4\ 100\ \text{N m}}$$

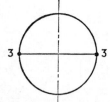

Positions of crank No. 3 for
maximum primary couple

(a)

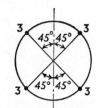

Positions of crank No. 3 for
maximum secondary couple

(b)

FIG. 7.8

The vector 05 is parallel with the line of stroke when the couple polygon has turned through 54°, 234°, 414° and 594° clockwise from the position shown. Since the secondary crank angles are twice the actual crank angles measured from the top dead centre, the actual cranks must be turned through 27°, 117°, 207° and 297° clockwise from the position shown in Fig. 7.5(a) for the maximum secondary couple to occur, i.e. crank No. 3 is at 45° on either side of the line of stroke, Fig. 7.8(b).

2. *The six cylinders of a single-acting, two-stroke cycle Diesel engine are pitched 1 m apart and the cranks are spaced at 60° intervals. The crank length is 300 mm and the ratio of connecting rod to crank is 4·5. The reciprocating mass per line is 1·35 Mg and the rotating mass is 1 Mg. The speed is 200 rev/min.*

Show with regard to primary and secondary balance, that the firing order 1—5—3—6—2—4 gives unbalance in primary moment only, and the order 1—4—5—2—3—6 gives secondary moment unbalance only. Compare the maximum values of these moments, evaluating with respect to the central plane of the engine. (U. Lond.)

Using the given data, the following table is compiled, taking the central plane of the engine as reference, Fig. 7.9.

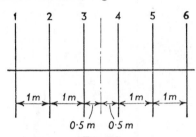

FIG. 7.9

Plane	R (kg)	M (kg)	r (m)	Rr (kg m)	$(R + M)r$ (kg m)	x (m)	Rrx (kg m²)	$(R + M)rx$ (kg m²)
1	1350	1000	0·3	405	705	−2·5	−1012·5	−1762·5
2	1350	1000	0·3	405	705	−1·5	−607·5	−1057·5
3	1350	1000	0·3	405	705	−0·5	−202·5	−352·5
4	1350	1000	0·3	405	705	+0·5	+202·5	+352·5
5	1350	1000	0·3	405	705	+1·5	+607·5	+1057·5
6	1350	1000	0·3	405	705	+2·5	+1012·5	+1762·5

The rotating mass M at each crankpin is included with the reciprocating mass R only for primary forces and couples.

Firing Order 1—5—3—6—2—4. The positions of the primary and secondary cranks are as shown in Fig. 7.10, crank No. 1 coinciding with the line of stroke.

Primary cranks

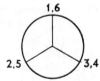

Secondary cranks

FIG. 7.10

From the crank symmetry, the primary and secondary forces are balanced. From the primary couple polygon, Fig. 7.11, using figures from the $(R + M)rx$ column, the closing side, 06, represents 2430 kg m².

∴ maximum unbalanced primary couple

$$= 2430 \times \left(200 \times \frac{2\pi}{60}\right)^2$$

$$= 1\ 065\ 000\ \text{N m} = \underline{1{\cdot}065\ \text{MN m}}*$$

The secondary couple polygon, Fig. 7.12, is a closed figure, so that there is no unbalanced secondary couple.

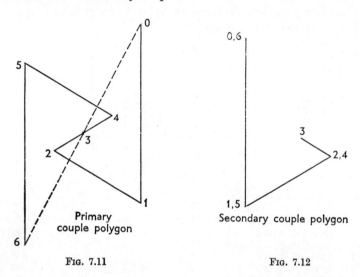

Primary couple polygon

Fig. 7.11

Secondary couple polygon

Fig. 7.12

Firing Order 1—4—5—2—3—6. The positions of the primary and secondary cranks are as shown in Fig. 7.13 and again the primary and secondary forces are balanced.

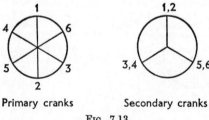

Primary cranks   Secondary cranks

Fig. 7.13

The primary couple polygon, Fig. 7.14, is a closed figure, so that there is no unbalanced primary couple.

From the secondary couple polygon, Fig. 7.15, using the figures from the Rrx column, the closing side, 06, represents 2820 kg m².

∴ maximum unbalanced secondary couple

$$= 2820 \times \left(200 \times \frac{2\pi}{60}\right)^2 \times \frac{1}{4\cdot5}$$

$$= 275\ 000 \text{ N m} = \underline{\underline{275 \text{ kN m}}}$$

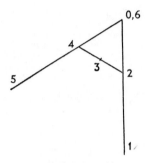

Primary couple polygon

Fig. 7.14

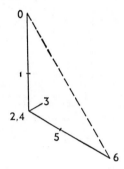

Secondary couple polygon

Fig. 7.15

---

* For any position of the cranks other than that for which the closing side is parallel with the line of stroke, the out-of-balance couple is made up of (a) a constant couple in the direction of the vector due to the rotating mass M and (b) the component in the line of stroke of part of this vector due to the reciprocating mass R

3. *Some particulars of a five-cylinder air compressor are given below. Cylinders 2 and 4 (which are the high-pressure cylinders) are to have their cranks at 180°. Determine a crank arrangement to satisfy the following conditions : (a) primary forces balanced, and (b) primary couple to have the least value compatible with condition (a). For this arrangement find the magnitude of the unbalanced primary and secondary couples about No. 3 cylinder line. Speed 250 rev/min, crank radius 0·3 m, connecting rod 1·5 m, axial spacing cylinders 0·825 m.*

Cylinder No.	Reciprocating mass (kg)	Crank angle
1	250	
2	125	0°
3	150	
4	125	180°
5	200	

(U. Glas.)

Using the given data, the following table is compiled, taking the plane of the cylinder No. 3 as reference, Fig. 7.16.

Plane	R (kg)	r (m)	Rr (kg m)	x (m)	Rrx (kg m²)
1	250	0·3	75	−1·65	−123·75
2	125	0·3	37·5	−0·825	−30·94
3	150	0·3	45	0	0
4	125	0·3	37·5	+0·825	+30·94
5	200	0·3	60	+1·65	+99·0

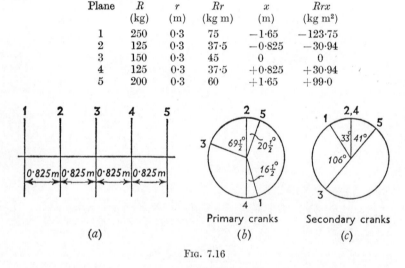

(a)   Primary cranks (b)   Secondary cranks (c)

Fig. 7.16

The primary forces due to cranks 2 and 4 balance, hence, for condition (a), the primary forces due to cranks 1, 3 and 5 must also balance. The relative positions of these cranks must therefore be as shown in Fig. 7.17.

From the couple polygon for these cranks, Fig. 7.18, the closing side, 05, represents 211 kg m² and is inclined at 20½° to the direction of crank No. 5. The couple due to cranks 2 and 4, represented by 61·88 kg m², acts vertically downwards, and hence for least resultant couple, the cranks 1, 3 and 5 must be rotated anticlockwise through $159\frac{1}{2}°$ so that the couple due to cranks 1, 3 and 5 is in opposition to that due to cranks 1 and 2. The relative position of all the cranks is then as shown in Fig. 7.16(b).

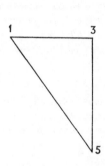

Primary force polygon

Fig. 7.17

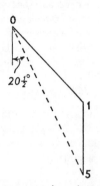

Primary couple polygon

Fig. 7.18

$\therefore$ maximum unbalanced primary couple

$$= (211 - 61 \cdot 88) \times \left(250 \times \frac{2\pi}{60}\right)^2$$

$$= 102\ 500 \text{ N m} = \underline{102 \cdot 5 \text{ kN m}}$$

The secondary crank positions are shown in Fig. 7.16(c). From the secondary couple polygon, Fig. 7.19, the closing side, 05, represents $137 \cdot 5$ kg m^2.

$\therefore$ maximum unbalanced secondary couple

$$= 137 \cdot 5 \times \left(250 \times \frac{2\pi}{60}\right)^2 \times \frac{0 \cdot 3}{1 \cdot 5}$$

$$= 18\ 840 \text{ N m} = \underline{18 \cdot 84 \text{ kN m}}$$

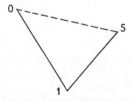

Secondary couple polygon

Fig. 7.19

4. *A two-cylinder vertical engine is direct-coupled to a single-cylinder compressor. The crank angles are to be arranged to give balanced primary force. Find the angles required and, for this arrangement, find the unbalanced primary and secondary effects, specifying magnitude and phase. Couples should be specified with reference to the middle cylinder.*

Reciprocating masses: engine 12 kg per cylinder, compressor 9 kg.

Ratios of connecting rods to cranks: engine 4, compressor 6.

Crank radii: engine 105 mm, compressor 60 mm.

Cylinder spacing: engine 300 mm, between compressor and nearer engine cylinder 390 mm.

Speed: 1600 rev/min.   (U. Glas.)

Using the given data, the following table is compiled, taking the plane of cylinder No. 2 as the reference plane. Since the value of n is different for the engine and compressor, the secondary forces and couples must be tabulated separately.

Plane	R	r	Rr	x	Rrx	$\dfrac{Rr}{n}$	$\dfrac{Rrx}{n}$
	(kg)	(m)	(kg m)	(m)	(kg m²)	(kg m)	(kg m²)
1	12	0·105	1·26	0·3	0·378	0·315	0·0945
2	12	0·105	1·26	0	0	0·315	0
C	9	0·06	0·54	−0·39	−0·211	0·09	−0·0351

Using the values from the Rr column, the primary force polygon, Fig. 7.21 gives the relative positions of the cranks; these are shown in Fig. 7.20(*b*).

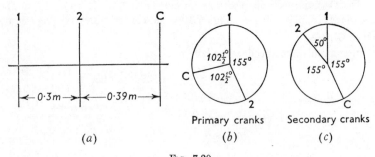

Primary cranks   Secondary cranks
(*b*)   (*c*)

(*a*)

Fɪɢ. 7.20

From the primary couple polygon, Fig. 7.22, using the values from the Rrx column, the closing side, OC, represents 0·472 kg m².

$$\therefore \text{ maximum unbalanced primary couple} = 0\cdot472 \times \left(1600 \times \frac{2\pi}{60}\right)^2$$

$$= 13\,250 \text{ N m} = \underline{13\cdot25 \text{ kN m}}$$

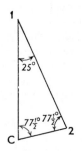

Primary force polygon

FIG. 7.21

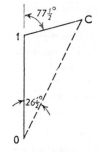

Primary couple polygon

FIG. 7.22

This occurs when crank No. 1 has turned through $26\tfrac{1}{2}°$ and $206\tfrac{1}{2}°$ anticlockwise from the position shown in Fig. 7.20(b).

The secondary crank positions are shown in Fig. 7.20(c). From the secondary force polygon, Fig. 7.23, using the values from the Rr/n column, the closing side, OC, represents 0·48 kg m.

$$\therefore \text{ maximum unbalanced secondary force} = 0·48 \times \left(1600 \times \frac{2\pi}{60}\right)^2$$

$$= 13\,500 \text{ N} = \underline{13·5 \text{ kN}}$$

This occurs when crank No. 1 has rotated through $12\tfrac{1}{2}°$, $102\tfrac{1}{2}°$, $192\tfrac{1}{2}°$ and $282\tfrac{1}{2}°$ clockwise from the position shown in Fig. 7.20(b).

From the secondary couple polygon, Fig. 7.24, using values from the Rrx/n column, the closing side, OC represents 0·1275 kg m².

$\therefore$ maximum unbalanced secondary couple

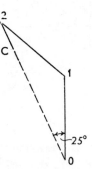

Secondary force polygon

FIG. 7.23

Secondary couple polygon

FIG. 7.24

$$= 0·1275 \times \left(1600 \times \frac{2\pi}{60}\right)^2 = 3580 \text{ N m} = \underline{3·58 \text{ kN m}}$$

This occurs when the cranks have rotated through $3\tfrac{1}{2}°$, $93\tfrac{1}{2}°$, $183\tfrac{1}{2}°$ and $273\tfrac{1}{2}°$ clockwise from the position shown in Fig. 7.20(b).

5. *An eight-cylinder vertical in-line marine diesel engine has a crank arrangement as follows: 0°—1 and 8; 90°—4 and 7; 180°—3 and 6; 270°—2 and 5. Calculate the unbalanced effects, up to the sixth order, from the following data. Reciprocating masses per cylinder, 24·5 kg; crank radius, 105 mm; axial pitch of cylinders, 240 mm; speed, 350 rev/min; harmonic coefficients $A_2 = 0·221$; $A_4 = -0·0027$; $A_6 = -0·000\ 035$.*

A simple interchange of crank angles would effect a considerable improvement in balance; demonstrate this. (U. Glas.)

Up to the sixth order,

inertia force $= R\omega^2 r(\cos \theta + A_2 \cos 2\theta + A_4 \cos 4\theta + A_6 \cos 6\theta)$

$$R\omega^2 r = 24·5 \times \left(350 \times \frac{2\pi}{60}\right)^2 \times 0·105 = 3460 \text{ N}$$

∴ inertia force

$$= 3460(\cos \theta + 0·221 \cos 2\theta - 0·0027 \cos 4\theta + 0·000\ 035 \cos 6\theta)$$

The positions of the secondary, fourth and sixth order cranks are obtained by multiplying the actual crank angles from the t.d.c. by 2, 4 and 6 respectively. The positions of these cranks are shown in Fig. 7.25, crank No. 1 coinciding with the cylinder centre-lines.

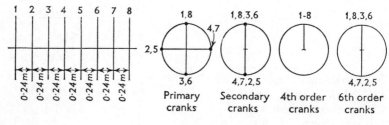

Fig. 7.25

From the symmetry of the cranks, the primary, secondary and sixth order forces are balanced. The maximum unbalanced force due to the fourth harmonic

$$= R\omega^2 r \times A_4 \times 8$$

$$= 3460 \times 0·0027 \times 8 = 74·7 \text{ N}$$

Taking the central plane of the engine as reference, the couples produced by the inertia forces are proportional to the distances of their lines of action from this plane, as shown in the following table:

Plane .	1	2	3	4	5	6	7	8
Couple	−0·84	−0·6	−0·36	−0·12	0·12	0·36	0·6	0·84

From the primary couple polygon, Fig. 7.26, the closing side, 08, represents 0·96 m.

∴ maximum unbalanced primary couple = 3460 × 0·96 = 3325 N m

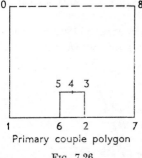

Primary couple polygon

Fig. 7.26

The secondary, fourth and sixth order couple polygons are closed figures. Since the second and sixth order forces are balanced and the corresponding couples are balanced for the central plane, they must be in balance for all planes. The fourth order forces, however, are unbalanced and hence the fourth order couple depends upon the choice of reference plane, e.g., the fourth order couple about the plane of crank No. 1 is 62·7 N m.

By interchanging cranks 2 and 4 or 5 and 7, the primary couples will be balanced, the fourth order unbalanced force remains unchanged and all the other forces and couples remain balanced.

6. *A twin-cylinder V-engine has the cylinders set at an angle of 45° with both pistons connected to a single crank. The crank radius is 75 mm and the connecting rods are 330 mm long. The reciprocating mass is 1·5 kg per line and the total rotating mass is equivalent to 2 kg at the crank radius. A balance mass is fitted opposite to the crank equivalent to 2·25 kg at a radius of 105 mm.*

Determine, for an engine speed of 1800 rev/min, the maximum and minimum values of the primary and secondary frame forces, due to inertia of the reciprocating and rotating masses. (U. Lond.)

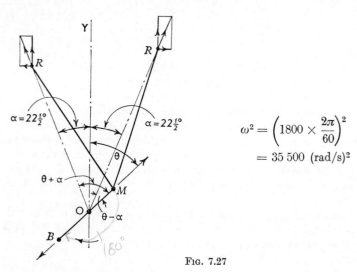

$$\omega^2 = \left(1800 \times \frac{2\pi}{60}\right)^2$$
$$= 35\,500 \ (\text{rad/s})^2$$

Fig. 7.27

Primary forces. For a crank angle θ, Fig. 7.27, measured from the axis of symmetry OY, the primary inertia force due to the reciprocating masses along the line of stroke of the L.H. piston is

$$R\omega^2 r \cos(\theta + \alpha)$$

and for the R.H. piston is

$$R\omega^2 r \cos(\theta - \alpha)$$

∴ net vertical force, including the effect of the rotating mass M at the crankpin and the balance mass B,

$$V_1 = R\omega^2 r \cos(\theta + \alpha) \cos\alpha + R\omega^2 r \cos(\theta - \alpha)\cos\alpha$$
$$+ M\omega^2 r \cos\theta - B\omega^2 b \cos\theta$$
$$= (2Rr\cos^2\alpha + Mr - Bb)\omega^2 \cos\theta$$
$$= (2\times1\cdot5\times0\cdot075\times0\cdot924^2 + 2\times0\cdot075 - 2\cdot25\times0\cdot105)\times35\,500\cos\theta$$
$$= 3770\cos\theta \ \text{N}$$

Net horizontal primary force,

$$H_1 = R\omega^2 r \cos(\theta - \alpha)\sin\alpha - R\omega^2 r \cos(\theta + \alpha)\sin\alpha$$
$$+ M\omega^2 r \sin\theta - B\omega^2 b \sin\theta$$

$$= (2Rr\sin^2\alpha + Mr - Bb)\omega^2 \sin\theta$$

$$= (2\times1.5\times0.075\times0.382^2 + 2\times0.075 - 2.25\times0.105)\times35\,500\sin\theta$$

$$= -1905\sin\theta \text{ N}$$

Resultant primary force, $\quad R_1 = \sqrt{(3770^2\cos^2\theta + 1905^2\sin^2\theta)}$

This is a maximum when $\cos\theta = 1$ and a minimum when $\cos\theta = 0$

Hence, maximum primary force = $\underline{3770 \text{ N}}$

and $\quad$ minimum primary force = $\underline{1905 \text{ N}}$

Secondary forces. The rotating masses have no effect on the secondary forces, which are due only to the second harmonic in the piston acceleration. The secondary inertia forces for the L.H. and R.H. pistons are respectively

$$R\omega^2 \frac{r}{n}\cos 2(\theta + \alpha) \quad \text{and} \quad R\omega^2 \frac{r}{n}\cos 2(\theta - \alpha)$$

Net vertical secondary force,

$$V_2 = R\omega^2 \frac{r}{n}\cos 2(\theta + \alpha)\cos\alpha + R\omega^2 \frac{r}{n}\cos 2(\theta - \alpha)\cos\alpha$$

$$= 2R\omega^2 \frac{r}{n}\cos\alpha\cos 2\alpha\cos 2\theta$$

$$= \frac{2\times1.5\times35\,500\times0.075}{330/75}\times0.924\times0.707\cos 2\theta = 1187\cos 2\theta$$

Net horizontal secondary force,

$$H_2 = R\omega^2 \frac{r}{n}\cos 2(\theta - \alpha)\sin\alpha - R\omega^2 \frac{r}{n}\cos 2(\theta + \alpha)\sin\alpha$$

$$= 2R\omega^2 \frac{r}{n}\sin\alpha\sin 2\alpha\sin 2\theta$$

$$= \frac{2\times1.5\times35\,500\times0.075}{330/75}\times0.382\times0.707\sin 2\theta = 491\sin 2\theta$$

Resultant secondary force, $\quad R_2 = \sqrt{(1187^2\cos^2 2\theta + 491^2\sin^2 2\theta)}$

This is a maximum when $\cos 2\theta = 1$ and a minimum when $\cos 2\theta = 0$

Hence, maximum secondary force = $\underline{1187 \text{ N}}$

and $\quad$ minimum secondary force = $\underline{491 \text{ N}}$

7. *Investigate the out-of-balance forces of an eight-cylinder V-engine consisting of two banks of cylinders, each having four cylinders in line and both working upon one four-throw crankshaft. The centre-lines of the two banks are inclined at angles $\frac{1}{2}\phi$ on each side of the vertical plane. The relative positions of the four cranks are $0°$, $180°$, $180°$ and $0°$ and two connecting rods work on each crank. Find the maximum values of the horizontal and vertical forces acting on the engine, in terms of the angle ϕ, the angular velocity ω of the crankshaft, the crank radius r, the connecting rod length l, and the reciprocating mass M per cylinder. State the nature and amount of the total force, (a) when $\phi = 90°$, (b) when $\phi = 60°$.* (U. Lond.)

Fig. 7.28 shows the arrangement of the cylinders and cranks; suffices 1 and 2 refer to the L.H. and R.H. banks respectively.

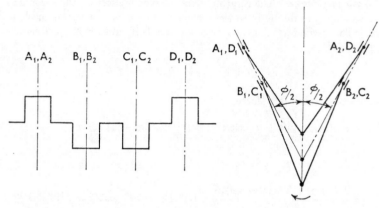

Fig. 7.28

Primary forces. The primary direct and reverse cranks are shown in Fig. 7.29(a) and (b).

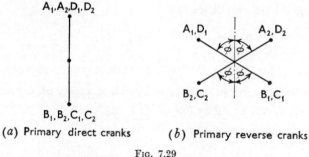

(a) Primary direct cranks   (b) Primary reverse cranks

Fig. 7.29

Due to the symmetry of the direct and reverse cranks, the primary forces are balanced.

Secondary forces. The secondary direct and reverse cranks are shown in Fig. 7.30(*a*) and (*b*).

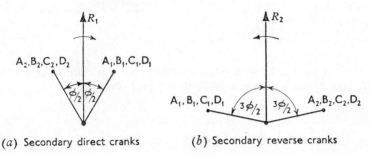

(*a*) Secondary direct cranks   (*b*) Secondary reverse cranks

FIG. 7.30

Resultant force due to direct cranks, $\qquad R_1 = 8 \times \dfrac{M}{2}\omega^2 \dfrac{r}{n} \cos \dfrac{\phi}{2}$

Resultant force due to reverse cranks, $\qquad R_2 = 8 \times \dfrac{M}{2}\omega^2 \dfrac{r}{n} \cos \dfrac{3\phi}{2}$

The maximum secondary vertical force occurs when the cranks are vertical, i.e. when R_1 and R_2 are parallel, Fig. 7.31,

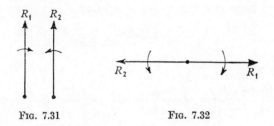

FIG. 7.31   FIG. 7.32

i.e. maximum vertical force

$$= 8 \times \frac{M}{2}\omega^2 \frac{r}{n} \cos \frac{\phi}{2} + 8 \times \frac{M}{2}\omega^2 \frac{r}{n} \cos \frac{3\phi}{2}$$

$$= \frac{8M\omega^2 r^2}{l} \cos \phi \cos \frac{\phi}{2}$$

The maximum secondary horizontal force occurs when the cranks are at 45° to the vertical, i.e. when R_1 and R_2 are in opposition, Fig. 7.32. i.e. maximum horizontal force

$$= 8 \times \frac{M}{2} \omega^2 \frac{r}{n} \cos \frac{\phi}{2} - 8 \times \frac{M}{2} \omega^2 \frac{r}{n} \cos \frac{3\phi}{2}$$

$$= \frac{8M\omega^2 r^2}{l} \sin \phi \sin \frac{\phi}{2}$$

When the cranks have rotated through an angle θ from the vertical R_1 and R_2 are inclined at 2θ to the vertical, Fig. 7.33.

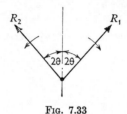

FIG. 7.33

$\therefore$ vertical force $= (R_1 + R_2) \cos 2\theta$

$$= \frac{8M\omega^2 r^2}{l} \cos \phi \cos \frac{\phi}{2} \cos 2\theta$$

and horizontal force $= (R_1 - R_2) \sin 2\theta$

$$= \frac{8M\omega^2 r^2}{l} \sin \phi \sin \frac{\phi}{2} \sin 2\theta$$

$\therefore$ resultant force,

$$R = \frac{8M\omega^2 r^2}{l} \sqrt{\left\{ \left(\cos \phi \cos \frac{\phi}{2} \cos 2\theta \right)^2 + \left(\sin \phi \sin \frac{\phi}{2} \sin 2\theta \right)^2 \right\}}$$

When $\phi = 90°$ $\qquad R = \frac{4\sqrt{2} M\omega^2 r^2}{l} \sin 2\theta$

When $\phi = 60°$, $\qquad R = \frac{2\sqrt{3} M\omega^2 r^2}{l}$

In the case when $\phi = 90°$, R varies sinusoidally as the cranks rotate, but when $\phi = 60°$, R is constant.

8. *An engine has three cylinders A, B and C in line, the distance between adjacent cylinders being a. Viewed from A towards C, the engine rotates clockwise with angular velocity ω rad/s, crank B following 120° behind crank A, and crank C 120° behind B. For each cylinder the mass of the reciprocating parts is m, the crank radius r, the length of the connecting rod nr. By considering only primary and secondary effects show that the engine is completely balanced in respect of forces and deduce an expression for the total unbalanced couple, at the instant when crank A has rotated through an angle φ from its inner dead centre.*

A V-6 engine consists of two such sets of cylinders, the plane containing the centre lines of one set, comprising cylinders A, B and C, being inclined at 45° anticlockwise, and the plane for the other set, comprising cylinders D, E and F, being inclined at 45° clockwise from the upward vertical through the single crankshaft. The connecting rod for D works on the same crankpin as that for A; similarly for E and B, F and C. Obtain expressions for the horizontal and vertical components of the total unbalanced couple acting on the engine, at an instant when the crank for cylinders A and D has rotated through an angle θ from the upward vertical. (U. Lond.)

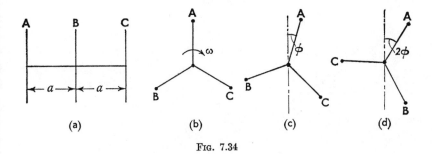

Fig. 7.34

When crank A has turned through an angle ϕ, the positions of the primary and secondary cranks are as shown in Fig. 7.34(c) and (d), respectively, and from symmetry the primary and secondary forces are balanced.

Taking the central plane as reference, the primary and secondary couples for plane A are respectively $-mra$ and $-\dfrac{mra}{n}$ and for plane B, they are $+mra$ and $+\dfrac{mra}{n}$. The couple polygons are shown in Fig. 7.35(a) and (b) and the total out-of-balance couple for the given position (i.e. the sum of the components of the closing sides in the vertical plane) is given by

$$\left\{ 2mra \cos 30° \cos (30° - \phi) + \frac{2mra}{n} \cos 30° \cos (30° + 2\phi) \right\} \times \omega^2$$

$$= \sqrt{3}\, m\omega^2 ra \left\{ \cos (30° - \phi) + \frac{\cos (30° + 2\phi)}{n} \right\}$$

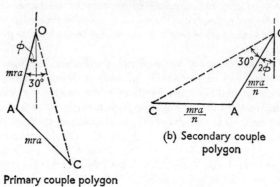

(a) Primary couple polygon

(b) Secondary couple polygon

FIG. 7.35

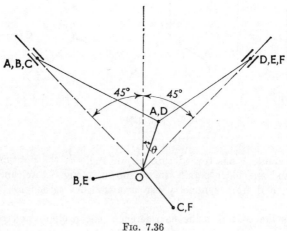

FIG. 7.36

For the V-6 engine, the arrangement is shown in Fig. 7.36.

For the L.H. bank, $\phi = 45° + \theta$

and for the R.H. bank, $\phi = -(45° - \theta)$

$$= \theta - 45°$$

Thus the couple in the plane of the L.H. bank due to A, B and C is given by

$$C_1 = \sqrt{3}\, m\omega^2 ra \left\{ \cos(30° - [45° + \theta]) + \frac{\cos(30° + [2\theta + 90°])}{n} \right\}$$

$$= \sqrt{3}\, m\omega^2 ra \left\{ \cos(-15° - \theta) + \frac{\cos(120° + 2\theta)}{n} \right\}$$

and the couple in the plane of the R.H. bank due to D, E and F is given by

$$C_2 = \sqrt{3}\, m\omega^2 ra \left\{ \cos(30° - [\theta - 45°]) + \frac{\cos(30° + [2\theta - 90°])}{n} \right\}$$

$$= \sqrt{3}\, m\omega^2 ra \left\{ \cos(75° - \theta) + \frac{\cos(-60° + 2\theta)}{n} \right\}$$

Therefore, the couple in the vertical plane,

$$V = (C_1 + C_2)\cos 45°$$

$$= \sqrt{\frac{3}{2}}\, m\omega^2 ra \left\{ \cos(-15° - \theta) + \cos(75° - \theta) \right.$$

$$\left. + \frac{\cos(120° + 2\theta) + \cos(-60° + 2\theta)}{n} \right\}$$

$$= \sqrt{3}\, m\omega^2 ra \cos(30° - \theta)$$

The couple in the horizontal plane,

$$H = (C_1 - C_2)\sin 45°$$

$$= \sqrt{\frac{3}{2}}\, m\omega^2 ra \left\{ \cos(-15° - \theta) - \cos(75° - \theta) \right.$$

$$\left. + \frac{\cos(120° + 2\theta) - \cos(-60° + 2\theta)}{n} \right\}$$

$$= \sqrt{3}\, m\omega^2 ra \left\{ \sin(30° - \theta) - \frac{\sqrt{2}}{n}\sin(30° + 2\theta) \right\}$$

9. A three-cylinder engine has the cranks spaced at equal angular intervals of 120°. Each crank is 150 mm long and each connecting rod is 625 mm long. The pitch of the cylinders is 450 mm and the speed is 500 rev/min. If the reciprocating parts per cylinder have a mass of 70 kg, find the maximum unbalanced primary and secondary effects of the reciprocating parts. (*I. Mech. E.*)

(*Ans.*: 22·35 kN m; 5·38 kN m)

10. The masses of the reciprocating parts of a four-cylinder engine are R_1, 300, 450 and R_4 kg and the centre-lines of the cylinders are 0·6, 1·2 and 0·9 m apart respectively. If the angle between the two inner cranks is 80°, find the magnitudes of R_1 and R_4 and the angular positions of the corresponding cranks for the primary forces and couples to balance.

If the crank radius is 0·3 m, connecting rod 1·5 m long, find the unbalanced secondary force at a speed of 120 rev/min. (*U. Lond.*)

(*Ans.*: R_1, 300 kg; 210° from crank 2; R_4, 318·5 kg; 248° from crank 2; 6·26 kN)

11. The spacing of the four cylinders A, B, C and D of a vertical ' in-line ' engine is 650 mm, 500 mm and 650 mm. The reciprocating masses of the inner cylinders B and C are 80 kg, and their cranks are at 60° to one another ; the stroke is 325 mm and the connecting rods are 600 mm long.

Find the magnitudes of the reciprocating masses for the outer cylinders A and D and the relative angular positions of all the cranks if all primary forces and couples are to be balanced.

What will be the maximum unbalanced secondary force acting on the base when the engine is run at 375 rev/min? (*U. Lond.*)

(*Ans.:* A, 70·4 kg ; D, 70·4 kg ; B, 0° ; C, 60° ; A, 201° ; D, 219° ; 14·34 kN)

12. In a four-cylinder in-line engine the relative angular positions of the cranks, taken in order, are 0°, 90°, 180° and 270°. For each cylinder the crank radius is 75 mm and the length of the connecting rod is 350 mm. The mass of the reciprocating parts, for each cylinder, is 7 kg. The distances between the centre-lines of adjacent cylinders, in order, are 200, 250 and 200 mm. Show that the engine is balanced in regard to primary and secondary forces.

Rotating masses are provided in the planes of No. 1 and No 4 cylinders, at 37·5 mm radius, to balance one-third of the primary couple. Find the magnitude of these masses and their angular positions. Calculate the amounts of the unbalanced primary and secondary couples at 400 rev/min. (*U. Lond.*)

(*Ans.:* 2·75 kg in plane of No. 1, at 45° to cranks 3 and 4 ; 2·75 kg in plane of No. 4, at 45° to cranks 1 and 2 ; 391 N m ; 78·75 N m)

13. The centre-lines of the cylinders of a four-crank reciprocating engine taken in order are 1·2, 1·5 and 0·9 m apart, and the reciprocating masses are x, 325, 350 and 200 kg. The stroke of each piston is 0·6 m. Find the value of x and the crank angles referred to the first crank, in order that the primary forces and couples may be in balance. If the connecting rods are each five times the length of the crank, find the secondary unbalanced force when the speed of the engine is 120 rev/min. (*U. Lond.*)

(*Ans.:* 165 kg ; 201·5° ; 66° ; 268° ; 4·675 kN m)

14. An air compressor has four vertical cylinders 1, 2, 3, 4, in line and the driving cranks at 90° interval reach their uppermost positions in this order. The cranks are 120 mm radius, the connecting rods 400 mm long, and the cylinder centre-lines 310 mm apart. The reciprocating parts for each cylinder have a mass of 20 kg and the speed of rotation is 400 rev/min. Show that there are no out-of-balance primary and secondary forces and determine the corresponding couples, indicating the positions of No. 1 crank for maximum values. The central plane of the machinery may be taken as reference plane. (*I. Mech. E.*)

(*Ans.:* Primary couple, 3675 N m, 45° and 225° to uppermost position ; secondary couple, 780 N m, 0°, 90° 180° and 270° to uppermost position)

15. A four-stroke engine has five identical cylinders with their centre-lines in one plane and spaced at equal intervals of 150 mm. The reciprocating parts per cylinder have a mass of 1·5 kg, the pistons have a stroke of 100 mm and the connecting rods are 175 mm long between centres. The cylinders are numbered consecutively from one end of the engine and the firing order is 1—4—5—3—2 at equal intervals. The engine speed is 600 rev/min.

Show that the engine is in complete balance with respect to primary and secondary forces. Find the maximum primary couple and the maximum secondary couple acting on the engine and state all the positions of crank No. 1 from its inner dead-centre position at which these maximum values occur.

Determine the total couple acting on the engine when crank No. 1 has turned 10° from its inner dead-centre position. Graphical solutions are acceptable. (*U. Lond.*)

(*Ans.:* 69·2 N m; 28° and 208° to i.d.c.; 60·6 N m; 20°, 110°, 200° and 290° to i.d.c.; 126 N m)

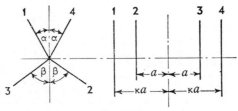

Fig. 7.37

16. The crank arrangement of a vertical four-cylinder in-line engine is shown in Fig. 7.37. The reciprocating mass and crank radius for each cylinder is given in the table below:

Cylinder number	.	.	1	2	3	4
Reciprocating mass (kg)	.	15	18	18	15
Crank radius (mm)	.	.	80	85	85	80

The ratio $\dfrac{\text{connecting rod length}}{\text{crank length}}$ is the same for all cylinders.

Find the values of α, β and κ for the engine to be balanced for primary and secondary forces and primary moments. (*U. Lond.*)

(*Ans.:* 37° 10′; 51° 15′; 1·65)

17. The centre-lines of successive cylinders of a four-cylinder in-line recipro-cating machine are spaced as follows: A to B ' *l* ' mm, B to C 250 mm, and C to D ' *l* ' mm. All the cranks have a radius of 75 mm. If the reciprocating masses at each of the outer cylinders A and D are 12·5 kg, whilst those at each of the inner cylinders are 20 kg, determine the angular disposition of the cranks for the primary and secondary forces to be in balance.

Find the value of ' *l* ' if the primary couples are to be in balance. (*U. Lond.*)

(*Ans.:* 26° 30′; 56°; 247 mm)

18. In a vertical five-cylinder engine with equal reciprocating masses the cylinders are arranged in line and their centre-lines are symmetrical about the centre cylinder 3. The distance between cylinders 1 and 5 is $2a$ and between cylinders 2 and 4 it is $2b$. When crank 3 is at the top dead centre position the crank angles taken in order from 1 to 5 are 144°, 288°; 0°, 72° and 216° respec-tively. Show that the primary and secondary forces are in balance and that the primary couples are in balance when $a/b = 1\cdot618$.

Determine the magnitude of the maximum unbalanced secondary couple for this value of a/b when the crank speed is ω, each reciprocating mass is R, and crank radius is r and the length of the connecting rod is l. (*U. Lond.*)

$$\left(Ans.:\ 4\cdot244\frac{R\omega^2 r^2 b}{l}\right)$$

19. A five-cylinder in-line engine has similar reciprocating parts at equal centre distances and the cranks are successively 72° apart. Show that the primary and secondary disturbing forces balance for all positions of the crankshaft.

If each reciprocating mass is 4 kg, each crank is 50 mm and each connecting rod is 175 mm long, and the cylinder centre distances are 100 mm, determine the maximum values of (a) the primary and (b) the secondary couples when the speed is 120 rev/min and state the positions of the central crank when these maxima occur. (*U. Lond*).

(*Ans.*: 13·35 N m ; 90°, 270° ; 2·365 N m ; 45°, 135°, 225°, 315°)

20. The six cylinders of a single-acting two-stroke diesel engine are in line and are symmetrically spaced on either side of the central plane. The cranks are spaced at 60° intervals and the centre-lines of cylinders 1 and 6 are 4·8 m apart. Cylinders 2 and 5 are 3 m apart and cylinders 3 and 4 are 1·2 m apart. The reciprocating mass per cylinder is 800 kg, the crank radius is 0·3 m, the speed is 180 rev/min and the connecting rod is 1·35 m long.

Show that primary and secondary forces are in balance for any order of firing in the cylinders and investigate the out-of-balance couple effects (primary and secondary) when the firing order is 1, 4, 5, 2, 3, 6, giving maximum values. (*U. Lond.*)

(*Ans.*: Primary couple, 25·4 kN m ; secondary couple, 127·5 kN m)

21. A four-cylinder oil engine has an additional scavenging-pump cylinder at each end of the crankshaft. The crank angles for the engine cylinders are to be, in order : 0°, 90°, 270°, 180°. Determine the crank angles for the pump cylinders to give balanced primary forces and the least possible unbalanced primary couple for the whole engine, stating the value of the couple. Determine also the magnitude and phase of the unbalanced secondary effects at 520 rev/min. Reciprocating masses per cylinder, 60 kg for the engine and 40 kg for the pumps. Crank radius, 225 mm for the engine and 175 mm for the pumps. Connecting rod 900 mm for all cylinders. Axial spacing 500 mm for all cylinders. (*U. Glas.*)

(*Ans.*: For pump crank adjacent to cylinder 4, 18° 25′ from crank 1 and between cranks 1 and 2, the other pump crank being diametrically opposite. Primary couple, 11·34 kN m ; secondary couple, 10·1 kN m ; secondary force, 8·1 kN ; engine crank 1 at 72°, 162°, 252° and 342° for maximum secondary effects)

22. An air-compressing plant consists of a four-cylinder petrol engine and a two-cylinder air compressor, coupled together with all the cylinders vertical. The relative angular positions of the six cranks and their axial distances are given in the following table :

Crank No.	Angular position (degrees)	Axial distance (mm)
Engine cranks 1	0	0
2	180	120
3	180	240
4	0	360
Compressor cranks 5	90	640
6	270	800

In each engine cylinder the crank radius is 50 mm, the connecting rod length is 200 mm and the reciprocating mass is 2 kg. The corresponding values for each compressor cylinder are 60 mm, 240 mm and 3 kg. The speed of the crankshaft is 900 rev/min.

Find, for the complete unit, the maximum values of the primary and secondary forces and moments due to the inertia of the reciprocating masses, and state the angular positions of the crankshaft at which these maximum values occur. In calculating moments, take as the reference plane the plane of crank No. 2, (*U. Lond.*)

(*Ans.:* Primary forces balanced ; secondary force, 88·6 N when crank 1 is in a vertical or horizontal position ; primary couple, 255 N m when crank 1 is in either horizontal position ; secondary couple, 425 N m when crank 1 is in a vertical or horizontal position)

23. A single-cylinder horizontal steam engine operates a reciprocating pump in a vertical line by means of a crank set opposite to the engine crank. The particulars are as follows :

	Engine	Pump
Crank radius (m)	0·3	0·15
Connecting rod length (m) . . .	1·5	0·675
Reciprocating mass (kg) . . .	75	30
Rotating mass at crank radius (kg) .	40	15

Determine, for a speed of 250 rev/min, the primary and secondary frame forces due to the inertia of the given masses. What part of the primary force could be directly balanced by a mass attached to the shaft, and what would be the nature and magnitude of the primary force then remaining ? (*U. Lond.*)

(*Ans.:* Maximum primary, 22·15 kN ; minimum primary, 3·59 kN ; maximum secondary, 3·15 kN ; minimum secondary, zero ; minimum primary force is always present and could be balanced directly leaving a force fluctuating from zero to 18·47 kN)

24. In a twin cylinder V-engine the cylinders are inclined at an angle x on either side of a vertical centre-line, and the two connecting rods are attached to the same crankpin. The mass of the reciprocating parts for each cylinder is 1 kg, the crank radius is 40 mm, and the connecting rod is 150 mm long. The crankshaft speed is 3000 rev/min.

Find the maximum values of the resultant primary and secondary forces, the directions in which they act and the crank position at which they occur, for the two cases :

(*a*) $x = 45°$ (90° V) ; (*b*) $x = 90°$ (opposed twin)   (*U. Lond.*)

(*Ans.:* (*a*) Primary force 3·93 kN, constant, acting along crank ; secondary force 1·484 kN, horizontal, when crank is at 45°, 135°, 225° and 315° to axis of symmetry, (*b*) Primary force 7·86 kN, horizontal, when crank is horizontal ; secondary forces balanced)

25. An engine has two cylinders arranged in the form of a V, the centre-lines of the cylinders being in one plane and inclined at 45° on either side of a central vertical line. The two connecting rods work on the same crank. The mass of the reciprocating parts for each cylinder is 0·5 kg, the crank radius is 35 mm, and the connecting rod length is 130 mm.

Show that the vertical force on this engine due to secondary inertia forces is zero, and that if suitable balance masses are attached to the crankshaft the primary inertia forces can also be reduced to zero. For this value of the balance masses, find the greatest out-of-balance force acting on the engine in the horizontal direction when the crankshaft speed is 3000 rev/min. (*U. Lond.*)

(*Ans.:* 657 N)

26. An engine has four cylinders arranged in two Vee banks, all being mounted symmetrically above the main shaft. Each bank consists of two cylinders whose centre-lines are inclined at 90° to one another and whose connecting rods act on a single crankpin. The cranks of the two banks are set 180° apart.

G

Investigate the state of balance of the engine with respect to primary and secondary forces and couples and determine the magnitude of any unbalanced forces or couples when the engine is running at 1500 rev/min. The reciprocating mass per cylinder is 5 kg, the crank radius 80 mm, the connecting rods 310 mm long and the banks are spaced 210 mm apart.

Could the balance of the engine be improved by adding masses to the crank-shaft? If so, find the magnitude and position of the balance masses to be added at 100 mm radius. (*U. Lond.*)

Ans.: Primary couple, 2·07 kN m; secondary force, 7·18 kN; 4 kg in each bank opposite crank.

27. A three-cylinder compressor is arranged as shown in Fig. 7.38 with the three pistons operated from a common crank and angles of 60° between the axes of the cylinders. The reciprocating masses are 9 kg for cylinders 1 and 3, and

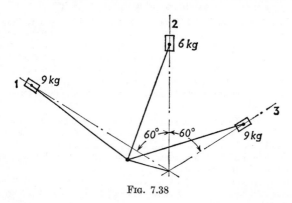

Fig. 7.38

6 kg for cylinder 2. The crank radius is 40 mm; and the ratio of connecting rod to crank is 5 for each cylinder. The speed is 960 rev/min. Find the primary unbalanced force at this speed, and obtain a suitable balance mass at crank radius for vertical primary balance. Find also the unbalanced secondary force. (*U. Glas.*) (*Ans.:* 4·24 kN to 5·45 kN; 10·5 kg; 0·121 kN to 1·09 kN)

28. An air compressor has three cylinders. The central cylinder is vertical, the other two cylinders being disposed symmetrically with their centre-lines at 60° from the vertical. The outer cylinders are in the same plane, with their connecting rods working on the same crank, but the central cylinder is displaced from this plane by an amount sufficient to allow its connecting rod to work on a separate crank. The two cranks are 180° apart. The reciprocating mass for each cylinder is 2 kg, the crank radius is 50 mm and the connecting rods are 190 mm long. Find the greatest force exerted on the machine in the vertical direction due to the inertia of the reciprocating parts at a speed of 500 rev/min. The calculation of inertia moments is not required. (*U. Lond.*) (*Ans.:* 172·6 N)

29. An air compressor has three cylinders, the centre-line of cylinder No. 2 being vertical and in a plane X and the centre-lines of cylinders No. 1 and No. 3 being inclined at 30° from the vertical, as shown in Fig 7.39, and in a plane Y. Planes X and Y are parallel and at a sufficient distance apart to permit crankshaft O to have two cranks at 180° to one another. The connecting rods for cylinders No. 1 and No. 3 operate on crankpin P while the connecting rod for cylinder No. 2 operates on crankpin Q.

The reciprocating parts for each cylinder have a mass m, each crank radius is r and each connecting rod is of length l between centres. The crankshaft speed is ω.

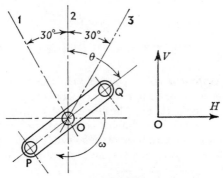

FIG. 7.39

Show that when the crank for cylinder No. 2 has turned through angle θ from its top-dead-centre position, the vertical and horizontal forces V and H acting on the air compressor due to primary and secondary unbalance can be expressed as

$$V = A \cos \theta + B \cos 2\theta$$
$$H = C \sin \theta + D \sin 2\theta$$

and determine the values of the constants A, B, C and D. (*U. Lond.*)

(*Ans.*: $A = -\dfrac{m}{2}\omega^2 r$; $B = 1\cdot866\ m\omega^2\ \dfrac{r^2}{l}$; $C = -\dfrac{m}{2}\omega^2 r$; $D = 0\cdot866 m\omega^2\dfrac{r^2}{l}$)

CHAPTER 8

FRICTION CLUTCHES

8.1 Plate clutches. In a plate clutch, the torque is transmitted by friction between one or more pairs of co-axial annular faces maintained in contact by an axial thrust. Both sides of each plate are normally effective, so that a *single-plate* clutch has two pairs of surfaces in contact.

Fig. 8.1 shows a simplified form of *multi-plate* clutch. The inner plates are free to slide axially in grooves connected to the driving shaft and

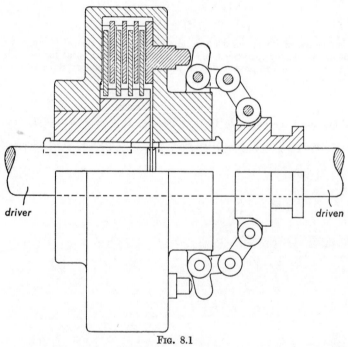

driver

driven

Fig. 8.1

the outer plates are free to slide axially in grooves connected to the driven shaft. The axial force exerted on the plates by the toggle mechanism is transmitted through each plate and so, in a clutch having n pairs of surfaces in contact, the torque transmitted is n times that for a single pair.

188

Consider two flat annular surfaces, Fig. 8.2, maintained in contact by an axial thrust W.

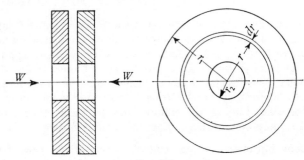

FIG. 8.2

Let $\qquad T$ = torque transmitted,

p = intensity of pressure between surfaces,

r_1, r_2 and R = outer, inner and mean radii of faces respectively

Normal force on elementary ring = $p \times 2\pi r \,.\, dr$

$$\therefore \text{ total axial force, } W = 2\pi \int_{r_2}^{r_1} pr \, dr$$

Friction force on ring = $\mu p \times 2\pi r \, dr$

$\therefore$ moment of friction force about axis = $\mu p \times 2\pi r^2 \, dr$

$$\therefore \text{ torque transmitted, } T = 2\pi\mu \int_{r_2}^{r_1} pr^2 \, dr$$

If it is assumed that the pressure is uniform over the contact area, then p is a constant, so that

$$W = p \times \pi(r_1{}^2 - r_2{}^2)$$

and
$$T = \tfrac{2}{3}\pi\mu p(r_1{}^3 - r_2{}^3)$$

$$= \tfrac{2}{3}\mu W \frac{r_1{}^3 - r_2{}^3}{r_1{}^2 - r_2{}^2}^* \qquad . \qquad . \qquad . \qquad (8.1)$$

If it is assumed that the wear is uniform over the contact area, then since

$$\text{wear} \propto \text{pressure} \times \text{velocity}$$

$$\propto \text{pressure} \times \text{radius}$$

i.e. $\qquad\qquad\quad pr = \text{constant} \;(= C)$

$$\therefore \; W = 2\pi C(r_1 - r_2)$$

* NOTE: As $r_2 \to r_1$, $T \to \mu W R$.

and
$$T = \pi\mu C(r_1{}^2 - r_2{}^2)$$
$$= \mu W \frac{(r_1 + r_2)}{2} = \mu W R \quad . \qquad . \qquad . \quad (8.2)$$

The assumption applicable to a particular clutch depends upon its condition ; for a new clutch the pressure will be approximately uniform, but for a worn clutch the uniform wear theory is more appropriate. Since the uniform pressure theory gives a higher friction torque than the uniform wear theory the latter theory should always be used, unless otherwise stated. Conversely, in calculating the power lost in friction at a pivot, footstep or collar bearing, the uniform pressure theory should be used. In this way, any error in assumption is on the safe side.

8.2 Cone clutches. A cone clutch consists of one pair of friction faces only, the area of contact being a frustum of a cone, Fig. 8.3.

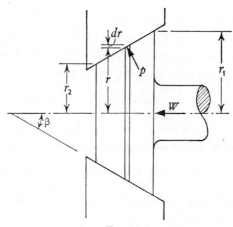

Fig. 8.3

If p is the normal pressure between the surfaces,

normal force on elementary ring $= p \times 2\pi r\, dr \operatorname{cosec} \beta$

Axial component of this force $= p \times 2\pi r\, dr \operatorname{cosec} \beta \times \sin \beta$
$$= 2\pi p r\, dr$$

$\therefore$ total axial force, $W = 2\pi \displaystyle\int_{r_2}^{r_1} pr\, dr$

Friction force on ring $= \mu p \times 2\pi r\, dr \operatorname{cosec} \beta$

$\therefore$ moment of force about axis $= \mu p \times 2\pi r^2\, dr \operatorname{cosec} \beta$

$\therefore$ torque transmitted, $T = 2\pi\mu \operatorname{cosec} \beta \displaystyle\int_{r_2}^{r_1} pr^2\, dr$

If p is assumed constant,
$$T = \tfrac{2}{3}\mu W \frac{r_1{}^3 - r_2{}^3}{r_1{}^2 - r_2{}^2} \operatorname{cosec} \beta \quad . \quad (8.3)$$

If pr is assumed constant,
$$T = \frac{\mu W}{2}(r_1 + r_2) \operatorname{cosec} \beta$$
$$= \mu W R \operatorname{cosec} \beta \quad . \quad . \quad (8.4)$$

8.3 Centrifugal clutches.

A centrifugal clutch consists of a number of shoes which can move in radial guides and bear on the inside of an annular rim, Fig. 8.4. The outer surfaces of the shoes are covered with

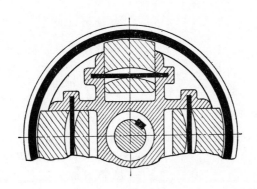

FIG. 8.4

a friction material, and as the speed rises the centrifugal force on the shoes causes them to transmit power by friction to the rim.

Springs are fitted to keep the shoes clear of the rim at low speeds and thus allow the motor to gain speed before taking up the load.

Let n = number of shoes,

F = centrifugal force on each shoe,

P = inward force on each shoe exerted by spring,

R = inside radius of rim,

μ = coefficient of friction between shoe and rim.

Then net radial force between each shoe and rim = $F - P$

$\therefore$ friction force on rim = $\mu(F - P)$

$\therefore$ friction torque on rim = $\mu R(F - P)$

$\therefore$ total friction torque = $n\mu R(F - P)$ . . (8.5)

1. *A shaft F is connected to a co-axial shaft G by a single-plate clutch, with two pairs of friction surfaces whose outer and inner diameters are 120 mm and 70 mm respectively. The total axial load on the clutch is 450 N and the coefficient of friction 0·35. Shaft G carries a pinion P gearing with a spur wheel S on a parallel shaft H. The masses and radii of gyration of the three shafts, F, G and H—with attached masses—are 12·5 kg, 80 mm; 20 kg, 70 mm; 37·5 kg, 120 mm respectively.*

Determine the minimum time in which the speed of H can be raised from 500 rev/min to 1500 rev/min by a torque applied to shaft F, and the required gear ratio between shafts G and H. Assume 'uniform wear' on the clutch surfaces.

(U. Lond.)

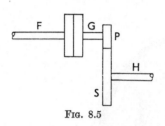

Fig. 8.5

Let the gear ratio, $\dfrac{N_S}{N_P} = n$, Fig. 8.5.

Then total equivalent moment of inertia of G

$$= I_G + n^2 I_H$$
$$= 20 \times 0·07^2 + n^2 \times 37·5 \times 0·12^2$$
$$= 0·098 + 0·54n^2 \text{ kg m}^2$$

Torque transmitted by clutch $= \mu W R \times 2$ from equation (8.2)
$$= 0·35 \times 450 \times 0·0475 \times 2$$
$$= 14·95 \text{ N m}$$

This is the maximum torque which can be transmitted by the clutch, so that

maximum acceleration of G, $\alpha_G = \dfrac{T}{I} = \dfrac{14·95}{0·098 + 0·54n^2}$ rad/s²

∴ maximum acceleration of H, $\alpha_H = n \times \alpha_G = \dfrac{14·95n}{0·098 + 0·54n^2}$ rad/s²

For α_H to be a maximum, $\dfrac{d}{dn}(\alpha_H) = 0$

from which
$$n = 0·426 \quad \text{or} \quad \dfrac{1}{2·345}$$

$$\therefore \alpha_H = \dfrac{14·95 \times 0·426}{0·098 + 0·54 \times 0·426^2} = 32·5 \text{ rad/s}^2$$

$$\therefore \text{ minimum time required} = \dfrac{2\pi}{60}\left(\dfrac{1500 - 500}{32·5}\right) = \underline{3·22 \text{ s}}$$

NOTE: provided that a sufficient torque is available at all times to transmit 14·95 N m to shaft G and to accelerate shaft F, the moment of inertia of shaft F is irrelevant.

2. *A machine is driven by a variable speed motor which develops a constant torque of 35 N m. The rotating parts of the machine have a moment of inertia of 3·5 kg m² and those for the motor, 2 kg m². On the shaft between the machine and the motor there is a friction clutch which can transmit a torque of 55 N m but no more.*

The motor and machine are running at a uniform speed of 600 rev/min when the clutch is accidentally disengaged for 4 s and then let into engagement again.

(a) Find the time during which the clutch slips after re-engagement.

(b) Find the loss of energy due to clutch slip, showing the effect of this loss on the energy supplied by the motor and on that supplied to the machine.

(U. Lond.)

(a) After disengagement, the motor torque of 35 N m causes an acceleration of the motor and the motor speed, N_1, at the end of the 4 s is given by

$$35 = 2 \times \frac{2\pi}{60} \frac{(N_1 - 600)}{4}$$

from which $\qquad N_1 = 1268$ rev/min

At the normal speed, the resisting torque on the machine is equal to the motor torque, i.e. 35 N m. After disengagement, this torque causes a deceleration of the machine and the machine speed, N_2, at the end of the 4 s is given by

$$35 = 3·5 \times \frac{2\pi}{60} \frac{(600 - N_2)}{4}$$

from which $\qquad N_2 = 218$ rev/min

After re-engagement, the torque retarding the motor is $55 - 35 = 20$ N m and the torque accelerating the machine is $55 - 35 = 20$ N m.

If N is the common speed when clutch slip ceases and t is the time of clutch slip, then

$$\text{for the motor, } 20 = 2 \times \frac{2\pi}{60} \frac{(1268 - N)}{t} \qquad . \qquad . \qquad (1)$$

and $\qquad$ for the machine, $20 = 3·5 \times \dfrac{2\pi}{60} \dfrac{(N - 218)}{t} \qquad . \qquad . \qquad (2)$

Hence, from equations (1) and (2),

$$N = 600 \text{ rev/min} \quad \text{and} \quad t = \underline{7 \text{ s}}$$

Alternatively, since equal torques are applied to both motor and machine during disengagement and re-engagement, speed when slipping ceases = 600 rev/min and time taken = $\frac{35}{20} \times 4 = \underline{7 \text{ s}}$.

(b) During acceleration and deceleration periods,

$$\text{average speed of motor} = \frac{2\pi}{60}\left(\frac{1268 + 600}{2}\right) = 97\cdot8 \text{ rad/s}$$

During deceleration and acceleration periods,

$$\text{average speed of machine} = \frac{2\pi}{60}\left(\frac{218 + 600}{2}\right) = 42\cdot8 \text{ rad/s}$$

After re-engagement,

$$\text{angle turned through by motor, } \theta_1 = 97\cdot8 \times 7 \text{ rad}$$

and　　　　　angle turned through by machine, $\theta_2 = 42\cdot8 \times 7$ rad

$$\therefore \text{ energy lost in slipping} = T(\theta_1 - \theta_2) = 55(97\cdot8 - 42\cdot8) \times 7$$
$$= \underline{21\ 200 \text{ J}}$$

Normal speed of motor and machine

$$= \frac{2\pi}{60} \times 600 = 62\cdot8 \text{ rad/s}$$

$\therefore$ work normally done by motor in 11 s

$$= 35 \times 62\cdot8 \times 11 \text{ J}$$

Work done during acceleration and deceleration period

$$= 35 \times 97\cdot8 \times 11 \text{ J}$$

$\therefore$ increase in energy supplied by motor

$$= 35(97\cdot8 - 62\cdot8) \times 11$$
$$= \underline{13\ 470 \text{ J}}$$

Work normally done on machine in 11 s

$$= 35 \times 62\cdot8 \times 11 \text{ J}$$

Work done during deceleration and acceleration period

$$= 55 \times 42\cdot8 \times 7 \text{ J}$$

$\therefore$ reduction in energy supplied to machine

$$= 35 \times 62\cdot8 \times 11 - 55 \times 42\cdot8 \times 7$$
$$= \underline{7700 \text{ J}}$$

3. *Two co-axial shafts, A and B, carrying masses of moment of inertia, respectively, 3·6 and 0·6 kg m², are coupled together by a hydraulic clutch. Initially no torque is being transmitted and both shafts revolve at 2000 rev/min. A steady driving torque of 100 N m is then applied to A and, simultaneously, a resisting torque of the same magnitude acts on B. If the torque transmitted by the clutch is given by $2(\omega - \omega')^2$ N m, in which ω and ω' are the respective instantaneous speeds of A and B in rad/s, find the final steady speeds of the two shafts and the power transmitted.* (U. Lond.)

Torque transmitted by the clutch, $T_c = 2(\omega - \omega')^2$

Accelerating torque on shaft A $= 100 - T_c$

and   decelerating torque on shaft B $= 100 - T_c$

Since these torques are equal at all times, the changes in momentum of A and B must be equal. If Ω and Ω' are the final steady speeds,

$$I_a\left(\Omega - 2000 \times \frac{2\pi}{60}\right) = I_b\left(2000 \times \frac{2\pi}{60} - \Omega'\right)$$

i.e. $3·6(\Omega - 209·4) = 0·6(209·4 - \Omega')$

or $6\Omega + \Omega' = 1465·8$.   .   (1)

When equilibrium has been attained, the clutch transmits 100 N m,

i.e. $100 = 2(\Omega - \Omega')^2$

or $\Omega - \Omega' = 7·071$.   .   .   (2)

∴ from equations (1) and (2), $\Omega = 210·41$ rad/s

and $\Omega' = 203·34$ rad/s

i.e. $N = \underline{2009·6 \text{ rev/min}}$ and $N' = \underline{1942·2 \text{ rev/min}}$

Power transmitted $= T_c \times \Omega' = 100 \times 203·34 = 20\ 334$ W

$= \underline{20·334 \text{ kW}}$

Alternative solution :

Applying $T = I\alpha$ to each shaft, then

for shaft A, $100 - 2(\omega - \omega')^2 = 3·6\dfrac{d\omega}{dt}$

and for shaft B, $2(\omega - \omega')^2 - 100 = 0·6\dfrac{d\omega'}{dt}$

Solving these simultaneous differential equations and applying the conditions that when $t = 0$, $\omega = \omega' = 2000 \times (2\pi/60)$ rad/s, the values of ω and ω' are obtained after any time t. Substituting $t = \infty$, ω and ω' are found to be 210·41 and 203·34 rad/s respectively, as before.

4. *In Fig. 8.6 part of a centrifugal brake is shown. The housing B, concentric with the fixed brake drum D, rotates about the axis Q and carries*

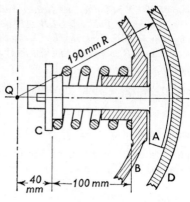

FIG. 8.6

a number of brake shoes A ; concentric with each shoe spindle a close coiled helical spring exerts an inward radial thrust on a collar C secured to each spindle.

The mass of each shoe assembly is 2 kg, and its mass centre is 150 mm from the centre Q when the shoe facing is in contact with the brake drum as shown. The spring of total mass 1·2 kg has a stiffness of 50 kN/m and a free length of 110 mm. The coefficient of friction between shoe and drum is 0·3.

*For a housing speed of 60 rad/s, find the braking torque due to one shoe
(a) neglecting the effect of centrifugal acceleration on the mass of the spring,
(b) allowing for the reduction in the spring force on the collar C due to the centrifugal effect of the spring mass.*   (U. Lond.)

(*a*) During engagement,

$$\text{compressive force in spring} = (0\cdot11 - 0\cdot1) \times 50 \times 10^3$$

$$= 500 \text{ N}$$

$$\text{Centrifugal force on shoe} = 2 \times 60^2 \times 0\cdot15$$

$$= 1080 \text{ N}$$

∴ net outward force $= 1080 - 500 = 580$ N

∴ tangential force $= 0\cdot3 \times 580 = 174$ N

∴ torque per shoe $= 174 \times 0\cdot19 = \underline{\underline{33 \text{ N m}}}$

(b) Let the compressive force in the spring at a radius r be P, allowing for the centrifugal force on the spring, Fig. 8.7. If this force increases

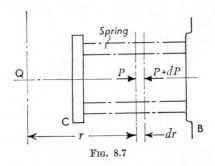

Fig. 8.7

to $P + dP$ as the radius increases to $r + dr$ and the centrifugal force on the element of the spring is dF, then

$$P + dP = P + dF$$

i.e.
$$dP = dF = \left(\frac{dr}{0 \cdot 1} \times 1 \cdot 2 \right) \times 60^2 \times r$$

$$= 43\,200r\,dr$$

$$\therefore P = 21\,600r^2 + C$$

$$\text{Change in spring force} = P - 500 \text{ N}$$

$$\therefore \text{ change in length of element} = K(P - 500)\,dr$$

where K is a constant depending upon the properties of the spring.

$$\therefore \text{ total change in length of spring} = K \int_{0 \cdot 04}^{0 \cdot 14} (P - 500)\,dr$$

But the length of the spring remains unaltered,

$$\therefore \int_{0 \cdot 04}^{0 \cdot 14} (21\,600r^2 + C - 500)\,dr = 0$$

from which
$$C = 307$$

$\therefore$ when $r = 0 \cdot 04$ m
$$P = 21\,600 \times 0 \cdot 04^2 + 307$$

$$= 341 \cdot 5 \text{ N}$$

$$\therefore \text{ net outward force} = 1080 - 341 \cdot 5 = 738 \cdot 5 \text{ N}$$

$$\therefore \text{ torque per shoe} = \frac{738 \cdot 5}{580} \times 33 = \underline{\underline{42 \text{ N m}}}$$

5. An electric motor drives a co-axial rotor through a single-plate clutch which has two pairs of driving surfaces each of 275 mm external and 200 mm internal diameter; the total spring load pressing the plates together is 550 N. The mass of the motor armature and shaft is 800 kg and its radius of gyration is 260 mm; the rotor has a mass of 1400 kg and its radius of gyration is 220 mm.

The motor is brought up to a speed of 1250 rev/min; the current is then switched off and the clutch suddenly engaged. Determine the final speed of the motor and rotor, and find the time taken to reach that speed and the kinetic energy lost during the period of slipping. How long would slipping continue if a constant torque of 55 N m were maintained on the armature shaft? Co-efficient of friction = 0.35. (*U. Lond.*)

(*Ans.* : 555 rev/min; 86·1 s; 257·5 kJ; 154·6 s)

6. A rotor A on a shaft X is connected through light gearing G with a shaft Y carrying a rotor B; the moments of inertia of the two rotors are 5 and 2 kg m² respectively. Between A and G is a single-plate clutch with two pairs of friction surfaces, of inner and outer diameters 150 mm and 250 mm; the total thrust of the springs is 700 N. The coefficient of friction on the friction surfaces is 0·35 and the speed ratio from X to Y is 1 : 4 when slipping has ceased.

The rotor A is given a speed of 2500 rev/min with the clutch disengaged; if the clutch is then engaged determine the final speeds of the two rotors and the time of slipping, stating clearly any assumptions made. The inertia of the clutch, the shafts and the gearing may be neglected. (*U. Lond.*)

(*Ans.* : 338 rev/min; 1352 rev/min; 23·1 s)

7. A shaft A carries a rotor at one end and the internal element of a cone clutch at the other; the total mass is 240 kg and the radius of gyration is 150 mm. The clutch has a mean radius of 125 mm and a half cone angle of 12°. It is lined with material for which the limiting coefficient of friction is 0·3 and the axial thrust of the operating spring is 550 N. The external element of the clutch is fixed to a gear wheel, of diameter 600 mm, mass 48 kg and radius of gyration 200 mm, which gears with a pinion of 150 mm diameter on a shaft B. The rotating parts on B have a mass of 28 kg, and a radius of gyration of 75 mm. Initially A is rotating at 300 rev/min and the gear wheel and shaft B are at rest. The gear is then clutched in. Determine the speeds of the shafts when slip ceases; and, during slipping, the time, the energy lost, and the tangential force at the gear teeth. (*U. Lond.*)

(*Ans.* : 164·5 rev/min; 658 rev/min; 0·773 s; 1210 J; 187·5 N)

8. Two co-axial machine rotors having moments of inertia I_1 and I_2 and running at speeds ω_1 and ω_2 respectively are engaged by a friction clutch. Show that when the final common speed is attained, the energy lost in the clutch is proportional to the square of the difference between the initial speeds and is given by

$$\frac{1}{2}\frac{I_1 I_2}{I_1 + I_2}(\omega_1 - \omega_2)^2$$

Show that for the particular case of a relatively small rotor being started from rest by engagement with a large rotor, the energy lost in the clutch is approximately equal to that given to the rotor. (*I. Mech. E.*)

9. Two shafts A and B are connected by a friction clutch. Keyed to shaft B is a spur wheel of 90 teeth gearing with a pinion of 30 teeth on the driving motor shaft. With the clutch disengaged and power switched off from the motor the shaft A is at rest and the motor pinion is rotating at 900 rev/min. If the clutch is suddenly engaged determine the common speed of A and B when slipping has ceased. Details of the component parts are :

	Mass (kg)	Radius of gyration (mm)
Shaft A assembly . . .	200	300
Shaft B assembly . . .	90	150
Motor pinion and armature .	22	75

(*U. Lond.*)   (*Ans. :* 44·6 rev/min)

10. Two co-axial shafts A and B carry rotors with moments of inertia of 7·2 and 1·2 kg m² respectively. The shafts are connected by a hydraulic clutch in which the torque is $2(\omega_A - \omega_B)^2$ N m where ω_A and ω_B are the angular velocities of the shafts in rad/s. Initially both shafts are rotating at 2000 rev/min and no torque is transmitted. A constant driving torque of 100 N m is then applied to shaft A and an equal resisting torque to shaft B. Find the steady speeds of the two shafts and the efficiency of the drive. (*U. Lond.*)

(*Ans. :* 2009·6 rev/min; 1942·2 rev/min; 96·6%)

11. Two flywheels A and B are keyed to two shafts in line which can be coupled together by a friction clutch, during the engagement of which the friction moment increases from zero uniformly with the time. The moments of inertia of the flywheels are A = 1·2 kg m², B = 2·0 kg m², and initially A revolves freely at 600 rev/min while B is at rest. The clutch is then engaged and slipping ceases after 3 s.

Find (*a*) the final speed of revolution ; (*b*) the kinetic energy lost by slipping ; (*c*) the number of revolutions of slipping ; (*d*) the friction of the clutch at the final instant of slipping. (*U. Lond.*)

(*Ans. :* 225 rev/min; 1·15 kJ; 20; 31·4 N m)

12. A motor drives a machine through a friction clutch which slips when the torque on it reaches 40 N m. The moment of inertia of the motor armature is 1·6 kg m² and that of the rotating part of the machine is 3·0 kg m². The torque developed by the motor is 27 N m assumed constant at all speeds and when the clutch is engaged the steady speed of motor and machine is 500 rev/min. At a given instant the clutch is disengaged and remains so for 4 s and then it is re-engaged. Find the time of slipping after re-engagement and determine how much energy is lost during slipping. (*U. Lond.*)

(*Ans. :* 8·3 s; 17·18 kJ)

CHAPTER 9

BELT DRIVES AND SHOE BRAKES

9.1 Centrifugal and driving tensions. Consider a flat belt partly wound round a pulley of radius r, Fig. 9.1. If the mass of the belt per unit length is m and its speed is v, the centrifugal force, F, acting on an element subtending an angle, $d\theta$

$$= mr \, d\theta \cdot \frac{v^2}{r} = mv^2 \, d\theta$$

If T_c is the tension in the belt due to this centrifugal force, then, resolving forces on the element radially:

$$mv^2 \, d\theta = 2T_c \frac{d\theta}{2}$$

$$\therefore T_c = mv^2 \qquad . \qquad . \qquad . \qquad (9.1)$$

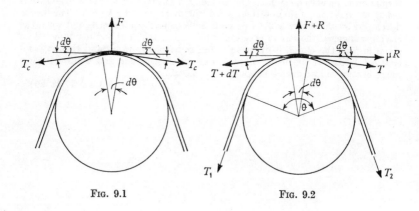

FIG. 9.1   FIG. 9.2

If the belt is transmitting power, let T_1 and T_2 be the *total* tight and slack side tensions respectively when slipping is about to occur and θ be the angle of lap, Fig. 9.2. An element subtending an angle $d\theta$ is subjected to tensions T and $T + dT$ at its two ends and also to the reaction, R, between the belt and pulley and the centrifugal force, F.

200

Resolving forces radially :

$$(T + dT)\frac{d\theta}{2} + T\frac{d\theta}{2} = R + F$$

$$= R + mv^2\,d\theta$$

$$\therefore R = (T - mv^2)\,d\theta = (T - T_c)\,d\theta$$

Resolving forces tangentially :

$$(T + dT) - T = \mu R$$

$$\therefore dT = \mu(T - T_c)\,d\theta$$

$$\therefore \int_{T_2}^{T_1}\frac{dT}{T - T_c} = \int_0^\theta \mu\,d\theta$$

from which

$$\frac{T_1 - T_c}{T_2 - T_c} = e^{\mu\theta} \qquad . \qquad . \qquad . \qquad . \qquad (9.2)$$

$T_1 - T_c$ and $T_2 - T_c$ are the *effective* driving tensions.

If the diameters of the driving and driven pulleys are unequal, the belt will slip first on the pulley having the smaller angle of lap, i.e. on the smaller pulley.

$$\text{The power transmitted} = (T_1 - T_2)v \qquad . \qquad . \qquad . \qquad (9.3)$$

$$= \{(T_1 - T_c) - (T_2 - T_c)\}v$$

$$= (T_1 - T_c)\left(1 - \frac{1}{e^{\mu\theta}}\right)v \qquad . \qquad (9.4)$$

For given values of T_1, μ and θ, the velocity at which the power transmitted is a maximum is given by

$$\frac{d}{dv}\{(T_1 - T_c)v\} = 0$$

i.e.

$$\frac{d}{dv}\{T_1 v - mv^3\} = 0$$

i.e.

$$T_1 - 3mv^2 = 0$$

or

$$T_c = \frac{T_1}{3} \qquad . \qquad . \qquad . \qquad (9.5)$$

The maximum power is then obtained by substituting this value of T_c and the corresponding value of v in equation (9.4).

If the mass or velocity of the belt is negligible, equations (9.2) and (9.4) reduce to

$$\frac{T_1}{T_2} = e^{\mu\theta} \qquad . \qquad . \qquad . \qquad . \qquad . \qquad (9.6)$$

and

$$\text{power} = T_1\left(1 - \frac{1}{e^{\mu\theta}}\right)v \qquad . \qquad . \qquad . \qquad (9.7)$$

9.2 Modification for V-grooved pulley. For a V-grooved pulley, the normal force between the belt or rope and the pulley is increased since the radial component of this force must equal R. Thus, if the *semi-angle* of the groove is β, Fig. 9.3,

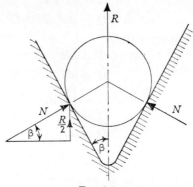

FIG. 9.3

$$N = \frac{R}{2} \operatorname{cosec} \beta$$

$$\therefore \text{ frictional resistance} = 2\mu N$$

$$= \mu R \operatorname{cosec} \beta$$

The friction force is therefore increased in the ratio $\operatorname{cosec} \beta : 1$, so that the V-grooved pulley is equivalent to a flat pulley having a coefficient of friction of

$$\mu \operatorname{cosec} \beta \qquad . \qquad . \qquad . \qquad . \qquad (9.8)$$

9.3 Initial tension. The belt is assembled with an initial tension, T_0. When power is being transmitted, the tension in the tight side increases from T_0 to T_1 and on the slack side decreases from T_0 to T_2. If the belt is assumed to obey Hooke's law and its length to remain constant, then the increase in length of the tight side is equal to the decrease in length of the slack side,

i.e. $$T_1 - T_0 = T_0 - T_2$$

since the lengths and cross-sectional areas of the belt are the same on each side.

Hence $$T_1 + T_2 = 2T_0 \qquad . \qquad . \qquad . \qquad . \qquad (9.9)$$

9.4 Belt creep. Due to the elasticity of the material, the belt will be longer per unit mass on the tight side than on the slack side and therefore, since a constant mass of belt passes any section in a given time, the velocity is slightly higher on the tight side than on the slack side. The peripheral velocity of the driver corresponds to the tight side velocity

and that of the follower to the slack side velocity, and in passing over
each pulley, the change in velocity is accounted for by a slight movement
of the belt relative to the pulley—this is termed 'belt creep'. The
angle γ over which creep takes place corresponds to that in which the
tension changes from T_1 to T_2, Fig. 9.4, and is given by

$$\frac{T_1}{T_2} = e^{\mu\gamma} \qquad . \qquad . \qquad . \qquad . \quad (9.10)$$

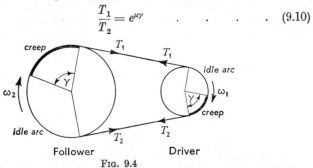

Fig. 9.4

The angle γ is measured from the point at which the belt leaves the
pulley. When γ becomes equal to θ, general slipping is about to occur.

If e_1 is the strain in the tight side and e_2 the strain in the slack side
and m is the mass per unit length of *unstrained* belt then the mass of
belt passing a given point per unit time

$$= \frac{mv_1}{1 + e_1} = \frac{mv_2}{1 + e_2}$$

$$\therefore \frac{v_2}{v_1} = \frac{1 + e_2}{1 + e_1} = (1 + e_2)(1 + e_1)^{-1}$$

$$\simeq (1 + e_2)(1 - e_1) \simeq 1 - e_1 + e_2$$

$$= 1 - \frac{(T_1 - T_2)}{aE} \qquad . \qquad . \qquad . \qquad . \quad (9.11)$$

where a and E are the cross-sectional area and modulus of elasticity
respectively. The effect of creep is therefore to reduce the ratio of the
pulley speeds, $\dfrac{\omega_2}{\omega_1}$, from $\dfrac{r_1}{r_2}$ to $\dfrac{r_1}{r_2}\left\{1 - \dfrac{(T_1 - T_2)}{aE}\right\}$

9.5 External and internal shoe brakes. The brake shoe is pivoted
at a fixed point and the other end is subjected to a force which presses
it in contact with the drum. The shoe is faced with a friction material
and may be applied either externally or internally to the brake drum.
If a point on the drum surface first makes contact with the shoe at the
end nearest the pivot, the shoe is termed a *trailing* shoe ; if it first makes
contact at the other end, the shoe is termed a *leading* shoe, the latter
giving a higher braking torque than the former for a given braking force.

The two cases of external and internal brakes are shown in Fig. 9.5 and Fig. 9.6 respectively; for the directions of rotation shown, the shoe in Fig. 9.5 is a trailing shoe and that in Fig. 9.6 is a leading shoe. In each case, let the radius of the drum be r and the distance from the pivot Q to the centre of rotation O be l. If the shoe rotates through a small angle about Q, the radial movement of any point on the arc of contact is proportional to the perpendicular distance of the radius at this point from the pivot. If the material of the brake lining is assumed to obey

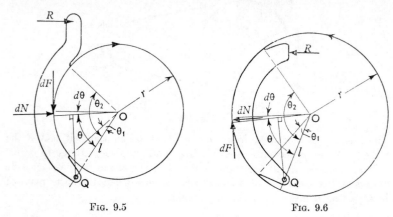

FIG. 9.5   FIG. 9.6

Hooke's Law, the pressure at this point will also be proportional to this perpendicular distance,

i.e. pressure on element at angle θ from OQ $= c \times l \sin \theta$

where c is a constant.

$$\therefore \text{ normal force on element, } dN = cl \sin \theta \times r \, d\theta$$

$$\therefore \text{ friction force on element, } dF = clr \sin \theta \, d\theta \times \mu$$

$$\therefore \text{ moment of friction force about O} = clr\mu \sin \theta \, d\theta \times r$$

$$\therefore \text{ friction torque on drum, } T = clr^2\mu \int_{\theta_1}^{\theta_2} \sin \theta \, d\theta \quad (9.12)$$

The forces exerted *by* the drum *on* the shoe, Figs. 9.7 and 9.8, are opposite in direction to those shown in Figs. 9.5 and 9.6. If R is the force applied at the end of the shoe at a perpendicular distance d from the pivot, then, taking moments about Q:

$$R \times d = \int_{\theta_1}^{\theta_2} dN \times l \sin \theta - \int_{\theta_1}^{\theta_2} dF(r - l \cos \theta)$$

$$= cl^2r \int_{\theta_1}^{\theta_2} \sin^2 \theta \, d\theta - clr\mu \int_{\theta_1}^{\theta_2} \sin \theta (r - l \cos \theta) \, d\theta. \quad (9.13)$$

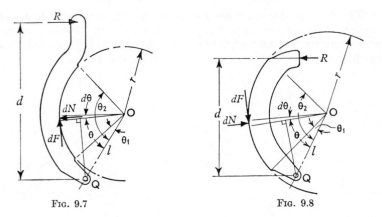

Fig. 9.7   Fig. 9.8

The constant c is determined from equations (9.12) or (9.13) and the unknown torque, T, or reaction, R, is then evaluated.

If the direction of rotation is reversed, the direction of dF is reversed; equation (9.12) is unaffected but the sign of dF changes in equation (9.13).

9.6 Graphical method. Divide the arc of contact into a number of equal parts, Fig. 9.9. Then the normal force on each element, assumed acting at its mid-point, is proportional to the perpendicular distance of its line of action from the pivot. If these forces are drawn to a convenient scale (their actual magnitudes being unknown), the direction and point of application of their resultant, N, can be obtained from the polygon of forces, this force passing through the centre O.

The friction force on each element is tangential to the surface and is equal to μ times the normal force. These forces are shown in Fig. 9.10 and if the polygon of forces is constructed, the resultant F will be perpendicular to N and of magnitude μN.

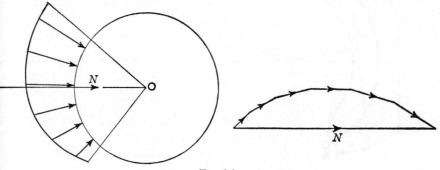

Fig. 9.9

To obtain the line of action of F, take a pole X and draw a funicular polygon across the lines of action of the friction forces. The intersection of the first and last lines of the funicular polygon then gives a point on the line of action of F and the torque on the drum, $T = Fh$.

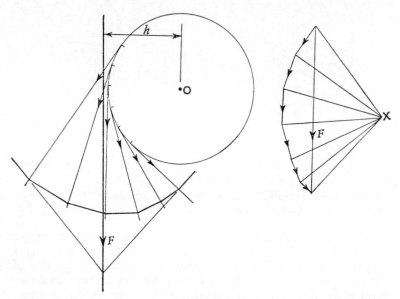

Fig. 9.10

The forces acting on the shoe are as shown in Fig. 9.11. Taking moments about Q,

$$R \times d = N \times a - F \times b$$
$$\therefore R = \frac{N(a - \mu b)}{d}$$

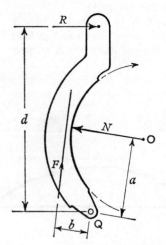

Fig. 9.11

1. *A pulley of 150 mm effective diameter running at 1500 rev/min drives a follower of 750 mm diameter, the two shafts being parallel, 1 m apart, and the free parts of the belt considered straight. The belt has a mass of 0·4 kg/m and the maximum tension is to be 720 N. If $\mu = 0\cdot4$, estimate the maximum tension difference allowing for the inertia of the belt. If the belt has a cross-sectional area of 320 mm² and E for the material is 300 MN/m², estimate the speed of the driven pulley at the maximum condition and the power transmitted to it.* (I. Mech. E.)

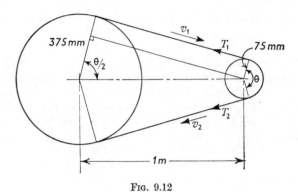

Fig. 9.12

Referring to Fig. 9.12,

$$\cos\frac{\theta}{2} = \frac{0\cdot375 - 0\cdot075}{1} = 0\cdot3$$

$$\therefore \frac{\theta}{2} = 72°\ 33' = 1\cdot266\ \text{rad}$$

∴ angle of lap on smaller pulley,

$$\theta = 2\cdot532\ \text{rad}$$

$$v = 1500 \times \frac{2\pi}{60} \times 0\cdot075 = 11\cdot78\ \text{m/s}$$

$$\therefore T_c = mv^2 \qquad . \qquad . \qquad . \qquad \text{from equation (9.1)}$$
$$= 0\cdot4 \times 11\cdot78^2 = 55\cdot6\ \text{N}$$

$$\frac{T_1 - T_c}{T_2 - T_c} = e^{\mu\theta} \qquad . \qquad . \qquad . \qquad \text{from equation (9.2)}$$

i.e. $$\frac{720 - 55\cdot6}{T_2 - 55\cdot6} = e^{0\cdot4 \times 2\cdot532} = 2\cdot754$$

from which $$T_2 = 296\cdot8\ \text{N}$$

$$\therefore T_1 - T_2 = 720 - 296\cdot8 = \underline{\underline{423\cdot2\ \text{N}}}$$

$$\frac{v_2}{v_1} = 1 - \frac{(T_1 - T_2)}{aE} \qquad \text{. from equation (9.11)}$$

$$\therefore v_2 = 11 \cdot 78\left(1 - \frac{423 \cdot 2}{320 \times 10^{-6} \times 300 \times 10^6}\right)$$

$$= 11 \cdot 728 \text{ m/s}$$

$$\therefore \text{ speed of follower} = \frac{11 \cdot 728}{0 \cdot 375} \times \frac{60}{2\pi} = \underline{298 \cdot 6 \text{ rev/min}}$$

$$\text{Power transmitted} = 423 \cdot 2 \times \left(298 \cdot 6 \times \frac{2\pi}{60} \times 0 \cdot 375\right) = \underline{4960 \text{ W}}$$

2. *The arrangement of a belt drive is shown in Fig. 9.13. The driving pulley on the shaft A is 300 mm diameter and runs at 750 rev/min. The driven pulley B is 900 mm diameter and the centre distance between A and B is 1200 mm, AB being horizontal. The jockey pulley, centre C, is 300 mm diameter and the length of the arm AC is 450 mm. The mass of the jockey pulley is 6 kg and the mass of the arm is 7·5 kg, its centre of gravity being 200 mm from A. The belt, 100 mm wide and 6 mm thick, has a mass of 0·6 kg/m. The coefficient of friction between belt and pulleys is 0·26 and the maximum permissible stress in the belt is 2·2 MN/m².*

Find (a) the greatest power that the drive can transmit,

(b) the torque required on the arm AC under these conditions.

A partly graphical solution will be accepted and the thickness of the belt may be neglected for geometrical purposes.   *(U. Lond.)*

(a) By calculation or drawing,

angle of lap on pulley $A = 197 \cdot 5° = 3 \cdot 45$ rad

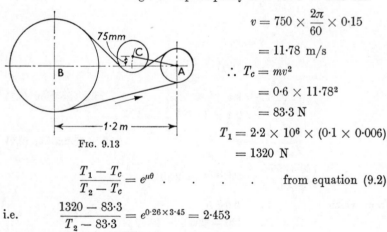

FIG. 9.13

$$v = 750 \times \frac{2\pi}{60} \times 0 \cdot 15$$

$$= 11 \cdot 78 \text{ m/s}$$

$$\therefore T_c = mv^2$$

$$= 0 \cdot 6 \times 11 \cdot 78^2$$

$$= 83 \cdot 3 \text{ N}$$

$$T_1 = 2 \cdot 2 \times 10^6 \times (0 \cdot 1 \times 0 \cdot 006)$$

$$= 1320 \text{ N}$$

$$\frac{T_1 - T_c}{T_2 - T_c} = e^{\mu\theta} \qquad . \qquad . \qquad . \qquad \text{from equation (9.2)}$$

i.e. $$\frac{1320 - 83 \cdot 3}{T_2 - 83 \cdot 3} = e^{0 \cdot 26 \times 3 \cdot 45} = 2 \cdot 453$$

from which $\qquad T_2 = 587 \cdot 3$ N

$$\therefore \text{ power} = (T_1 - T_2)v$$
$$= (1320 - 587 \cdot 3) \times 11 \cdot 78 = \underline{8630 \text{ W}}$$

(b) In Fig. 9.14, let R be the resultant force acting on C due to the driving tension, $T_2 - T_c$.

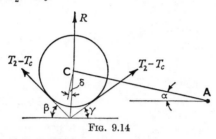

Fig. 9.14

By calculation or drawing,

$$\alpha = 9° \ 36', \quad \beta = 46° \ 28'$$
$$\delta = 5° \ 21' \quad \text{and} \quad \gamma = 32° \ 12'$$
$$\therefore \ R = 2 \times (587 \cdot 3 - 83 \cdot 3) \times \cos \left\{ \frac{180° - (\gamma + \beta)}{2} \right\}$$
$$= 639 \text{ N}$$

Component of R normal to AC $= 639 \cos \delta = 635$ N

Moment of this force about A $= 635 \times 0 \cdot 45 = 286$ N m

$\therefore$ torque required on AC $= 286 - (6 \times 0 \cdot 45 + 7 \cdot 5 \times 0 \cdot 2) \times 9 \cdot 81 \cos \alpha$
$$= \underline{254 \cdot 4 \text{ N m}}$$

3. *Explain with the aid of a sketch the creeping action of the belt round the two pulleys of a belt drive.*

A leather belt 200 mm wide and 10 mm thick embraces one-half of the circumference of the driving pulley. Assuming that the leather has a density of 1 Mg/m³, that the coefficient of friction is 0·3 and that the greatest permissible stress in the leather is 1·7 MN/m², find the speed of the belt at which the power transmitted is a maximum, the maximum power and the stress in the slack side of the belt.

If at this speed the power transmitted is 15 kW and the tension in the slack side is 1·3 MN/m², find the angle subtended at the centre of the driving wheel by the portion of the belt which creeps on it. (U. Lond.)

$$\theta = \pi \text{ rad}$$
$$m = 1000 \times 0 \cdot 2 \times 0 \cdot 01 \times 1 = 2 \text{ kg/m}$$
$$T_1 = 1 \cdot 7 \times 10^6 \times 0 \cdot 2 \times 0 \cdot 01 = 3400 \text{ N}$$

For maximum power, $mv^2 = \dfrac{T_1}{3} = 1133 \cdot 3$ N .   from equation (9.5)

$$\therefore v = \sqrt{\dfrac{1133 \cdot 3}{2}} = \underline{23 \cdot 8 \text{ m/s}}$$

$$\dfrac{T_1 - T_c}{T_2 - T_c} = e^{\mu\theta} \quad . \quad . \quad . \quad \text{from equation (9.2)}$$

i.e. $\dfrac{3400 - 1133 \cdot 3}{T_2 - 1133 \cdot 3} = e^{0 \cdot 3\pi} = 2 \cdot 57$

from which $T_2 = 2014 \cdot 3$

$\therefore$ maximum power $= (T_1 - T_2)v$

$$= (3400 - 2014 \cdot 3) \times 23 \cdot 8 = \underline{33\ 000 \text{ W}}$$

Stress in slack side $= \dfrac{2014 \cdot 3}{0 \cdot 2 \times 0 \cdot 01} = \underline{1 \cdot 007 \times 10^6 \text{ N/m}^2}$

Let T_1' and T_2' be the tight and slack side tensions when transmitting 15 kW.

Then $T_2' = 1 \cdot 3 \times 10^6 \times 0 \cdot 2 \times 0 \cdot 01 = 2600$ N

and $T_1' - T_2' = \dfrac{15 \times 10^3}{23 \cdot 8} = 630$ N

$$\therefore T_1' = 3230 \text{ N}$$

$$\therefore \dfrac{3230 - 1133 \cdot 3}{2600 - 1133 \cdot 3} = e^{0 \cdot 3\gamma}$$

$$\therefore \gamma = \underline{68 \cdot 3°}$$

4. *A leather belt connects a 1·2-m diameter pulley on a shaft running at 250 rev/min with another pulley running at 500 rev/min, the angle of lap on the latter being 175°; the maximum permissible load in the belt is 1·35 kN, and the coefficient of friction between belt and pulley surface is 0·25. If the initial tension in the belt may have any value between 900 N and 1100 N, what is the maximum power the belt should transmit?*

(U. Lond.)

$$\theta = 175° = 3 \cdot 05 \text{ rad}$$

$$v = 250 \times \dfrac{2\pi}{60} \times 0 \cdot 6 = 15 \cdot 71 \text{ m/s}$$

(a) Neglecting the possibility of slipping, the maximum power transmissible depends on the initial tension and the maximum permissible tension.

$$T_1 + T_2 = 2T_0$$
$$\therefore T_2 = 2T_0 - T_1$$
$$\therefore T_1 - T_2 = 2(T_1 - T_0) = 2(1350 - T_0)$$
$$\therefore \text{power} = (T_1 - T_2)v$$
$$= 2(1350 - T_0) \times 15\cdot71 = 31\cdot42(1350 - T_0) \text{ W} \quad (1)$$

(b) Neglecting the possibility of overloading the belt, the **maximum power** transmissible depends on belt slip, so that

$$\frac{T_1}{T_2} = e^{\mu\theta} = e^{0\cdot25 \times 3\cdot05} = 2\cdot144$$
$$\therefore T_1 + T_2 = 3\cdot144T_2 = 2T_0$$
$$\therefore T_2 = 0\cdot637T_0 \quad \text{and} \quad T_1 = 1\cdot365T_0$$
$$\therefore T_1 - T_2 = 0\cdot728T_0$$
$$\therefore \text{power} = (T_1 - T_2)v$$
$$= 0\cdot728T_0 \times 15\cdot71 = 11\cdot45T_0 \text{ W} \qquad . \qquad . \qquad (2)$$

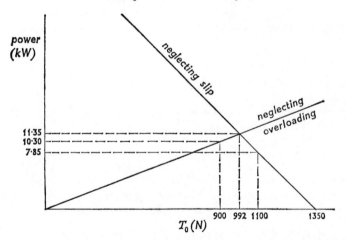

Fig. 9.15

The graphs of equations (1) and (2) are shown in Fig. 9.15. The optimum value of T_0 is 992 N, at which the power transmissible is 11·35 kW. If the initial tension is 900 N, the belt will slip before becoming overloaded and the maximum power is 10·3 kW, as obtained from equation (2). If the initial tension is 1100 N, the belt will be overloaded before slipping occurs and the maximum power is 7·85 kW, as obtained from equation (1). Assuming that the initial tension cannot be determined within this range, the maximum power which the drive should be expected to transmit is therefore 7·85 kW.

5. *In a belt drive, the radius of the driving pulley is r, the angle of lap of the belt on the pulley is θ rad, the coefficient of friction between the belt and the pulley is μ, and the mass of the belt is m per unit length. The drive is initially stationary and a torque is then applied to the driving pulley, giving it an acceleration of α rad/s². If the applied torque is such that slipping is about to commence when the belt speed has reached v, determine the difference between the tight-side tension T_1 and the slack side tension T_2 in terms of T_1, r, v, θ, α, m and μ.*

From this relationship, discuss the extent to which the difference of the tensions when slipping is about to commence is affected by the acceleration of the belt as compared with the speed of the belt.   (U. Lond.)

Referring to Fig. 9.16, a short length of the belt, subtending an angle $d\theta$ at the centre, is in equilibrium under

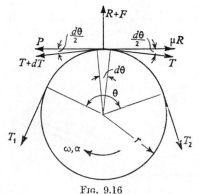

FIG. 9.16

(*a*) the reaction of the pulley, R,

(*b*) the friction force, μR,

(*c*) the centrifugal force, F,

(*d*) the inertia force, P,

(*e*) the belt tensions at the ends of the element, T and $T + dT$.

$$F = mr\, d\theta \,.\, \frac{v^2}{r} = mv^2\, d\theta$$

$$P = mr\, d\theta \,.\, \alpha r = mr^2\alpha\, d\theta$$

Resolving forces radially :

$$(T + dT)\frac{d\theta}{2} + T\frac{d\theta}{2} = R + mv^2\, d\theta$$

i.e.
$$R = T\, d\theta - mv^2\, d\theta. \qquad . \qquad (1)$$

Resolving forces tangentially :

$$(T + dT) + mr^2\alpha\, d\theta = T + \mu R$$

i.e.
$$\mu R = dT + mr^2\alpha\, d\theta . \qquad . \qquad (2)$$

Eliminating R between equations (1) and (2) :

$$dT = \mu(T - mv^2)\, d\theta - mr^2\alpha\, d\theta$$

$$\therefore \int_{T_2}^{T_1} \frac{dT}{T - mv^2 - \dfrac{mr^2\alpha}{\mu}} = \int_0^\theta \mu \, d\theta$$

$$\therefore \frac{T_1 - m\left(v^2 + \dfrac{r^2\alpha}{\mu}\right)}{T_2 - m\left(v^2 + \dfrac{r^2\alpha}{\mu}\right)} = e^{\mu\theta}$$

from which $\qquad T_1 - T_2 = \underline{\left\{T_1 - m\left(v^2 + \dfrac{r^2\alpha}{\mu}\right)\right\}(1 - e^{-\mu\theta})}$

The extent to which the difference in tensions when slipping is about to commence is affected by the acceleration and the speed of the belt depends upon the relative magnitudes of v^2 and $r^2\alpha/\mu$. The latter term is, however, likely to be relatively unimportant, except when starting.

6. *A heavy belt hangs over a wheel of radius a, and the ends, of lengths l_1 and l_2, hang vertically downwards. Prove that when slipping is about to occur,*

$$l_2 = l_1 e^{\mu\pi} + \frac{2a\mu}{1 + \mu^2}(1 + e^{\mu\pi})$$

(U. Lond.)

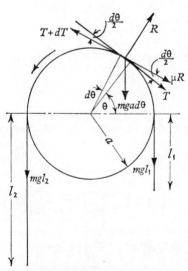

Fig. 9.17

From the given equation, it is evident that l_2 is greater than l_1 so that slipping will take place in the direction shown in Fig. 9.17.

A short length of the belt, subtending an angle $d\theta$ at the centre, is in equilibrium under

(a) its weight, $mga\,d\theta$, where m is the mass per unit length,

(b) the reaction of the pulley, R,

(c) the friction force, μR, and

(d) the belt tensions at the ends of the element, T and $T + dT$.

Resolving forces radially :

$$R = T\frac{d\theta}{2} + (T + dT)\frac{d\theta}{2} + mga\,d\theta\,\sin\theta$$

$$= T\,d\theta + mga\,\sin\theta\,d\theta \qquad . \qquad . \qquad . \qquad . \qquad (1)$$

Resolving forces tangentially :

$$\mu R + T + mga\,d\theta\,\cos\theta = (T + dT)$$

i.e. $$\mu R = dT - mga\,d\theta\,\cos\theta \qquad . \qquad (2)$$

Eliminating R between equations (1) and (2) :

$$\frac{dT}{d\theta} - \mu T = mga(\cos\theta + \mu\sin\theta)$$

The solution of this equation is

$$T = A\,e^{\mu\theta} - \frac{mga}{1 + \mu^2}\{2\mu\cos\theta - (1 - \mu^2)\sin\theta\}$$

When $\theta = 0$, $T = mgl_1$,

i.e. $$mgl_1 = A - \frac{mga \times 2\mu}{1 + \mu^2} \qquad \therefore A = mgl_1 + \frac{2mga\mu}{1 + \mu^2}$$

When $\theta = \pi$, $T = mgl_2$,

i.e. $$mgl_2 = \left\{mgl_1 + \frac{2mga\mu}{1 + \mu^2}\right\}e^{\mu\pi} - \frac{mga}{1 + \mu^2}(-2\mu)$$

$$\therefore l_2 = l_1\,e^{\mu\pi} + \frac{2a\mu}{1 + \mu^2}(1 + e^{\mu\pi})$$

7. *The brake mechanism shown in Fig. 9.18 is operated by the action of the spring at A, which is in compression.*

(a) Assuming that the material of the brake lining obeys Hooke's law when in compression, show, graphically or otherwise, that the resultant normal thrusts of the shoes on the drum are inclined at $6\frac{1}{2}°$ to the horizontal centre line.

(b) Assuming the coefficient of friction between drum and linings to be 0·3, but ignoring friction elsewhere, find the spring force necessary to produce a braking torque of 340 N m (U. Lond.)

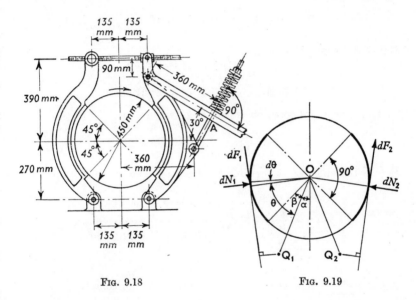

FIG. 9.18   FIG. 9.19

Referring to Fig. 9.19,

$$\alpha = \tan^{-1} \frac{0\cdot135}{0\cdot27} = 26° \ 34'$$

$$\therefore \beta = 45° - 26° \ 34' = 18° \ 26'$$

Pressure on element of L.H. shoe

$$= c_1 \times OQ_1 \sin \theta \quad \text{where } c_1 \text{ is a constant*}$$

$$= 0\cdot3018c_1 \sin \theta \ \text{N/m}$$

---

* The constants of proportionality are not the same for the two sides since the trailing edge of one shoe corresponds to the leading edge of the other shoe.

∴ normal force on element,

$$dN_1 = 0.3018c_1 \sin \theta \times 0.225 \, d\theta$$

$$= 0.0679c_1 \sin \theta \, d\theta \text{ N}$$

Vertical component of dN_1

$$= dN_1 \cos (\theta + \alpha)$$

$$= 0.0679c_1 \sin \theta \cos (\theta + 26° \, 34') \, d\theta \text{ N}$$

∴ total vertical force on shoe,

$$V = 0.0679c_1 \int_{18° \, 26'}^{108° \, 26'} \sin \theta \cos (\theta + 26° \, 34') \, d\theta \text{ N}$$

Horizontal component of dN_1

$$= dN_1 \sin (\theta + \alpha)$$

$$= 0.0679c_1 \sin \theta \sin (\theta + 26° \, 34') \, d\theta \text{ N}$$

∴ total horizontal force on shoe,

$$H = 0.0679c_1 \int_{18° \, 26'}^{108° \, 26'} \sin \theta \sin (\theta + 26° \, 34') \, d\theta \text{ N}$$

If γ is the angle of the resultant normal force to the horizontal, then

$$\tan \gamma = \frac{V}{H} = \frac{\displaystyle\int_{18° \, 26'}^{108° \, 26'} \sin \theta \cos (\theta + 26° \, 34') \, d\theta}{\displaystyle\int_{18° \, 26'}^{108° \, 26'} \sin \theta \sin (\theta + 26° \, 34') \, d\theta}$$

$$= -0.11$$

∴ $\gamma = \underline{-6° \, 18'}$ (i.e. inclined downwards from the horizontal)

Friction force on element, $dF_1 = 0.0679c_1 \sin \theta \, d\theta \times 0.3$

$$= 0.0204c_1 \sin \theta \, d\theta \text{ N}$$

∴ moment of dF_1 about O $= 0.0204c_1 \sin \theta \, d\theta \times 0.225$

$$= 0.00458c_1 \sin \theta \, d\theta \text{ N m}$$

∴ total moment about O for L.H. shoe $= 0.00458c_1 \int_{18° \, 26'}^{108° \, 26'} \sin \theta \, d\theta$

$$= 0.0058c_1 \text{ N m}$$

Similarly,

total moment about O for R.H. shoe $= 0.0058c_2 \text{ N m}$

∴ total torque on drum, $T = 0.0058(c_1 + c_2) \text{ N m}$

Taking moments about Q_1, Fig. 9.20, of the forces acting on the L.H. shoe :

$$0 \cdot 66R = \int_{18^\circ\,26'}^{108^\circ\,26'} dN_1 \times 0 \cdot 3018 \sin \theta$$

$$- \int_{18^\circ\,26'}^{108^\circ\,26'} dF_1 \times (0 \cdot 225 - 0 \cdot 3018 \cos \theta)$$

$$= 0 \cdot 020\,52c_1 \int_{18^\circ\,26'}^{108^\circ\,26'} \{\sin^2 \theta - 0 \cdot 3 \times (0 \cdot 746 \sin \theta - \sin \theta \cos \theta)\}\, d\theta$$

$$\therefore R = \frac{0 \cdot 020\,52c_1}{0 \cdot 66}(1 \cdot 0855 - 0 \cdot 163) = 0 \cdot 0287c_1 \text{ N}$$

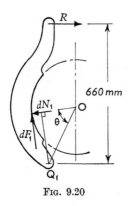

FIG. 9.20

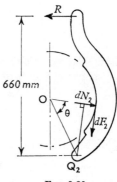

FIG. 9.21

Taking moments about Q_2, Fig. 9.21, of the forces acting on the R.H. shoe and lever* :

$$0 \cdot 66R = \int_{18^\circ\,26'}^{108^\circ\,26'} dN_2 \times 0 \cdot 3018 \sin \theta$$

$$+ \int_{18^\circ\,26'}^{108^\circ\,26'} dF_2 \times (0 \cdot 225 - 0 \cdot 3018 \cos \theta)$$

$$\therefore R = \frac{0 \cdot 020\,52c_2}{0 \cdot 66}(1 \cdot 0855 + 0 \cdot 163) = 0 \cdot 0389c_2 \text{ N}$$

* The only *external* force acting on the R.H. assembly is R and the reaction at Q_2; hence the moment about Q_2 is $0 \cdot 66R$ N m. Alternatively, the forces acting on the shoe alone are as shown dotted in Fig. 9.22, so that moment about Q_2

$$= \left(R + \frac{S}{2}\right) \times 0 \cdot 57 + S \times 0 \cdot 075$$

$$= 0 \cdot 57R + 0 \cdot 36S$$

$$= 0 \cdot 66R \quad \text{since } S = \frac{R}{4}$$

H

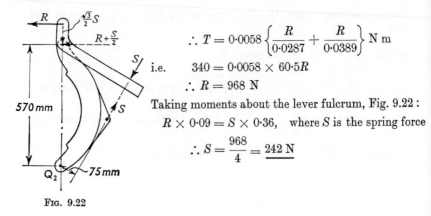

$$\therefore T = 0.0058 \left\{ \frac{R}{0.0287} + \frac{R}{0.0389} \right\} \text{ N m}$$

i.e. $\quad 340 = 0.0058 \times 60.5R$

$$\therefore R = 968 \text{ N}$$

Taking moments about the lever fulcrum, Fig. 9.22 :

$$R \times 0.09 = S \times 0.36, \quad \text{where } S \text{ is the spring force}$$

$$\therefore S = \frac{968}{4} = \underline{242 \text{ N}}$$

FIG. 9.22

Graphical solution. The arc of contact for the L.H. shoe is divided into six equal parts, Fig. 9.23, and the polygon of forces is constructed

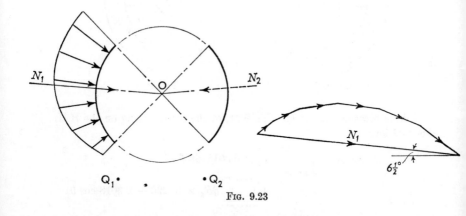

FIG. 9.23

for the normal forces ; it will be found that the resultant normal force, N_1, is inclined at approximately $6\frac{1}{2}°$ to the horizontal.

A similar polygon for the friction forces is drawn in Fig. 9.24 and a funicular polygon is then constructed across the lines of action of these forces to give a point on the line of action of the resultant friction force, F_1.

The forces exerted by the R.H. shoe are similar and symmetrical, except that the direction of the friction force, F_2, is reversed.

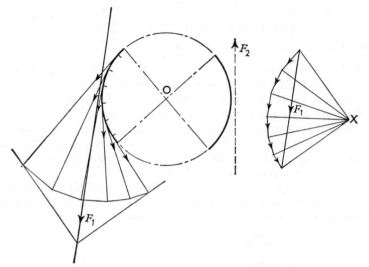

Fig. 9.24

Fig. 9.25 shows the forces acting on the shoes.

The braking torque,

$$T = F_1 h_1 + F_2 h_2 = \mu(N_1 h_1 + N_2 h_2)$$

By symmetry,

$$h_1 = h_2 = 0 \cdot 246 \text{ m}$$

$$\therefore 340 = 0 \cdot 3 \times 0 \cdot 246 (N_1 + N_2)$$

$$\therefore N_1 + N_2 = 4600 \text{ N}$$

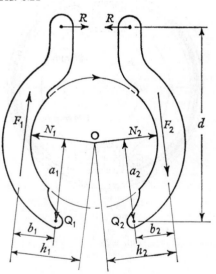

Fig. 9.25

Taking moments about Q_1 and Q_2 respectively,

$$R \times d = N_1 a_1 - F_1 b_1$$
$$= N_1(a_1 - \mu b_1)$$

and

$$R \times d = N_2 a_2 + F_2 b_2$$
$$= N_2(a_2 + \mu b_2)$$

Hence $\quad R \times d\left(\dfrac{1}{a_1 - \mu b_1} + \dfrac{1}{a_2 + \mu b_2}\right) = N_1 + N_2$

By symmetry, $\quad\quad a_1 = a_2 = 0 \cdot 284 \text{ m}$

and $\quad\quad\quad\quad\quad b_1 = b_2 = 0 \cdot 138 \text{ m}$

$$\therefore R \times 0{\cdot}66\left(\frac{1}{0{\cdot}284 - 0{\cdot}3 \times 0{\cdot}138} + \frac{1}{0{\cdot}284 + 0{\cdot}3 \times 0{\cdot}138}\right) = 4600$$

from which $\qquad\qquad\qquad R = \underline{968\ N}$

8. *The arrangement of an internal expanding two leading shoe brake is shown in Fig. 9.26. When the brake is operated, fluid under pressure enters the chambers marked A and the pistons force the shoes against the inside of the brake drum which is 288 mm internal diameter. The pistons are both 30 mm diameter and the wheel to which the brake drum is fitted is 830 mm diameter. The normal reaction between the wheel and the road is 2·25 kN, the coefficient of friction between the tyre and road 0·75 and the coefficient of friction between drum and brake lining 0·3.*

Determine the hydraulic pressure required to lock the wheel for the direction of drum rotation shown and express this as a percentage of the pressure required when the direction of rotation is reversed. (U. Lond.)

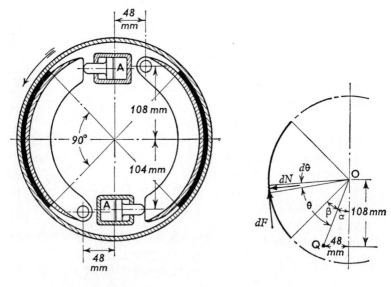

FIG. 9.26   FIG. 9.27

$$\alpha = \tan^{-1}\frac{0{\cdot}048}{0{\cdot}108} = 23^{\circ}\ 58', \text{ Fig. 9.27}, \qquad \therefore \beta = 45^{\circ} - 23^{\circ}\ 58' = 21^{\circ}\ 2'$$

Pressure on element $= c \times \text{OQ} \sin \theta = 0{\cdot}1182c \sin \theta$ N/m

$\therefore$ normal force on element, $dN = 0{\cdot}1182c \sin \theta \times 0{\cdot}144\ d\theta$

$$= 0{\cdot}017c \sin \theta\ d\theta\ \text{N}$$

$\therefore$ friction force on element, $dF = 0.017c \sin\theta \, d\theta \times 0.3$
$$= 0.0051c \sin\theta \, d\theta \text{ N}$$

$\therefore$ moment of dF about O $= 0.0051c \sin\theta \, d\theta \times 0.144$
$$= 0.000\,734c \sin\theta \, d\theta \text{ N m}$$

$\therefore$ total torque on drum, $T = 2\displaystyle\int_{21°\,2'}^{111°\,2'} 0.000\,734c \sin\theta \, d\theta$
$$= 0.0019c \text{ N m}$$

Friction force on road $= 2250 \times 0.75$ N

$\therefore$ torque on wheel $= 2250 \times 0.75 \times 0.415 = 700$ N m

$$\therefore 700 = 0.0019c$$

$$\therefore c = 368\,500$$

Taking moments about Q of the forces acting on the shoe, Fig. 9.28

$$P \times 0.212 = \int_{21°\,2'}^{111°\,2'} dN \times 0.1182 \sin\theta - \int_{21°\,2'}^{111°\,2'} dF \times (0.144 - 0.1182 \cos\theta)$$

$$= c\int_{21°\,2'}^{111°\,2'} \{0.002\,01 \sin^2\theta - 0.000\,734 \sin\theta + 0.000\,602 \sin\theta \cos\theta\}d\theta$$

$$= 368\,500 \times 0.001\,525$$

$$\therefore P = 2650 \text{ N}$$

$$\therefore p = \frac{2650}{\dfrac{\pi}{4} \times 0.03^2} = 3\,750\,000 \text{ N/m}^2$$

$$= 3.75 \text{ MN/m}^2$$

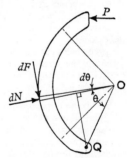

Fig. 9.28

When the direction of rotation is reversed, the direction of dF becomes reversed,

$$\therefore P \times 0.212 = \int_{21°\,2'}^{111°\,2'} dN \times 0.1182 \sin\theta + \int_{21°\,2'}^{111°\,2'} dF \times (0.144 - 0.1182 \cos\theta)$$

$$= c\int_{21°\,2'}^{111°\,2'} \{0.002\,01 \sin^2\theta + 0.000\,734 \sin\theta - 0.000\,602 \sin\theta \cos\theta\}\,d\theta$$

$$= 368\,500 \times 0.002\,98$$

$$\therefore P = 5170 \text{ N}$$

$$\therefore p = \frac{5170}{\dfrac{\pi}{4} \times 0.03^2} = 7\,320\,000 \text{ N/m}^2$$

$$\frac{3.75}{7.32} = 0.511 \quad \text{or} \quad \underline{51.1\%}$$

9. A short vertical rope drive is required to transmit power from a pulley of 1·15 m effective diameter. The ropes have a mass of 1·2 kg/m run, the groove angle is 50° and the angle of lap is 170°. The coefficient of friction is 0·3.

(a) With an initial or static tension of 900 N in each rope, what is the maximum power which can be transmitted per rope ? What will then be the load in the rope and its linear speed ?

(b) If the permissible load is 1·6 kN/rope and this is to be fully utilized for maximum power, what then should be the initial tension in the rope ?

Hooke's Law may be assumed to apply to the rope. (*U. Lond.*)

(*Ans. :* 14 kW ; 1·273 kN ; 18·8 m/s ; 1·13 kN)

10. A pulley 300 mm diameter running at 1500 rev/min transmits 15 kW to an elastic belt driving a similar pulley. Find the required width of belt and the uniform tension when stationary if the maximum total tension under load is 17·5 kN/m of width. It may be assumed that the sum of the tensions in the two sides remains constant, $e^{\mu\theta} = 2 \cdot 5$, where μ is the coefficient of belt friction and θ the angle of lap, and the mass of the belt $m = 7$ kg/m of width per metre of length.

If the belt is 5 mm thick and $E = 200$ MN/m², find also the fractional loss of speed of the driven pulley. (*I. Mech. E.*)

(*Ans.:* 77·8 mm ; 1·043 kN ; 0·817%)

11. An elastic belt makes contact with a pair of similar flat pulleys over arcs of 180°, the coefficient of friction being 0·45. The initial mean tension is 400 N and it may be assumed that the sum of the tensions in the two sides remains constant. Find the difference in the tensions at which the belt slips on the pulleys.

If the belt is 24 mm wide and 6 mm thick and E for the material is 200 MN/m², estimate the effect on the maximum tension difference caused by the pulley centres moving together by 3 mm if they are initially 0·9 m apart.

(*I. Mech. E.*)   (*Ans. :* 486·5 N ; 23·9%)

12. In an experiment on power transmission by a belt running on flat pulleys of equal diameter, the driving motor was mounted so that a constant mean tensioning force could be applied. Both the driving motor and the dynamometer were pivoted to enable the corresponding torques to be measured, and a differential device enabled differences in speed to be measured.

One set of readings was as follows : mean belt tension = 340 N, speed of motor = 1000 rev/min, motor torque = 6·8 N m, speed loss of dynamometer = 5 rev in 40 s, dynamometer torque = 6·5 N m, pulley diameter = 85 mm, belt = 24 mm wide × 4 mm thick.

Estimate the percentage loss of speed due to creep, the efficiency of the transmission, the tension in the tight side, the tension ratio, the effective value of Young's Modulus for the belt material and the necessary friction coefficient between belt and pulley to prevent slip. (*I. Mech. E.*)

(*Ans.:* 0·75% ; 94·8% ; 420 N ; 1·615 ; 222·5 MN/m² ; 0·153)

13. In the belt drive shown in Fig. 9.29, an idler pulley is supported on an arm which is free to turn about the shaft carrying the small driving pulley. A deadweight on the arm causes the idler to press against the slack side of the belt and increase the angle of lap. The portion of belt between the top of the small pulley and the bottom of the idler pulley is to be horizontal.

The belt is 150 mm wide and has a mass of 1·2 kg/m. Its speed is 1200 m/min and the coefficient of friction between it and the pulleys is 0·3. If the maximum tension per metre width of belt is 14 kN, find the power which can be transmitted and, neglecting the mass of the arm, find the necessary size of deadweight required, the idler pulley having a mass of 18 kg.

If the total work expended in bending the belt is 750 J/m² of belt passing over the pulleys, what will be the efficiency of the transmission, neglecting friction at the bearings ? (*U. Lond.*)

(*Ans.:* 21·35 kW ; 26·87 kg ; 89·47%)

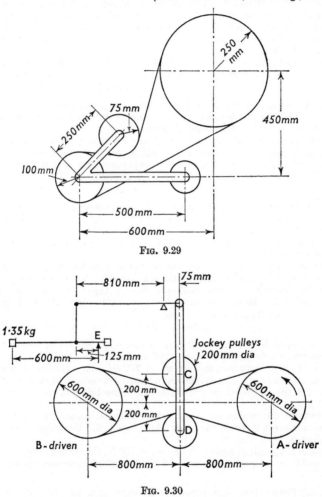

Fɪɢ. 9.29

Fɪɢ. 9.30

14. In the transmission dynamometer shown in Fig. 9.30, the driving pulley A and the driven pulley B each have a diameter of 600 mm, their centres being equidistant from the jockey pulley centre-lines and 1·6 m apart. The jockey pulleys C and D are 200 mm diameter and their spindles are 400 mm apart. The masses of the jockey pulleys and levers are counterbalanced and friction at the jockey pulleys is negligible. Calculate the power transmitted if the 1·35-kg mass maintains balance when moved 600 mm from the fulcrum E and the speed of the pulleys A and B is 210 rev/min. (*U. Lond.*)

(*Ans.:* 9·54 kW)

15. A leather belt transmitting power from a pulley of 750 mm diameter, and running at 420 rev/min, has a minimum arc of contact of 175° ; the maximum permissible load in the belt is 1·6 kN, and the coefficient of friction between shaft and pulley is 0·25. Determine the maximum power which can be transmitted when the belt is on the point of slipping, and the necessary initial (or static) load in the belt.

If the initial load in the belt were 1·1 kN, what would be the maximum power which could be transmitted under the above conditions ? (*U. Lond.*)

(*Ans.:* 14·1 kW ; 1·173 kN ; 13·2 kW)

16. A leather belt, 6 mm thick, is to transmit power from a 0·9-m diameter pulley running at 250 rev/min to one running at 450 rev/min, the distance between the pulley centres being 3·15 m.

If the initial tension can only be adjusted within ± 55 N of its calculated value, determine the maximum power which can be transmitted when the co-efficient of friction is 0·25 and the maximum permissible load in the belt is 1·1 kN. The effect of centrifugal tension may be neglected. (*U. Lond.*)

(*Ans.:* 5·6 kW)

17. A leather belt 6 mm thick connects two pulleys of 0·95 m and 0·525 m diameters, carried on parallel shafts at 3·45 m centres ; the speed of the larger pulley is 240 rev/min. The coefficient of friction is 0·25 and the maximum permissible load in the belt is 1150 N. Find the maximum power and the necessary initial tension.

Plot on a base of initial mean tension, curves showing the power which can be transmitted : (*a*) allowing for maximum loading but neglecting possibility of slip, (*b*) allowing for limitation due to slip but neglecting possibility of over-loading the belt. Hence determine the permissible range of initial tension if 5·5 kW is to be transmitted. (*U. Lond.*)

(*Ans.:* 7·37 kW ; 845·5 N ; 632 and 923 N)

18. An open belt, 6 mm thick and 90 mm wide, is used to transmit power between two shafts whose axes are 2·85 m apart ; the pulley on the driving shaft is 1·15 m in diameter, and rotates at 275 rev/min ; the driven shaft is to rotate at 400 rev/min. The maximum permissible tension in the belt is 12 kN/m. width, and the coefficient of friction between the belt and the pulley surface is 0·25.

Determine the maximum power which may be transmitted if the initial tension in the belt is within the range 730 N to 850 N. Centrifugal tension may be neglected. (*U. Lond.*)

(*Ans. :* If the initial tension can be set at 806 N, the maximum power = 9·05 kW ; at 730 N, maximum power = 8·2 kW ; at 850 N maximum power = 7·61 kW)

19. (*a*) Derive from first principles an expression for the ratio of speeds between two belt-driven pulleys transmitting power when allowance is made for ' creep ', stating clearly what assumptions are made.

(*b*) If $\dfrac{T_1}{T_2} = e^{\mu\theta}$, T_{max} = maximum permissible load in a belt and T_0 = the initial tension or mean of the driving tensions T_1 and T_2, derive expressions for the power transmitted by a belt running at V m/s: (*i*) in terms or T_0, T_{max}, and V, and (*ii*) in terms of T_0, $e^{\mu\theta}$, and V. (*U. Lond.*)

$$\left(Ans.:\ \frac{v_2}{v_1} = 1 - \frac{(T_1 - T_2)}{aE}\ ;\ \ 2(T_{max} - T_0)V\ ;\ \ 2T_0\!\left(\frac{e^{\mu\theta} - 1}{e^{\mu\theta} + 1}\right)\!V\right)$$

20. A small pulley of radius r_1 on a lineshaft drives a large pulley of radius r_2 on a machine vertically below it, the centre distance being d. Show that slipping is equally likely to occur at either pulley if the tension in the belt where it runs on to the large pulley is given by

$$T = \frac{e^{\mu(\pi - 2\theta)} - 1}{e^{\mu(\pi + 2\theta)} - e^{\mu(\pi - 2\theta)}}.mgd$$

where m = mass of belt per unit length and $\sin\theta = (r_2 - r_1)/d$. (*U. Lond.*)

21. Fig. 9.31 shows the arrangement of an internal-expanding friction brake, in which the brake shoe is pivoted at a fixed point C. The distance of this point from the centre O of the rotating drum is 75 mm. The internal radius of the drum is 100 mm. The friction lining extends over an arc AB, such that the angle AOC is 135° and BOC is 45°. The brake is applied by means of a force at Q, perpendicular to CQ, the distance CQ being 150 mm.

Assuming that the local rate of wear on the lining is proportional to the normal pressure, and taking the coefficient of friction as 0·40, calculate the force required at Q to produce a braking torque of 25 N m. Give separately the results for the two cases, (*a*) when the drum rotates in the clockwise direction, and (*b*) when the rotation is anticlockwise. (*U. Lond.*)

(*Ans.:* 450·5 N ; 117·2 N)

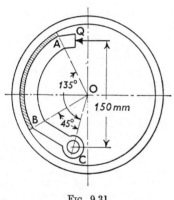

Fig. 9.31

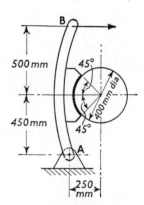

Fig. 9.32

22. Fig. 9.32 shows a brake shoe applied to a drum by a lever AB which is pivoted at a fixed point A. The shoe is rigidly fixed to the lever. The coefficient of friction between the brake lining and the drum is 0·35, and it may be assumed that the material of the brake lining obeys Hooke's law. The drum rotates clockwise.

Using either graphical or analytical methods, find the braking torque on the drum due to a horizontal force of 750 N applied at B. (If a graphical method is used, it is suggested that the arc of contact be divided into four sections. (*U. Lond.*)

(*Ans.:* 108 N m)

23. Fig. 9.33 shows a brake drum 330 mm diameter, acted on by two brake shoes which are mounted on a pin A, and pushed apart by two hydraulically operated pistons at B, each exerting a force of P N on the shoe on which it makes contact. The brake lining on each shoe extends 60° above to 60° below the horizontal centre line. The coefficient of friction is 0·2. The radial pressure between the lining and the drum is proportional to the rate of wear of the lining.

Find graphically or otherwise, the value of P to produce a braking torque of 180 N m   (*U. Lond.*)   (*Ans.:* 1·07 kN)

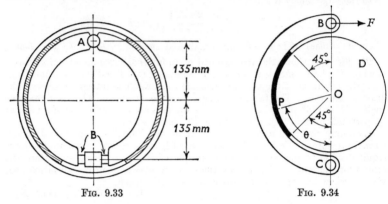

FIG. 9.33   FIG. 9.34

24. Fig. 9.34 shows a friction brake in which the curved lever BC is pivoted at the fixed point C and carries a friction lining which presses on the rotating drum D over an arc of 90°. The diameter of the drum is 250 mm. The distances OB and OC are each 150 mm. The pressure exerted by the friction lining on the drum at any point P is 350 sin θ kN/m². The lining is 50 mm wide and the coefficient of friction is 0·3.

Calculate the braking torque exerted on the drum, also the amount of the force F which is required at B to apply the brake. The drum rotates in a clockwise direction. (*U. Lond.*)   (*Ans.:* 116 N m; 1·02 kN)

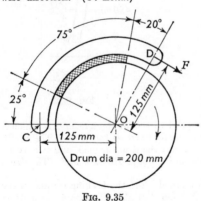

FIG. 9.35

25. The details of an externally applied brake are given in Fig. 9.35. The brake is pivoted about the fixed pin at C and is operated by the force F which acts at right angles to OD. The brake lining is of uniform width, it subtends an angle of 75° at the drum shaft O and it obeys Hooke's law in compression. The angle of friction between the lining and the drum is 20° and the drum has a diameter of 200 mm.

When the drum is rotating clockwise the brake is applied with the force $F = 135$ N. Determine graphically or otherwise (*a*) the force on the pin at C and (*b*) the braking torque on the drum. (*U. Lond.*)
(*Ans.:* 10·46 N m : 247 N)

CHAPTER 10

GYROSCOPES

10.1 Gyroscopic couple. Let a disc, of polar moment of inertia I, rotate about its axis OX, Fig. 10.1, at ω rad/s and let this axis be rotated (or *precessed*) in the horizontal plane XOZ about the axis OY at Ω rad/s. The angular momentum of the disc, $I\omega$, about the axis of spin may be represented, in the usual convention, by the vector Oa, the sense of the vector corresponding to the forward movement of a corkscrew turned in the direction of rotation of the disc.

When the disc has rotated about OY through an angle $d\theta$, the angular momentum of the disc, $I\omega$, is then represented by the vector Ob and the change in angular momentum by the vector ab,

i.e. change in angular momentum $= I\omega \, d\theta$

If this change takes place in time dt,

$$\text{rate of change of angular momentum} = I\omega \frac{d\theta}{dt}$$

i.e. gyroscopic couple, $C = I\omega\Omega$* . (10.1)

Applying the corkscrew rule to the vector ab, the couple required to cause precession is directed along the OZ axis, clockwise looking in the direction ZO. The reaction to this couple, i.e. the couple exerted by the shaft OX on its bearings, is opposite in direction to the applied couple.

It will be seen from Fig. 10.1 that the axes of spin, precession and couple are mutually perpendicular.

The disc also has angular momentum $I_d\Omega$ about the axis OY, I_d being the moment of inertia about the diameter, but this momentum is not being precessed, hence no gyroscopic couple is induced.

---

* Note the analogy between this derivation and that of centripetal force, $mv\omega$, Art. 1.4.

10.2 Effect of gyroscopic couple. Gyroscopic motion arises when-
ever the axis of a rotating body is caused to change direction. Examples
may be found in the rotor of a turbine which is pitching in a ship, in the
wheels of a vehicle turning round a bend and in gyroscopic instruments.
In every case, the couple which must be applied to the rotating body
to cause it to precess is determined from a vector diagram, such as in
Fig. 10.1, using the corkscrew rule, and the reaction which the body
applies to its bearings is then obtained. In the case of a ship's turbine
rotor, the reaction couple on the ship causes it to swing sideways ; in a
vehicle, the reaction couple caused by the precession of the wheels tends
to overturn the vehicle in the same way as the centrifugal force.

10.3 Alternative derivation of gyroscopic couple. Fig. 10.2
shows three views of the disc,
rotating about axes OX and OY
with angular velocities ω and
Ω rad/s respectively. A particle of mass dm at point P,
radius r, has a velocity ωr perpendicular to OP. This has
components

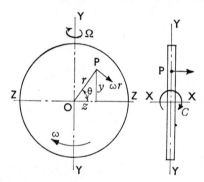

$\omega r \sin \theta$ (or ωy), parallel to OZ
and

$\omega r \cos \theta$ (or ωz), parallel to OY.

Thus, in the plan view, P
is moving to the right with
velocity ωy along a line parallel
with OZ which is rotating at Ω
rad/s and hence has an acceleration $2(\omega y)\Omega$* perpendicular to
the plane of the disc.

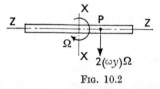

Fig. 10.2

Similarly, particles below the ZZ axis are moving to the left and therefore have an acceleration $2(\omega y)\Omega$ acting in the opposite direction ; the
accelerating forces on particles above and below ZZ then produce a couple
about ZZ.

$$\text{Force to accelerate particle} = dm \, . \, 2\omega y \Omega$$

$$\therefore \text{ moment of force about ZZ} = dm \, . \, 2\omega y^2 \Omega$$

$$\therefore \text{ total moment about ZZ} = 2\omega\Omega \int y^2 \, dm$$

i.e.
$$C = 2\omega\Omega I_{ZZ}$$

* See Art. 2.4.

But $2I_{ZZ} = I$, the polar moment of inertia of the disc,

$$\therefore \; C = I\omega\Omega$$

The projection of P on YY axis moves with S.H.M. but the force necessary to produce this motion has no moment about the ZZ axis.

The above result is only applicable to a body for which the value of I_{ZZ} remains constant as it revolves.

In the case of a thin rod rotating in the vertical plane about a hori-

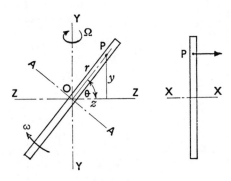

FIG. 10.3

zontal axis through its centre, Fig. 10.3, the accelerating force on a particle of mass dm at P has a moment about the axis AA of magnitude

$$dm \;.\; 2\omega y\Omega r \quad \text{or} \quad dm \;.\; 2\omega\Omega r^2 \sin\theta$$

$$\therefore \; \text{total moment about AA} = 2\omega\Omega \sin\theta \int r^2 \, dm$$

i.e.

$$C = 2\omega\Omega \sin\theta \, I_{AA}$$

$$= 2I\omega\Omega \sin\theta \quad . \quad . \quad (10.2)$$

since for a thin rod, $I_{AA} = I$, the polar moment of inertia (i.e. about the axis of rotation).

This couple may be resolved into components about the axes OY and OZ.

$$\text{Couple about OZ} = 2I\omega\Omega \sin^2\theta$$

This varies from 0 at $\theta = 0°$ and $180°$ to a maximum value of $2I\omega\Omega$ when $\theta = 90°$ and $270°$.

$$\text{Couple about OY} = 2I\omega\Omega \sin\theta \cos\theta$$

This varies from 0 at $\theta = 0°$, $90°$, $180°$ and $270°$ to a maximum value of $I\omega\Omega$ when $\theta = 45°$, $135°$, $225°$ and $315°$.

The thin rod is the idealized case of a two-bladed propeller.

10.4   General case. Consider a body which is rotating and accelerat-

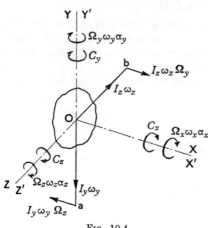

ing simultaneously about each of the *principal* axes of the body, OX, OY and OZ, which are themselves rotating with angular velocities Ω_x, Ω_y and Ω_z respectively relative to a set of axes OX′, OY′ and OZ′, fixed in space and, at the instant considered, coincident with OX, OY and OZ, Fig. 10.4. Let the moments of inertia of the body about OX, OY and OZ be I_x, I_y and I_z, the angular velocities ω_x, ω_y and ω_z and the angular accelerations α_x, α_y and α_z respectively.

FIG. 10.4

The angular momentum about OY is $I_y\omega_y$ and is represented by Oa. This vector is being precessed about OZ′ at a rate Ω_z so that the couple required is $I_y\omega_y\Omega_z$; its direction is parallel to the OX′ axis and the sense is the same as that of ω_x. The angular momentum about OZ is $I_z\omega_z$ and is represented by Ob. This vector is being precessed about OY′ at a rate Ω_y so that the couple required is $I_z\omega_z\Omega_y$; its direction is parallel to the OX′ axis and the sense is opposite to that of ω_x. There is, in addition, a couple $I_x\alpha_x$ required about OX to produce the acceleration α_x, so that the total couple about OX′,

$$C_x = I_x\alpha_x + I_y\omega_y\Omega_z - I_z\omega_z\Omega_y \qquad . \qquad . \quad (10.3)$$

Similarly, it can be shown that

$$C_y = I_y\alpha_y + I_z\omega_z\Omega_x - I_x\omega_x\Omega_z \qquad . \qquad . \quad (10.4)$$

and

$$C_z = I_z\alpha_z + I_x\omega_x\Omega_y - I_y\omega_y\Omega_x \qquad . \qquad . \quad (10.5)$$

If h_x, h_y and h_z represent the angular momenta, $I_x\omega_x$, $I_y\omega_y$ and $I_z\omega_z$ respectively, equations (10.3), (10.4) and (10.5) can be written

$$C_x = \frac{dh_x}{dt} + h_y\Omega_z - h_z\Omega_y \qquad . \qquad . \qquad . \quad (10.6)$$

$$C_y = \frac{dh_y}{dt} + h_z\Omega_x - h_x\Omega_z \qquad . \qquad . \qquad . \quad (10.7)$$

and

$$C_z = \frac{dh_z}{dt} + h_x\Omega_y - h_y\Omega_x \qquad . \qquad . \qquad . \quad (10.8)$$

These are known as Euler's Dynamical Equations.

1. *One of the driving axles of a locomotive, with its two wheels, has a moment of inertia of 350 kg m². The wheels are 1·85 m diameter. The distance between the planes of the wheels is 1·5 m. When travelling at 100 km/h, the locomotive passes over a defective rail which causes the right-hand wheel to fall 12 mm and rise again in a total time of 0·1 s, the vertical movement of the wheel being with S.H.M.*

Find the maximum gyroscopic torque caused and make a sketch showing the direction in which it acts when the wheel is falling. (U. Lond.)

Periodic time of the S.H.M. = 0·1 s, Fig. 10.5.

∴ angular velocity of vector generating the S.H.M., $p = \dfrac{2\pi}{0\cdot1}$ rad/s

Angular amplitude of motion, $\phi = \dfrac{0\cdot006}{1\cdot5} = 0\cdot004$ rad

∴ maximum velocity of precession, $\Omega = p\phi = \dfrac{2\pi}{0\cdot1} \times 0\cdot004$

$$= 0\cdot251 \text{ rad/s}$$

Angular velocity of wheels, $\omega = \dfrac{v}{r} = \dfrac{100}{3\cdot6 \times 0\cdot925}$

$$= 30 \text{ rad/s}$$

∴ maximum gyroscopic couple,

$$C = I\omega\Omega \qquad . \qquad . \qquad . \qquad \text{from equation (10.1)}$$

$$= 350 \times 30 \times 0\cdot251 = \underline{2640 \text{ N m}}$$

Fig. 10.6 shows a view of the wheels and axle. When the right-hand wheel falls, the momentum vector oa moves to ob, the change in momentum being represented by ab. Thus the couple required for precession

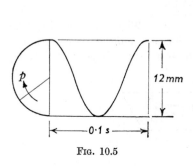

FIG. 10.5

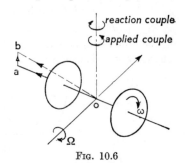

FIG. 10.6

is clockwise looking upward and the reaction couple exerted *by* the wheels *on* the locomotive is equal and opposite, i.e. anticlockwise looking upward. The locomotive thus tends to swing to the right.

2. *The turbine rotor of a ship has a mass of 30 t, a radius of gyration of 600 mm, and rotates at 2400 rev/min in a clockwise direction when viewed from aft. The ship pitches through a total angle of 15°, 7½° above and 7½° below the horizontal, the motion being simple harmonic and having a period of 12 s.*

Determine the maximum gyroscopic couple on the holding-down bolts on the turbine, and the direction of yaw as the bow rises.   (U. Lond.)

Angular velocity of vector generating the S.H.M.,

$$p = \frac{2\pi}{12} \text{ rad/s}$$

Angular amplitude of motion,

$$\phi = 7\tfrac{1}{2} \times \frac{\pi}{180} \text{ rad}$$

∴ maximum velocity of precession,

$$\Omega = p\phi = \frac{2\pi}{12} \times 7\tfrac{1}{2} \times \frac{\pi}{180} = 0.068\ 65 \text{ rad/s}$$

Angular velocity of rotor,

$$\omega = 2400 \times \frac{2\pi}{60} = 251.5 \text{ rad/s}$$

Moment of inertia of rotor,

$$I = 30 \times 10^3 \times 0.6^2 = 10\ 800 \text{ kg m}^2$$

∴ maximum gyroscopic couple,

$$C = I\omega\Omega$$
$$= 10\ 800 \times 251.5 \times 0.068\ 65$$
$$= 186\ 500 \text{ N m} = \underline{186.5 \text{ kN m}}$$

Fig. 10.7 shows a side view of the ship. When the bow rises, the momentum vector *oa* moves to *ob*, the change in momentum being represented by *ab*. Thus the couple required for precession is clockwise looking

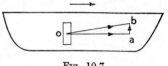

Fig. 10.7

upward and the reaction of the rotor on the ship is equal and opposite, i.e. anticlockwise looking upward. The ship therefore yaws to starboard, i.e. to the right looking forward.

3. *A rear-engine automobile is travelling around a track of 100 m mean radius. Each of the four road wheels has a moment of inertia of 1·6 kg m² and an effective diameter of 600 mm. The rotating parts of the engine have a moment of inertia of 0·85 kg m², the engine axis is parallel to the rear axle and the crankshaft rotates in the same sense as the road wheels. The gear ratio, engine to back axle, is 3 to 1. The vehicle has a mass of 1400 kg and its centre of gravity is 450 mm above road level. The width of the track of the vehicle is 1·5 m.*

Determine the limiting speed of the vehicle round the curve for all four wheels to maintain contact with the road surface, if this is not cambered.

(U. Lond.)

$$\text{Moment of inertia of wheels} = 4 \times 1\cdot6 = 6\cdot4 \text{ kg m}^2$$

$$\text{Moment of inertia of engine} = 0\cdot85 \text{ kg m}^2$$

If v is the speed of the vehicle, r and R the radii of the wheels and track respectively, then

$$\text{angular velocity of wheels} = \frac{v}{r} = \frac{v}{0\cdot3} \text{ rad/s}$$

$$\therefore \text{ angular velocity of engine} = \frac{3v}{0\cdot3} = 10v \text{ rad/s}$$

and $\qquad$ $$\text{angular velocity of precession} = \frac{v}{R} = \frac{v}{100} \text{ rad/s}$$

Since the wheels and engine rotate in the same direction, the gyroscopic couples are additive,

i.e. $\qquad$ $$C = \Sigma I\omega\Omega = \left(6\cdot4 \times \frac{v}{0\cdot3} + 0\cdot85 \times 10v\right) \times \frac{v}{100}$$

$$= 0\cdot2983v^2 \text{ N m}$$

Centrifugal couple $= m\dfrac{v^2}{R} \times h$

$$= 1400 \times \frac{v^2}{100} \times 0\cdot45$$

$$= 6\cdot3v^2 \text{ N m}$$

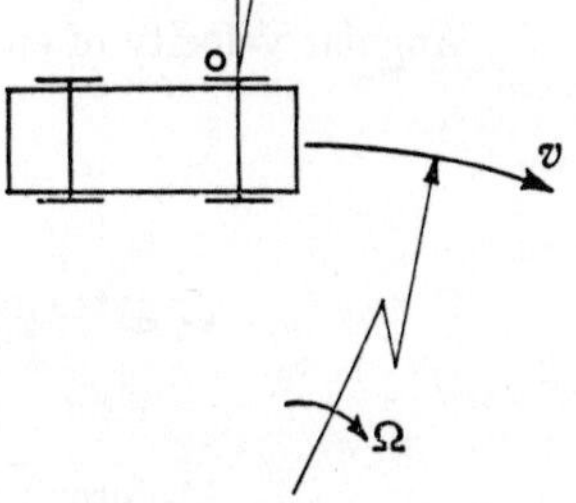

Fig. 10.8

Fig. 10.8 shows a plan view of the car. When turning to the right, the change in angular momentum is represented by the vector ab. The couple required for precession is therefore clockwise looking from the rear and the reaction is in the opposite direction, tending to overturn the car outwards.

The effects of the gyroscopic and centrifugal couples are therefore similar, so that the total overturning couple

$$= 0 \cdot 2983 v^2 + 6 \cdot 3 v^2$$
$$= 6 \cdot 6 v^2 \text{ N m}$$

$$\therefore \text{ lifting force on inside pair of wheels} = \frac{6 \cdot 6 v^2}{1 \cdot 5} = 4 \cdot 4 v^2 \text{ N}$$

The dead weight on the inside pair of wheels is $700 \times 9 \cdot 81$ N, so that the limiting speed occurs when

$$4 \cdot 4 v^2 = 700 \times 9 \cdot 81$$

from which $\qquad v = 39 \cdot 5 \text{ m/s} \quad \text{or} \quad \underline{142 \text{ km/h}}$

4. *A section of an electric rail track of gauge 1·5 m has a left-hand curve of radius 250 m, the superelevation of the outer rail being 250 mm. The approach to the curve is along a straight length of track, over the last 45 m of which there is a uniform increase in the elevation of the outer rail from level track to the superelevation of 250 mm. Each motor used for traction has a rotor of mass 500 kg and a radius of gyration of 240 mm. The motor shaft is parallel to the axes of the running wheels, it is supported in bearings 750 mm apart and runs at five times the wheel speed but in the opposite direction. The diameter of the running wheels is 1·1 m.*

Determine the forces on the bearings due to gyroscopic action when the train is travelling at 100 km/h, (a) on the last 45 m of approach track, (b) round the curve. The directions of the forces must be clearly shown on diagrams. (U. Lond.)

Moment of inertia of engine rotor, $I = 500 \times 0 \cdot 24^2 = 28 \cdot 8 \text{ kg m}^2$

Angular velocity of engine rotor, $\omega = 5 \times \dfrac{v}{r}$

$$= \frac{5 \times 100}{3 \cdot 6 \times 0 \cdot 55} = 253 \text{ rad/s}$$

$$\therefore \text{ gyroscopic couple, } C = I\omega\Omega = 28 \cdot 8 \times 253\Omega$$
$$= 7280\Omega \text{ N m}$$

$$\therefore \text{ force on bearings, } P = \frac{7280\Omega}{0 \cdot 75} = 9700\Omega \text{ N}$$

(a) Angle turned through by engine shaft in last 45 m $= \dfrac{0.25}{1.5} = \dfrac{1}{6}$ rad

Time taken to cover this distance $= \dfrac{45}{100/3.6} = 1.62$ s

∴ velocity of precession, $\Omega = \dfrac{1/6}{1.62} = 0.103$ rad s

∴ $P = 9700 \times 0.103 = \underline{1000 \text{ N}}$

As the right-hand bearing rises, the change in momentum is represented by ab, Fig. 10.9, and thus the couple required for precession is

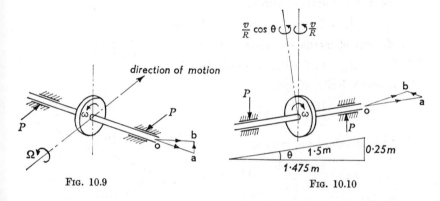

Fig. 10.9   Fig. 10.10

clockwise looking upward. The reaction couple is therefore anticlockwise looking upward and the forces exerted by the shaft on the bearings are in the directions shown.

(b) On the curve, the effective angular velocity of precession about the axis perpendicular to the axis of rotation

$$= \frac{100}{3.6 \times 250} \cos \theta = \frac{100}{3.6 \times 250} \times \frac{1.475}{1.5}$$

i.e. $\qquad \Omega = 0.1093$ rad/s

Effective angular velocity of spin $= \omega - \Omega \sin \theta$

$\qquad \simeq \omega$ since $\Omega \sin \theta$ is negligible compared with ω

∴ $P = 9700 \times 0.1093 = \underline{1060 \text{ N}}$

As the right-hand bearing moves forward relative to the left-hand bearing, the change in momentum is represented by ab, Fig. 10.10, and thus the couple required for precession is clockwise looking forward. The reaction couple is therefore anticlockwise looking forward and the forces exerted by the shaft on the bearings are in the direction shown.

5. *A solo motor cycle, rider and passenger together have a mass of 320 kg, the combined centre of gravity being 525 mm above ground level. The wheels of the machine each have a mass of 9 kg, a radius of gyration of 225 mm and an effective rolling radius of 300 mm. The rotating parts of the engine have a mass of 12 kg, a radius of gyration of 75 mm and rotate in the same sense as the road wheels. The gear ratio, engine to back wheel, is 3·5 : 1. The machine is travelling round a banked curve.*

Determine the angle of the banking necessary for the machine to ride normal to the track on a bend of 60 m radius at a speed of 160 km/h, allowing for gyroscopic effects. (U. Lond.)

Moment of inertia of wheels,

$$I_1 = 2 \times 9 \times 0.225^2 = 0.912 \text{ kg m}^2$$

Moment of inertia of engine parts,

$$I_2 = 12 \times 0.075^2 = 0.0675 \text{ kg m}^2$$

Linear velocity of cycle,

$$v = \frac{160}{3.6} = 44.44 \text{ m/s}$$

Angular velocity of wheels,

$$\omega_1 = \frac{v}{r} = \frac{44.44}{0.3} = 148 \text{ rad/s}$$

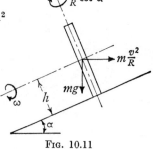

Fig. 10.11

Angular velocity of engine parts,

$$\omega_2 = 3.5 \times 148 = 518 \text{ rad/s}$$

If α is the inclination of the cycle to the vertical, Fig. 10.11, the velocity of precession in the plane of the wheel and engine axes (i.e. perpendicular to the axes of spin)

$$= \frac{v}{R} \cos \alpha$$

Since the engine and wheels rotate in the same sense, the gyroscopic couples are additive,

i.e.
$$C = (I_1\omega_1 + I_2\omega_2) \times \frac{v}{R} \cos \alpha$$

$$= (0.912 \times 148 + 0.0675 \times 518) \times \frac{44.44}{60} \cos \alpha$$

$$= 126 \cos \alpha \text{ N m}$$

As in Example 4, this couple tends to overturn the cycle outwards.

$$\text{Centrifugal couple} = \frac{mv^2}{R} \times h \cos \alpha$$

$$= 320 \times \frac{44 \cdot 44^2}{60} \times 0 \cdot 525 \ \cos \alpha$$

$$= 5530 \cos \alpha \ \text{N m}$$

$$\therefore \ \text{total overturning moment} = (126 + 5530) \cos \alpha$$

$$= 5656 \cos \alpha \ \text{N m}$$

$$\therefore \ \text{for equilibrium,} \quad 5656 \cos \alpha = 320 \times 9 \cdot 81 \times 0 \cdot 525 \sin \alpha$$

$$\therefore \ \tan \alpha = 3 \cdot 43$$

$$\therefore \ \alpha = \underline{73° \ 45'}$$

6. *A gyroscopic wheel of mass 0·25 kg and with radius of gyration 20 mm is mounted in a pivoted frame C, as shown in Fig. 10.12. The axis AB of the pivot passes through the centre of rotation of the wheel but the centre of gravity of the frame C is at a distance 10 mm below the axis. The frame C has a mass of 0·14 kg. The speed of rotation of the wheel is 3000 rev/min.*

The arrangement is mounted in a vehicle so that the axis AB is parallel to the direction of motion of the vehicle. If the vehicle travels at 15 m/s in a curve of 48 m radius, find the angle of inclination of the gyroscope from the vertical, (a) when the vehicle moves in the direction shown, and (b) when it moves in the opposite direction. (U. Lond.)

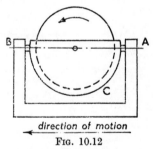

direction of motion

FIG. 10.12

Moment of inertia of wheel, $I = 0 \cdot 25 \times 0 \cdot 02^2 = 0 \cdot 0001 \ \text{kg m}^2$

Angular velocity of spin, $\omega = 3000 \times \dfrac{2\pi}{60} = 314 \cdot 2 \ \text{rad/s}$

If α is the inclination of the wheel to the vertical, Fig. 10·13, the velocity of precession in the plane of the wheel axis,

$$\Omega = \frac{v}{R} \cos \alpha = \frac{15}{48} \cos \alpha = 0 \cdot 3125 \cos \alpha \ \text{rad/s}$$

$$\therefore \ \text{gyroscopic couple,} \ C = I\omega\Omega = 0 \cdot 0001 \times 314 \cdot 2 \times 0 \cdot 3125 \cos \alpha$$

$$= 0 \cdot 009 \ 81 \cos \alpha \ \text{N m}$$

$$\text{Centrifugal couple} = \frac{mv^2}{R} \times \text{OG} \cos \alpha$$

$$= 0.14 \times \frac{15^2}{48} \times 0.01 \cos \alpha$$

$$= 0.006\,56 \cos \alpha \text{ N m}$$

$$\text{Couple due to weight of frame} = mg \times \text{OG} \sin \alpha$$

$$= 0.14 \times 9.81 \times 0.01 \sin \alpha$$

$$= 0.0137 \sin \alpha \text{ N m}$$

(a) When travelling in the direction shown in Fig. 10.13, the change in momentum is represented by the vector ab. Thus the couple exerted on the frame by the wheel is anticlockwise, as shown.

Therefore, for equilibrium,

$$0.009\,81 \cos \alpha = 0.006\,56 \cos \alpha + 0.0137 \sin \alpha$$

from which
$$\alpha = \underline{13° \ 20'}$$

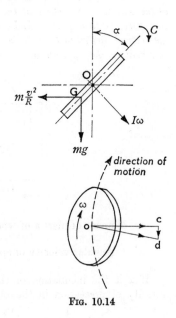

10.13

Fig. 10.14

(b) When travelling in the direction shown in Fig. 10.14, the change in momentum is represented by the vector cd. Thus the couple exerted on the frame by the wheel is clockwise, as shown.

Therefore, for equilibrium,

$$0 \cdot 009\ 81 \cos \alpha + 0 \cdot 006\ 56 \cos \alpha = 0 \cdot 0137 \sin \alpha$$

from which $\qquad \therefore \alpha = \underline{50° \, 5'}$

7. *A disc rotor has a mass of 30 kg and a radius of gyration about its axis of symmetry of 125 mm, while its radius of gyration about a diameter of the rotor at right angles to the axis of symmetry is 75 mm. This rotor is pressed on to a shaft but, due to incorrect boring, the angle between the axis of symmetry and the actual axis of rotation is 0·25°, though both these axes pass through the centre of gravity of the rotor. Assuming that the shaft is rigid and is carried between bearings 200 mm apart, determine the bearing forces due to the misalignment at a speed of 6000 rev/min.*

If a formula is used to determine the centrifugal couple, this should be proved. (U. Lond.)

Let $I =$ polar moment of inertia of disc,

$\quad I_d =$ moment of inertia of disc about diameter,

$\quad \omega =$ angular velocity of shaft.

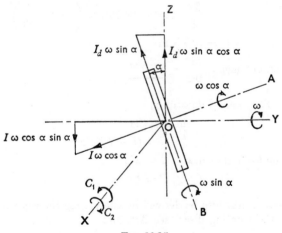

Fig. 10.15

Referring to Fig. 10.15,

$$\text{angular velocity of disc about OA} = \omega \cos \alpha$$

$$\therefore \text{angular momentum of disc about OA} = I\omega \cos \alpha$$

The component of this momentum perpendicular to OY

$$= I\omega \cos \alpha \sin \alpha$$

This component is being precessed about OY at ω rad/s, so that

$$\text{couple required, } C_1 = I\omega^2 \cos \alpha \sin \alpha$$

acting about OX in an anticlockwise direction.

$$\text{Angular velocity of disc about OB} = \omega \sin \alpha$$

$\therefore$ angular momentum of disc about OB $= I_d\omega \sin \alpha$

The component of this momentum perpendicular to OY

$$= I_d\omega \sin \alpha \cos \alpha$$

This component is being precessed about OY at ω rad/s, so that

$$\text{couple required, } C_2 = I_d\omega^2 \sin \alpha \cos \alpha$$

acting about OX in a clockwise direction.

$$\therefore \text{resultant couple about OX} = C_1 - C_2$$

$$= \frac{I - I_d}{2} \omega^2 \sin 2\alpha$$

$$= \frac{30}{2}(0.125^2 - 0.075^2) \times \left(6000 \times \frac{2\pi}{60}\right)^2 \times \sin \tfrac{1}{2}^\circ$$

$$= 515 \text{ N m}$$

$$\therefore \text{load on bearings} = \frac{515}{0.2} = \underline{2750 \text{ N}}$$

NOTE. For small values of α,

$$\sin 2\alpha \simeq 2\alpha$$

$$\therefore C = (I - I_d)\omega^2\alpha$$

For a uniform thin disc, $\quad I = 2I_d$

$$\therefore C = I_d\omega^2\alpha$$

This expression may also be derived by considering the centrifugal forces set up by the rotating disc; see Art. 15.2.

8. *The wheel shown in Fig. 10.16 rotates at 10 rad/s about its polar axis OX, the rectangular frame OX, OY, OZ being rotated at 5 rad/s about the vertical axis OW. This vertical axis is in the same plane as OX and OZ and bisects the angle XOZ. OY is horizontal.*

Determine the gyroscopic torques in N m to be applied about each of the axes OX, OY and OZ if I and I_d of the wheel are 0·17 and 0·09 kg m² respectively. (U. Lond.)

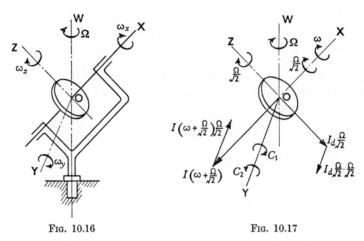

FIG. 10.16   FIG. 10.17

Referring to Fig. 10.17, the angular velocity of the frame about OW, Ω, can be resolved into components $\dfrac{\Omega}{\sqrt{2}}$ about each of the axes OX and OZ; thus the effective angular velocity of the wheel is $\left(\omega + \dfrac{\Omega}{\sqrt{2}}\right)$.

The angular momentum of the wheel about OX, $I\left(\omega + \dfrac{\Omega}{\sqrt{2}}\right)$, is being precessed about OZ at a rate $\dfrac{\Omega}{\sqrt{2}}$ and hence the gyroscopic couple about OY, C_1, is $I\left(\omega + \dfrac{\Omega}{\sqrt{2}}\right)\dfrac{\Omega}{\sqrt{2}}$.

The angular momentum of the wheel about OZ, $I_d\dfrac{\Omega}{\sqrt{2}}$, is being precessed about OX at a rate $\dfrac{\Omega}{\sqrt{2}}$ and hence the gyroscopic couple about OY, C_2 is $I_d\dfrac{\Omega}{\sqrt{2}}\cdot\dfrac{\Omega}{\sqrt{2}}$.

Hence, resultant couple about $OY = C_1 - C_2$

$$= I\left(\omega + \frac{\Omega}{\sqrt{2}}\right)\frac{\Omega}{\sqrt{2}} - I_d \frac{\Omega}{\sqrt{2}} \cdot \frac{\Omega}{\sqrt{2}}$$

$$= 0.17\left(10 + \frac{5}{\sqrt{2}}\right)\frac{5}{\sqrt{2}} - 0.09 \cdot \frac{5^2}{2}$$

$$= \underline{7.025 \text{ N m}}$$

The only changes in angular momentum which take place due to precession are parallel with the OY axis and hence no couples are required about OX or OZ.

Alternatively, from equations (10.6), (10.7) and (10.8), transposed to agree with the notation of the diagram,

$$C_x = \frac{dh_x}{dt} + h_z\Omega_y - h_y\Omega_z$$

$$C_y = \frac{dh_y}{dt} + h_x\Omega_z - h_z\Omega_x$$

$$C_z = \frac{dh_z}{dt} + h_y\Omega_x - h_x\Omega_y$$

$$h_x = I\left(\omega + \frac{\Omega}{\sqrt{2}}\right), \quad h_y = 0 \quad \text{and} \quad h_z = I_d\frac{\Omega}{\sqrt{2}}$$

$$\therefore C_x = \frac{d}{dt}\left\{I\left(\omega + \frac{\Omega}{\sqrt{2}}\right)\right\} + I_d\frac{\Omega}{\sqrt{2}} \times 0 - 0 \times \frac{\Omega}{\sqrt{2}} = 0$$

$$C_y = \frac{d}{dt}(0) + I\left(\omega + \frac{\Omega}{\sqrt{2}}\right)\frac{\Omega}{\sqrt{2}} - I_d\frac{\Omega}{\sqrt{2}} \cdot \frac{\Omega}{\sqrt{2}}$$

$$= I\left(\omega + \frac{\Omega}{\sqrt{2}}\right)\frac{\Omega}{\sqrt{2}} - I_d\frac{\Omega^2}{2}$$

$$C_z = \frac{d}{dt}\left(I\frac{\Omega}{\sqrt{2}}\right) + 0 \times \frac{\Omega}{\sqrt{2}} - I\left(\omega + \frac{\Omega}{\sqrt{2}}\right) \times 0 = 0$$

Note that the effective angular velocity about OX concerned with the angular momentum of the wheel about that axis is $\left(\omega + \frac{\Omega}{\sqrt{2}}\right)$ but the angular velocity about OX causing precession of the momentum $I_d\frac{\Omega}{\sqrt{2}}$ is merely $\frac{\Omega}{\sqrt{2}}$.

9. Fig. 10.18 shows a gyroscope mounted on a framework which is attached by an arm PQ to a wheel R which turns freely about a vertical spindle. The arm PQ is free to swing in a vertical plane about Q. The gyroscope is a uniform disc of diameter 125 mm, of mass 12 kg and having its centre of gravity 450 mm from the spindle axis. The framework has a mass of 5 kg and its centre of gravity G is 375 mm from the spindle axis.

If the disc spins at 15 000 rev/min in the direction shown and the arm PQ is horizontal, determine the angular velocity with which the system will precess about the verticle spindle.   (*Ans.:* 15·44 rev/min)

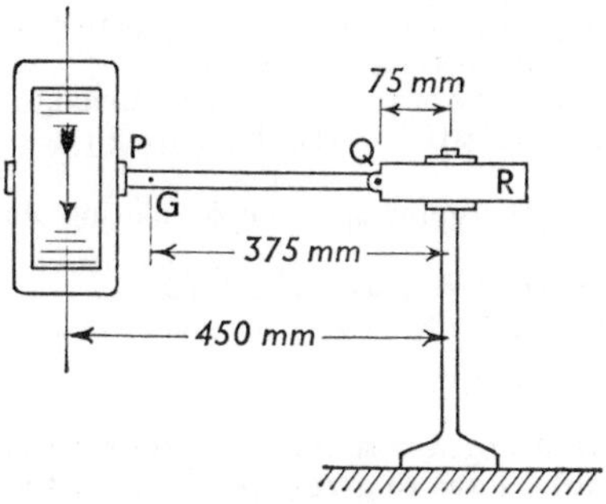

FIG. 10.18 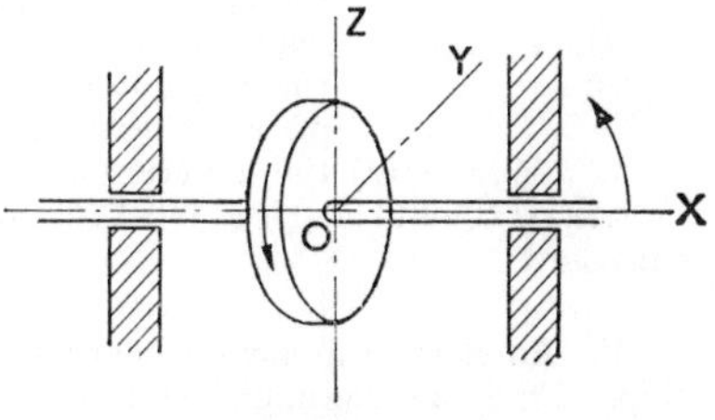   FIG. 10.19

10. Fig. 10.19 shows a uniform disc of 150 mm diameter and mass 5 kg mounted centrally in bearings which maintain its axle in a horizontal plane. The disc spins about its axle with a constant speed of 1000 rev/min while the axle precesses uniformly about the vertical, turning 1 rev in 1 s. The directions of rotation are as shown. If the distance between the bearings is 100 mm, find the resultant reaction at each (due to weight and gyroscopic effects) and show clearly on the sketch the points of contact between axle and bearing. (*I. Mech. E.*)

(*Ans.:* Right: 67·9 N, down: left: 117·1 N, up)

11. A pair of locomotive driving wheels with the axle have a moment of inertia of 380 kg m². The diameter of the wheel treads is 2 m and the distance between the wheel centres is 1·5 m. When the locomotive is travelling on a level track at 110 km/h, defective ballasting causes one wheel to fall 10 mm and to rise again in a total time of 0·1 s. If the displacement of the wheel takes place with simple harmonic motion, find the gyroscopic reaction on the locomotive. (*I. Mech. E.*)   (*Ans.:* 2·435 kN m)

12. A motor vehicle having a rigid rear axle is travelling along a straight level road at 160 km/h. The rear wheel on the left-hand side of the vehicle drops into a pot-hole which lowers the end of the axle by 80 mm. The track of the vehicle is 1·5 m, the rolling radius of the wheels is 0·3 m and the movement of the wheel into and out of the pot-hole may be regarded as approximating in form to a complete sine wave, 0·45 m long, having its mean 40 mm below the road level. The polar moment of inertia of the axle with its two wheels is 2·5 kg m².

Determine the maximum gyroscopic couple that acts on the vehicle and the effects of this on the course of the vehicle. Explain clearly, with the aid of diagrams, how you determine the direction of the gyroscopic couple. (*U. Lond.*)

(*Ans.:* 6·12 kN m, turning car to left)

13. A generating set is arranged on board ship with its axis parallel to the longitudinal centre-line of the ship. The revolving parts have a mass of 1400 kg, a radius of gyration of 400 mm and revolve at 420 rev/min. If the ship steams at 36 km/h, round a curve of 180 m radius, find the magnitude and sense of the gyroscopic couple transmitted to the ship. (*I. Mech. E.*)

(*Ans.:* 547 N m; tends to lift bow when turning to port if revolving clockwise looking forward)

14. The turbine rotor of a ship has a mass of 20 t, has a radius of gyration of 0·6 m and rotates at 3000 rev/min. The ship is pitching 10° above and 10° below the horizontal, the motion being simple harmonic and having a period of 12 s. The rotor turns in a clockwise direction when viewed from aft.

Determine : (*a*) the maximum angular velocity of the ship during pitching ;
(*b*) the maximum angular acceleration of the ship during pitching ;
(*c*) the maximum value of the gyroscopic couple, stating its plane of action ;
(*d*) the direction of yaw as the bow rises. (*U. Lond.*)

(*Ans.:* 0·0914 rad/s; 0·04775 rad/s²; 207 kN m in a horizontal plane; starboard)

15. An electric motor on board ship is arranged with its rotor athwart the ship. Find the maximum load on its bearings due to gyroscopic action if the ship rolls with S.H.M. 30° on each side of the vertical and the time for one complete roll is 3·4 s; the mass of the rotor is 220 kg, its radius of gyration is 215 mm, the bearings are 1·1 m apart and the speed of the rotor is 3000 rev/min clockwise viewed from the starboard (right) side.

Explain precisely with the aid of diagrams the effect this has on the ship's hull. (*U. Lond.*)

(*Ans.:* 2·81 kN; when the ship heels to port, it tends to turn to port)

16. A truck with four wheels, each of 750 mm diameter, travels on rails round a curve of 75 m radius at a speed of 50 km/h. The total mass of the truck is 5 t and its centre of gravity is midway between the axles, 1·05 m above the rails and midway between them. Each pair of wheels is driven by a motor rotating in the opposite direction to the wheels and at four times the speed. The moment of inertia of each pair of wheels is 15 kg m² and of each motor shaft 5 kg m². The rails lie on a horizontal plane and 1·45 m apart. Determine the load on each rail. (*U. Lond.*) (*Ans.:* outer, 33·87 kN; inner, 15·17 kN)

17. Using the Coriolis component of acceleration, derive an expression for the gyroscopic torque on a wheel of moment of inertia I rotating at an angular velocity ω and precessing with an angular velocity Ω.

A diesel electric locomotive has two axles 4 m apart. The wheels on each axle are 1·5 m apart, 1·2 m diameter and moment of inertia of each axle is 70 kg m². The rotating parts of the engine and generator have a moment of inertia of 60 kg m² and they rotate about a longitudinal axis of the locomotive at 2200 rev/min in a counter-clockwise direction when looking forward. When the speed of the locomotive is 40 km/h, it enters a left-hand curve of 150 m radius. Find the change in the vertical reactions on each wheel due to gyroscopic action only. (*U. Lond.*)

(*Ans.:* front outer, +192 N; front inner, +64 N; rear outer, −64 N; rear inner, −192 N)

18. The following particulars are given for a motor vehicle: total mass, 1·5 t; wheel base, 3·2 m; track width, 1·5 m; centre of gravity 1·8 m behind the front axle and 0·95 m above road level; moment of inertia of two front wheels, 10 kg m²; moment of inertia of two rear wheels, 15 kg m²; moment of inertia of parts turning at engine speed, 2 kg m²; wheel diameter, 0·64 m, gear ratio from engine to road wheels, 10 to 1. The engine turns in a clockwise direction when viewed from the front of the vehicle. The vehicle travels at a constant speed of 80 km/h and enters a right-hand curve of 150 m radius.

Determine: (*a*) the vertical load on each wheel, taking into account

> (*i*) gravitational effects;
> (*ii*) centrifugal effects;
> (*iii*) the gyroscopic effect due to the engine rotation;

> (*b*) the rolling couple acting on the vehicle due to the gyroscopic effect of the road wheels. (*U. Lond.*)

(*Ans.:* (*a*) front inner, +3·22, −1·565, −0·0322 kN; front outer, +3·22, +1·565, −0·0322 kN; rear inner, +4·14, −1·565, +0·0322 kN; rear outer, +4·14, +1·565, +0·0322 kN; (*b*) 257 N m)·

19. A solo motor cycle, complete with rider, has a mass of 225 kg, the centre of gravity being 0·6 m above ground level. The moment of inertia of each road wheel is 1 kg m² and the rolling diameter is 0·6 m. The engine crankshaft rotates, in the same sense as the wheels, at 5 times the speed of the wheels. The rotating parts of the engine are equivalent to a flywheel whose moment of inertia is 0·2 kg m².

Determine the heel-over angle required when the unit is travelling at 100 km/h in a curve of radius 60 m. (*U. Lond.*)   (*Ans.:* 54° 36′)

20. A racing motor cyclist travels at 140 km/h round a curve of 120 m radius (measured horizontally). The cycle and rider have mass 150 kg and their mass centre is 0·7 m above ground level when the machine is vertical. Each wheel is 0·6 m diameter and has moment of inertia about its axis of rotation 1·5 kg m². The engine has rotating parts whose moment of inertia about their axis of rotation is 0·25 kg m², and it rotates at five times the wheel speed in the same direction.

Find (*a*) the correct angle of banking of the track so that there will be no tendency to side slip (*b*) the correct angle of inclination of the cycle and rider to the vertical.

(*Ans.:* 52° 6′; 55° 36′)

21. A motor vehicle of all-up mass 900 kg has a track of 1·5 m and the centre of gravity is 0·38 m above ground level. The four road wheels have a total moment of inertia of 5 kg m² and a rolling radius of 0·3 m. The vehicle is travelling at a speed of 45 m/s in a circular path of radius 180 m.

Determine: (*a*) the total vertical component of the reaction on the two outer wheels, if the track is not banked;

> (*b*) the angle of banking necessary for there to be no tendency for the vehicle to sideslip;

> (*c*) the total component (perpendicular to the track surface) of the reaction on the two outer wheels, if the track is banked to the angle specified in (*b*) above. (*U. Lond.*)

(*Ans.:* 7·108 kN; 48° 57′; 6·8 kN)

22. Fig. 10.20 shows the basic construction of a rate of turn indicator for an aircraft. View A is obtained by looking at the instrument from the starboard (right-hand) side of the aircraft and view B is that observed by the pilot when looking straight ahead.

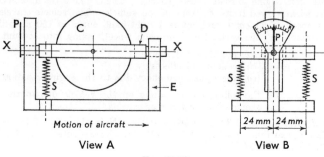

View A   View B

FIG. 10.20

The main frame E of the instrument is fixed to the aircraft while frame D can rotate relatively to E about axis XX, its angular position being indicated by the pointer P. The position of frame D relative to frame E is controlled by the two equal springs S which are attached to both frames as shown. These springs have sufficient initial stretch to ensure that they remain in tension as frame D tilts, and any slight inclination which they make with the vertical may be ignored. C is a thin rigid disc of uniform thickness having a mass of 0·35 kg and a radius of 40 mm. It rotates at 8000 rev/min on a spindle carried in frame D.

When the aircraft is in level flight and is altering course steadily at the rate of 90° in 15 s it is desired that the pointer should read 10° in the appropriate sense indicating the direction of turn. Calculate the necessary stiffness of the springs and determine the direction of spin of the disc, as seen in view A. Show the derivation from first principles of any formula used, indicating where any approximations are made, and explaining clearly how the direction of spin is determined. (*U. Lond.*)   (*Ans.* : 122·6 N/m ; anticlockwise)

23. A point P moves along a straight line OA with a uniform velocity. At the same time, OA rotates about O with a uniform angular velocity. Show that the acceleration of P is independent of its distance from O.

A gyroscope turn indicator for an aeroplane consists essentially of a uniform disc, 50 mm in diameter, which rotates at 3000 rev/min. Its spindle bearings are carried on a frame which is free to turn in trunnions, the centre-line of the trunnions being at right angles to, and 5 mm above, the axis of rotation of the disc. On a straight course, the plane of the disc is vertical and at right angles to the fore-and-aft centre-line of the machine. Find the angle through which the frame will tilt when the speed is 240 km/h and the course is altered to a circular arc of 300 m radius. (*U. Lond.*)   (*Ans.* : 24°)

24. A shaft carries a disc flywheel of mass 140 kg and diameter 750 mm. The disc axis is set out of alignment with the shaft axis by 1°. Find the couple on the bearings when the shaft rotates at 1800 rev/min. (*U. Lond.*)

                                  (*Ans.* : 3·055 kN m)

25. The inclined disc shown in Fig. 10.21 is mounted on a shaft and is supported in bearings at A and B. The material of the disc has a density of 7·8 Mg/m³. When the disc rotates at 300 rev/min, what forces are exerted on the bearings by the centrifugal action of the disc ? The diameter of the shaft may be taken as small compared with that of the disc. (*U. Lond.*) (*Ans.:* 20 N)

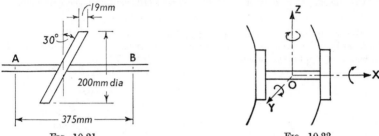

FIG. 10.21   FIG. 10.22

26. A pair of loco driving wheels of combined polar moment of inertia 400 kg m² travels at 30 m/s around a curve of mean radius 300 m. The wheels are 1·8 m in diameter and they are 1·5 m apart. Due to a weak point in the support of the outer rail, the outer wheel drops momentarily with a velocity of 0·35 m/s. Determine the gyroscopic torques about the axes OX, OY and OZ, Fig. 10.22. (*U. Lond.*) (*Ans.:* 0, −3·11 kN m, −1·333 kN m)

27. A solid circular steel disc, 250 mm diameter and 50 mm thick, is mounted with its polar axis on the line OX of the three perpendicular axes OX, OY and OZ. If, at a particular instant, this disc is spinning about OX, at +12 rad/s with an acceleration of +7 rad/s² and the frame is rotated at +5 rad/s and accelerated at +3 rad/s² about OY, determine the magnitudes and sense of the torques about each of the axes. Take the density of steel as 7·8 Mg/m³. (*U. Lond.*) (*Ans.:* 1·048, 0·237 and 8·96 N m)

28. The inertia about OX of the wheel shown in Fig. 10.23 is 0·018 kg m² and that about a diameter is 0·011 kg m², the diametral axis $y_1 y_2$ being parallel to and 200 mm away from OY. The mass of the wheel is 2·8 kg.

Determine, neglecting the torque due to the weight, the external torque required about each of the axes OX, OY and OZ, stating the directions, if the wheel is driven about OX at 100 rad/s (*a*) for steady precession at 20 rad/s about OZ in the direction shown, i.e. clockwise looking downwards, (*b*) if the

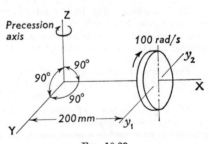

FIG. 10.23

precession is again 20 rad/s but is accelerated in the same direction at 10 rad/s². (*U. Lond.*) (*Ans.:* (*a*) 0, 36, 0 N m; (*b*) 0, 36, 1·23 N m)

29. A thin disc is mounted with its principal axis OA inclined at an angle α to the vertical co-ordinate axis OZ, as shown in Fig. 10.24. The disc rotates with uniform angular velocity ω about OA and precession also takes place with uniform angular velocity Ω about OZ, the angle α between OA and OZ remaining constant during the motion.

Resolve the angular velocity Ω about OZ into components about the axis OA and the principal diametral axis OB and, assuming the moments of inertia about OA and OB to be I_a and I_b respectively, obtain the angular momentum about each of these axes and hence about the horizontal co-ordinate axis OX which is situated in the plane AOB.

Considering OX to rotate with the plane AOB, establish the expression

$$\omega\Omega I_a \sin\alpha + \tfrac{1}{2}\Omega^2(I_a - I_b)\sin 2\alpha$$

for the torque which must be applied to the disc to maintain the motion, and indicate by a diagram the axis about which this torque must be applied and the direction of application about this axis. (*I. Mech. E.*)

(*Ans.:* Torque is applied about OY)

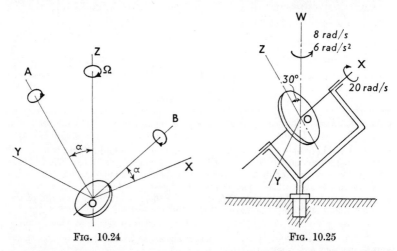

Fig. 10.24   Fig. 10.25

30. The wheel in Fig. 10.25 rotates at 20 rad/s about its polar axis OX whilst the rectangular frame OX, OY, OZ is rotated at 8 rad/s about OW in the direction shown and is also being accelerated in the same sense at 6 rad/s². The vertical axis OW is in the same plane as OX and OZ and the angle ZOW is 30°; OY is horizontal.

Determine the gyroscopic couples to be applied about each of the axes OX, OY and OZ if I_p and I_d are respectively 5 and 3 kg m². (*U. Lond.*)

(*Ans.:* 15 N m; $368\sqrt{3}$ N m; $9\sqrt{3}$ N m)

31. A uniform beam of mass 1·8 kg and length 0·6 m rotates in a vertical plane about its mid-point with an angular velocity of 65 rad/s. Its axis of rotation is precessed in a horizontal plane with an angular velocity of 10 rad/s.

(*a*) Calculate the value of the gyroscopic torque when the beam is at an angle θ rad to the horizontal. Show on a diagram the direction in which the torque acts.

(*b*) Sketch the graph connecting the gyroscopic torque and the angle θ for values of θ from 0 to 2π.

Expressions for the acceleration of a point on the beam need not be proved, but otherwise work from first principles. (*U. Lond.*)   (*Ans.:* $70\sin\theta$ N m)

CHAPTER 11

TOOTHED GEARING

11.1 Spur gears. Spur teeth are parallel to the axis of the wheel and are usually of involute profile. When two gears are in mesh, the larger is termed the *wheel* or *spur* and the smaller the *pinion*.

The *pitch circle diameters* are the diameters of discs which would transmit the same velocity ratio by friction as the gear wheels. If D_1 and D_2 are the pitch circle diameters, T_1 and T_2 the numbers of teeth and ω_1 and ω_2 the angular velocities of the pinion and wheel respectively, then

$$\frac{\omega_1}{\omega_2} = \frac{D_2}{D_1} = \frac{T_2}{T_1}$$

The *pitch point* is the point of contact of two pitch circles.

The *circular pitch* (p) is the distance between a point on one tooth and the corresponding point on an adjacent tooth, Fig. 11.1, measured along the pitch circle,

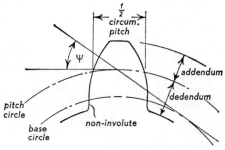

FIG. 11.1

i.e.
$$p = \frac{\pi D_1}{T_1} = \frac{\pi D_2}{T_2}$$

The *diametral pitch* (P) is the number of teeth per millimetre of p.c.d.,

i.e.
$$P = \frac{T_1}{D_1} = \frac{T_2}{D_2} = \frac{\pi}{p}$$

The *module* (m) is the number of millimetres of p.c.d. per tooth,

i.e.
$$m = \frac{D_1}{T_1} = \frac{D_2}{T_2} = \frac{1}{P}$$

The *base circle* is the circle from which the involute curves* forming the tooth profiles are drawn.

* See *Practical Geometry and Engineering Graphics*, W. Abbott, p. 36.

The *addendum* is the radial height of a tooth above the pitch circle.

The *dedendum* is the radial depth of a tooth below the pitch circle.

The *working depth* is the sum of the addenda of two mating teeth.

The *pressure angle* or *angle of obliquity* (ψ) is the angle between the common normal to two teeth in contact and the common tangent to the pitch circles.

The proportions recommended by the British Standards Institution in BSS 436–1940 are :

addendum = $1/P = m$ working depth = $2/P = 2m$

dedendum = $1\cdot25/P = 1\cdot25m$ pressure angle = $20°$

Standard modules are : $\frac{1}{2} \times \frac{1}{2} - 6\frac{1}{2}$, $7 \times 1 - 16$, 18, 20, 25, 30, 35, 40, 45, 50.

11.2   Condition for transmission of constant velocity ratio.   In Fig. 11.2, O_1 and O_2 are the centres of the pinion and wheel respectively, I_1I_2 is the common normal at the point of contact, C, and O_1I_1 and O_2I_2 are the perpendiculars from O_1 and O_2 respectively to the common normal.

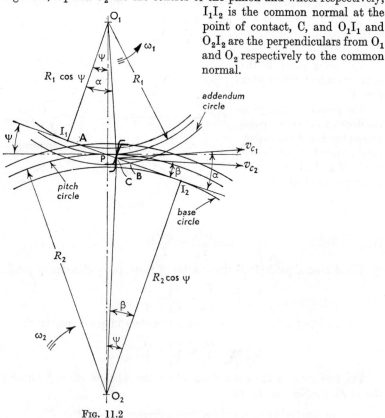

Fig. 11.2

Let v_{c_1} be the velocity of point C on the pinion and v_{c_2} be the velocity of point C on the wheel. If the teeth are to remain in contact, the components of these velocities along the common normal I_1I_2 must be equal,

i.e. $$v_{c_1} \cos \alpha = v_{c_2} \cos \beta$$

i.e. $$\omega_1 O_1 C \cos \alpha = \omega_2 O_2 C \cos \beta$$

i.e. $$\omega_1 O_1 I_1 = \omega_2 O_2 I_2$$

$$\therefore \frac{\omega_1}{\omega_2} = \frac{O_2 I_2}{O_1 I_1} = \frac{O_2 P}{O_1 P}$$

If ω_1/ω_2 is to be constant, P must be the pitch point for the two wheels, i.e. the common normal at the point of contact must pass through the pitch point. This condition is fulfilled by teeth of involute form, provided that the base circles from which the profiles are generated are tangential to the common normal. Since all points of contact lie on the common normal, it is called the *line of contact*. The force between two mating teeth acts along this line in the absence of friction, so that it is also called the *pressure line*.

11.3 Velocity of sliding. The velocity of sliding is the velocity of one tooth relative to its mating tooth along the common tangent at the point of contact. If the wheel is regarded as fixed, the pinion rotates instantaneously about the point P with a relative velocity $\omega_1 + \omega_2$. Hence,

$$\text{velocity of sliding at C} = (\omega_1 + \omega_2) \times \text{PC} . \qquad . \quad (11.1)$$

The maximum velocity of sliding thus occurs at the first or last point of contact.

11.4 Path of contact. Assuming the pinion to be the driver, the first and last points of contact are A and B, Fig. 11.3, where the addenda circles cut the common normal. The path of contact is AB, which is divided into the path of approach, AP, and the path of recess, PB. If R_1 and R_2 are the pitch circle radii and ρ_1 and ρ_2 the radii of the addenda circles, then

$$AP = AI_2 - PI_2 = \sqrt{(\rho_2{}^2 - R_2{}^2 \cos^2 \psi)} - R_2 \sin \psi \quad . \quad (11.2)$$

$$PB = BI_1 - PI_1 = \sqrt{(\rho_1{}^2 - R_1{}^2 \cos^2 \psi)} - R_1 \sin \psi \quad . \quad (11.3)$$

and $\quad AB = AP + PB$

$$= \sqrt{(\rho_2{}^2 - R_2{}^2 \cos^2 \psi)} + \sqrt{(\rho_1{}^2 - R_1{}^2 \cos^2 \psi)} - (R_1 + R_2) \sin \psi \quad (11.4)$$

11.5 Arc of contact and contact ratio. The arc of contact is the arc of the pitch circle EF, Fig. 11.3, between the positions of a tooth at the first and last points of contact with its mating tooth. It is divided into the arc of approach, EP, and the arc of recess, PF.

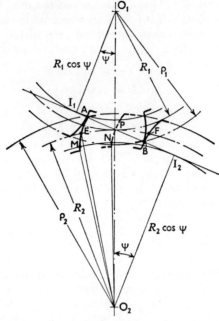

Fig. 11.3

$$\angle \, MO_2N = \angle \, EO_2P$$

i.e.
$$\frac{\text{arc MN}}{R_2 \cos \psi} = \frac{\text{arc EP}}{R_2}$$

$$\therefore \; \text{arc EP} = \frac{\text{arc MN}}{\cos \psi}$$

But, from the construction of the involute curve, arc MN = line AP

$$\therefore \; \text{arc of approach} = \frac{\text{path of approach}}{\cos \psi} \qquad . \qquad . \qquad (11.5)$$

Similarly,
$$\text{arc of recess} = \frac{\text{path of recess}}{\cos \psi} \qquad . \qquad . \qquad (11.6)$$

$$\therefore \; \text{arc of contact} = \frac{\text{path of contact}}{\cos \psi}. \qquad . \qquad . \qquad (11.7)$$

The number of pairs of teeth in contact, or *contact ratio*,

$$= \frac{\text{arc of contact}}{\text{circular pitch}} \qquad . \qquad . \qquad (11.8)$$

The maximum and minimum numbers of pairs of teeth in contact are the nearest whole numbers above and below this value.

11.6 Interference. For correct tooth action, the points of contact on two mating teeth must lie on the involute profiles. If the addendum of one tooth is too large, however, contact may occur between the tip of that tooth and the non-involute portion of the mating tooth between the base circle and the dedendum circle, Fig. 11.1. This causes under-cutting of the mating tooth and interference is said to occur.

For no interference between the teeth, the first and last points of contact must lie between the points of tangency, I_1 and I_2, Fig. 11.3, i.e. the addendum circles must cut the common tangent to the base circles between I_1 and I_2. The limiting case occurs when the addendum circles pass through these points, and since the limiting addendum for the pinion is larger than that for the wheel, it is usually interference between the tips of the wheel teeth and the flanks of the pinion teeth which has to be prevented. This limits either the maximum wheel addendum or the minimum number of teeth on the pinion.

For no undercutting of the pinion teeth, maximum permissible addendum radius of wheel

$$= O_2I_1 = \sqrt{(O_2I_2{}^2 + I_1I_2{}^2)}$$
$$= \sqrt{\{R_2{}^2 \cos^2 \psi + (R_1 \sin \psi + R_2 \sin \psi)^2\}}$$
$$= \sqrt{(R_2{}^2 + 2R_1R_2 \sin^2 \psi + R_1{}^2 \sin^2 \psi)}$$

$\therefore$ maximum wheel addendum

$$= \sqrt{(R_2{}^2 + 2R_1R_2 \sin^2 \psi + R_1{}^2 \sin^2 \psi)} - R_2$$

If the standard addendum, m, is used, then for no interference,

$$m = \frac{2R_1}{T_1} \leqslant \sqrt{(R_2{}^2 + 2R_1R_2 \sin^2 \psi + R_1{}^2 \sin^2 \psi)} - R_2$$

i.e. $\quad \dfrac{1}{T_1} \leqslant \dfrac{1}{2}\sqrt{(G^2 + 2G \sin^2 \psi + \sin^2 \psi)} - \dfrac{G}{2}$

where G is the gear ratio, R_2/R_1.

$$\therefore \; T_1 \geqslant \frac{2}{\sqrt{\{G^2 + \sin^2 \psi(1 + 2G)\}} - G} \qquad . \qquad . \qquad . \qquad . \quad (11.10)$$

11.7 Methods of avoiding interference. Interference between the tips of the wheel teeth and the flanks of the pinion teeth may be avoided by one of the following methods.

(a) Undercutting the flanks of the pinion teeth or otherwise modifying their profiles. This leads to a weakening of the tooth and complication of manufacture. Alternatively, the tips of the wheel teeth may be modified.

(b) Increasing the centre distance slightly. Correct tooth action is maintained but the pressure angle is increased, leading to higher tooth pressures for a given torque transmitted and backlash is increased.

(c) By tooth ' correction ', i.e. modifying the wheel and pinion addenda. The wheel addendum is decreased by moving the rack-cutter towards the wheel axis while generating the teeth and the pinion addendum is increased by moving the cutter away from the pinion axis. The reduction in the wheel addendum is usually made equal to the increase in the pinion addendum to maintain the same working depth. The pressure angle, centre distance and base circles remain unaltered but the thickness of the wheel tooth at the pitch circle becomes less than $p/2$ and that of the pinion becomes greater than $p/2$.

The amount by which the addendum is modified is known as the ' correction '. The *least* correction is the difference between the standard addendum and the maximum addendum for no interference.

11.8 Rack and pinion. A rack may be regarded as a wheel of infinite radius ; the teeth are straight-sided and are normal to the line of contact. Referring to Fig. 11.4 and assuming the pinion to be the driver,

path of approach, $AP = a \operatorname{cosec} \psi$ where a is the rack addendum

and   path of recess, $PB = BI - PI$

$$= \sqrt{(\rho_1{}^2 - R_1{}^2 \cos^2 \psi)} - R_1 \sin \psi$$

$\therefore$ path of contact, $AB = \sqrt{(\rho_1{}^2 - R_1{}^2 \cos^2 \psi)} - R_1 \sin \psi + a \operatorname{cosec} \psi$

$$(11.11)$$

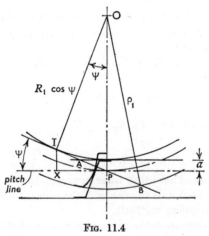

FIG. 11.4

For no interference between the rack tips and the pinion flanks,

maximum rack addendum $= IX = IP \sin \psi$

$$= R_1 \sin^2 \psi \qquad . \qquad . \qquad . \qquad . \quad (11.12)$$

Maximum path of contact $= IB = \sqrt{(\rho_1{}^2 - R_1{}^2 \cos^2 \psi)}$. (11.13)

If the standard addendum, m, is used for the rack, then

$$m = \frac{2R_1}{T_1} \leqslant R_1 \sin^2 \psi \quad . \qquad . \qquad . \quad (11.14)$$

$$\therefore \; T_1 \geqslant 2 \operatorname{cosec}^2 \psi \quad . \qquad . \qquad . \qquad . \quad (11.15)$$

If $\psi = 20°$ $T_1 \geqslant 18$

If $\psi = 14\frac{1}{2}°$ $T_1 \geqslant 32$

11.9   Internal teeth. Fig. 11.5 shows a pinion, of centre O_1, in mesh with an internally toothed wheel (annulus), of centre O_2. The

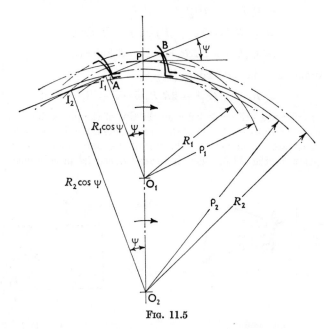

Fɪɢ. 11.5

profiles of the internal teeth are parts of involutes generated from a base circle tangential to the common normal at I_2; the addendum of the internal teeth is measured radially *inwards* from the pitch circle.

Assuming the pinion to be the driver, the first and last points of contact are A and B, where the addenda circles cut the common normal. Path of approach,

$$AP = PI_2 - AI_2 = R_2 \sin \psi - \sqrt{(\rho_2{}^2 - R_2{}^2 \cos^2 \psi)} . \qquad . \quad (11.16)$$

Path of recess,

$$PB = BI_1 - PI_1 = \sqrt{(\rho_1{}^2 - R_1{}^2 \cos^2 \psi)} - R_1 \sin \psi. \qquad . \quad (11.17)$$

$\therefore$ path of contact,

$$AB = AP + PB = \sqrt{(\rho_1{}^2 - R_1{}^2 \cos^2 \psi)}$$
$$- \sqrt{(\rho_2{}^2 - R_2{}^2 \cos^2 \psi)} + (R_2 - R_1) \sin \psi \quad . \quad (11.18)$$

The arc of contact and contact ratio are obtained as for external teeth (Art. 11.5).

For no interference to occur between the tips of the internal teeth and the flanks of the pinion teeth, the addendum circle of the annulus must cut the common normal between P and I_1. The maximum permissible addendum radius for the annulus

$$= O_2I_1 = \sqrt{(O_2I_2{}^2 + I_1I_2{}^2)}$$
$$= \sqrt{\{R_2{}^2 \cos^2 \psi + (R_2 \sin \psi - R_1 \sin \psi)^2\}}$$
$$= \sqrt{(R_2{}^2 - 2R_1R_2 \sin^2 \psi + R_1{}^2 \sin^2 \psi)}$$

$\therefore$ maximum annulus addendum

$$= R_2 - \sqrt{(R_2{}^2 - 2R_1R_2 \sin^2 \psi + R_1{}^2 \sin^2 \psi)} \quad . \quad (11.19)$$

No interference will occur between the tips of the pinion teeth and the flanks of the annulus teeth, but if the pinion addendum is too large, interference may later occur between the tips of the pinion and annulus teeth at the point X, Fig. 11.6. This is termed *trochoidal* interference.

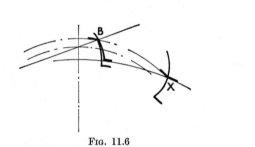
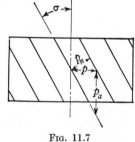

FIG. 11.6   FIG. 11.7

11.10 Helical gears. Helical gears are used in place of spur gears to give a smoother drive between parallel shafts, since engagement between helical teeth is gradual. The transverse section of a tooth is identical with that of an involute spur tooth but the teeth are inclined to the axis of the wheel.

The *spiral or helix angle* (σ) is the angle between the tooth axis and the plane containing the wheel axis, Fig. 11.7. The teeth of mating helical gears must have the same spiral angle and must be of opposite hands, as shown in Fig. 11.11.

The *lead* (*L*) is the distance which a point on the helix would advance parallel to the axis of the gear in moving one revolution round the helix.

Thus
$$\tan \sigma = \frac{\text{pitch circle circumference}}{\text{lead}} = \frac{\pi D}{L}$$

The *circular pitch* (*p*) is the distance between corresponding points on two adjacent teeth, measured along the pitch circle,

i.e.
$$p = \frac{\pi D}{T} \quad . \quad . \quad . \quad . \quad . \quad (11.20)$$

The *diametral pitch* (*P*) is the number of teeth per millimetre of p.c.d.,

i.e.
$$P = \frac{T}{D} = \frac{\pi}{p} \quad . \quad . \quad . \quad . \quad (11.21)$$

The *module* (*m*) is the number of millimetres of p.c.d. per tooth,

i.e.
$$m = \frac{D}{T} = \frac{p}{\pi} \quad . \quad . \quad . \quad . \quad (11.22)$$

The *normal circular pitch* (*p_n*) is the distance between corresponding points on two adjacent teeth, measured normal to the tooth helix,

i.e.
$$p_n = p \cos \sigma \quad . \quad . \quad . \quad . \quad (11.23)$$

The *normal diametral pitch* (*P_n*) corresponds to the normal circular pitch and is given by

$$P_n = \frac{\pi}{p_n}$$

or
$$P_n = P \sec \sigma \quad . \quad . \quad . \quad . \quad (11.24)$$

The *normal module* (*m_n*) is the reciprocal of the normal diametral pitch

i.e.
$$m_n = \frac{p_n}{\pi} = \frac{p \cos \sigma}{\pi} = m \cos \sigma \quad . \quad . \quad (11.25)$$

The *axial pitch* (*p_a*) is the distance between corresponding points on two adjacent teeth, measured parallel to the axis of the gear,

i.e.
$$p_a = p \tan \sigma \quad . \quad . \quad . \quad . \quad (11.26)$$

The *pressure angle* (*ψ*) is the angle between the common normal to two teeth in contact and the common tangent to the pitch circles, in a plane normal to the gear axis, Fig. 11.8.

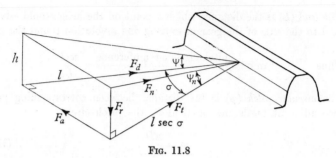

Fig. 11.8

The *normal pressure angle* (ψ_n) is the true pressure angle in a plane normal to the tooth axis. The relation between ψ and ψ_n is given by

$$h = l \tan \psi = l \sec \sigma \tan \psi_n$$

i.e.
$$\tan \psi = \sec \sigma \tan \psi_n \qquad . \qquad . \qquad . \qquad . \qquad (11.27)$$

11.11 Forces on bearings. The resultant force between the teeth, F_n (i.e. the bearing load), is assumed to act normal to the tooth surface at the centre of the face width. This may be resolved into the radial force on the tooth, F_r, and a force tangential to the pitch cylinder, normal to the tooth axis, F_t. The latter force has components F_a and F_d, parallel and perpendicular to the wheel axis, being the axial and driving forces on the wheel respectively.

If C is the torque on the wheel and R the pitch circle radius, then

$$F_d = \frac{C}{R} \qquad . \qquad . \qquad . \qquad . \qquad . \qquad . \qquad (11.28)$$

$$F_a = F_d \tan \sigma \qquad . \qquad . \qquad . \qquad . \qquad . \qquad (11.29)$$

$$F_t = F_d \sec \sigma \qquad . \qquad . \qquad . \qquad . \qquad . \qquad (11.30)$$

$$F_r = F_t \tan \psi_n = F_d \sec \sigma \tan \psi_n \qquad . \qquad . \qquad (11.31)$$

$$F_n = F_t \sec \psi_n = F_d \sec \sigma \sec \psi_n \qquad . \qquad . \qquad (11.32)$$

11.12 Equivalent spur wheel. The cross-section of a helical tooth is similar to that of a straight tooth and the milling cutter used in its manufacture must correspond to the tooth size on an imaginary spur wheel whose plane is perpendicular to the axis of the helical tooth. The equivalent spur wheel has a pitch circle radius equal to the radius of curvature of the pitch cylinder in a plane normal to the tooth axis and a circular pitch equal to the normal circular pitch of the helical teeth, Fig. 11.9.

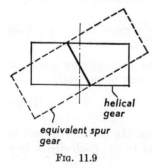

helical gear

equivalent spur gear

Fig. 11.9

Fig. 11.10 shows a helical gear cut by a plane XX through the pitch point normal to the helix. The intersection of this plane with the pitch cylinder is part of an ellipse, the semi-minor and semi-major axes of which are R and $R \sec \sigma$ respectively, and the radius of the equivalent spur gear corresponds to the radius of curvature of the ellipse at the end of the minor axis. This radius is given by*

$$\rho = \frac{(R \sec \sigma)^2}{R}$$
$$= R \sec^2 \sigma$$

so that the diameter of the equivalent spur gear,

$$D_e = 2\rho$$
$$= D \sec^2 \sigma \qquad . \qquad . \qquad . \qquad . \quad (11.33)$$

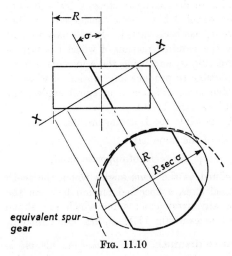

FIG. 11.10

equivalent spur gear

The module of the equivalent spur wheel is the same as the normal module of the helical wheel, so that the number of teeth on the equivalent spur wheel,

$$T_e = \frac{D_e}{m_n}$$

$$= \frac{D \sec^2 \sigma}{m \cos \sigma}$$

$$= T \sec^3 \sigma \qquad . \quad (11.34)$$

where T is the number of teeth on the helical wheel.

* The equation of an ellipse, of semi-major and minor axes a and b respectively, is given by

$$\frac{x^2}{a^2} + \frac{y^2}{b^2} = 1$$

from which

$$y = \frac{b}{a} \sqrt{(a^2 - x^2)}$$

$$\therefore \frac{dy}{dx} = -\frac{b}{a} \frac{x}{\sqrt{(a^2 - x^2)}}$$

and

$$\frac{d^2y}{dx^2} = -\frac{ab}{(a^2 - x^2)^{3/2}}$$

When $x = 0$ and $y = b$, $\dfrac{dy}{dx} = 0$ and $\dfrac{d^2y}{dx^2} = -\dfrac{b}{a^2}$

$$\therefore \rho = \frac{\left\{ 1 + \left(\dfrac{dy}{dx} \right)^2 \right\}^{3/2}}{\dfrac{d^2y}{dx^2}} = -\frac{a^2}{b}$$

11.13 Spiral gears. Helical gears used to connect non-parallel non-intersecting shafts are known as spiral, skew, screw or crossed-helical gears. Spiral gears give only theoretical point contact, instead of line contact as for helical gears connecting parallel shafts, and are therefore only suitable for light loads.

The shaft angle θ is defined as the angle through which one shaft would have to be rotated relative to the other about a line joining the wheel centres to bring the shafts parallel and rotating in opposite directions.

Fig. 11.11 shows various positions of wheel 1 in relation to wheel 2, the plan view showing the development of the top of the lower wheel and the bottom of the upper wheel. The complete range of shaft angles has been shown, with θ increasing or decreasing in steps of 30° and both left- and right-hand spirals for wheel 2 have been considered; σ_2 has been kept constant at 45° while σ_1 has been varied. It will be seen that with some tooth arrangements, the required motion of wheel 1 is impossible, but by increasing or decreasing σ_2 and using the appropriate hand for wheel 2, it is theoretically possible to produce any motion of wheel 1, although for cases which involve large differences in spiral angles, the drive would be very inefficient (see Art. 11.16).

It will be seen that, for the possible tooth arrangements,

$$\theta = \sigma_1 + \sigma_2 \quad \text{for teeth of the same hand}$$

and $\quad \theta = \sigma_1 - \sigma_2 \quad \text{or} \quad \sigma_2 - \sigma_1 \quad$ for teeth of opposite hands

and if the spiral angles are defined as the *acute* angles which the teeth make with their respective wheel axes, a specified motion between the wheels becomes impossible for any arrangements for which the above relations are not fulfilled (in such cases in Fig. 11.11, σ_1 has been omitted).

11.14 Gear ratio and centre distance. The *normal* module, m_n, is the same for each wheel but the modules m_1, m_2, in the planes of revolution depend on the spiral angles.

Thus $\qquad m_n = m_1 \cos \sigma_1 = m_2 \cos \sigma_2 \quad$ from equation (11.25)

Also $\qquad m_1 = \dfrac{D_1}{T_1} \quad \text{and} \quad m_2 = \dfrac{D_2}{T_2}$

$\therefore$ gear ratio, $G = \dfrac{\omega_1}{\omega_2} = \dfrac{T_2}{T_1} = \dfrac{m_1 D_2}{m_2 D_1}$

$$= \dfrac{D_2 \cos \sigma_2}{D_1 \cos \sigma_1} \qquad . \qquad . \qquad . \qquad . \qquad . \quad (11.35)$$

Centre distance, $C = \frac{1}{2}(D_1 + D_2) \qquad . \qquad . \qquad . \qquad . \qquad . \quad (11.36)$

$$= \dfrac{m_n}{2}(T_1 \sec \sigma_1 + T_2 \sec \sigma_2). \qquad . \qquad . \quad (11.37)$$

11.15 Velocity of sliding. The circumferential velocities of the spiral wheels at the pitch point are $\omega_1 r_1$ and $\omega_2 r_2$. The components of these velocities along the tooth helices are $\omega_1 r_1 \sin \sigma_1$ and $\omega_2 r_2 \sin \sigma_2$ respectively, and since these are in opposite directions, there is a velocity of sliding given by

$$v_s = \omega_1 r_1 \sin \sigma_1 + \omega_2 r_2 \sin \sigma_2 \quad . \quad . \quad . \ (11.38)$$

11.16 Efficiency of spiral gearing. Let wheel 1 be the driver[1] and wheel 2 the follower. Without friction, the reaction between the teeth in a plane tangential to the pitch cylinders is normal to the tooth axis but with friction, the reaction F is inclined at the friction angle ϕ to the normal,[2] Fig. 11.12.

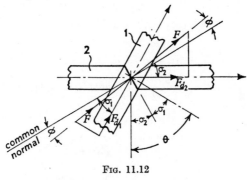

FIG. 11.12

If F_{d_1} and F_{d_2} are the driving forces exerted *by* wheel 1 and *on* wheel 2 respectively, then

$$F_{d_1} = F \cos(\sigma_1 - \phi)$$

and

$$F_{d_2} = F \cos(\sigma_2 + \phi)$$

The efficiency of the drive is the ratio of the work done on wheel 2 to the work done by wheel 1 in the same time,

i.e.

$$\eta = \frac{F_{d_2} \times \pi\omega_2 D_2}{F_{d_1} \times \pi\omega_1 D_1}$$

$$= \frac{F \cos(\sigma_2 + \phi) \cos \sigma_1}{F \cos(\sigma_1 - \phi) \cos \sigma_2} \quad \text{from equation (11.35)}$$

$$\qquad\qquad . \quad . \ (11.39)$$

$$= \frac{\cos \sigma_2 \cos \phi - \sin \sigma_2 \sin \phi}{\cos \sigma_1 \cos \phi + \sin \sigma_1 \sin \phi} \frac{\cos \sigma_1}{\cos \sigma_2}$$

$$= \frac{1 - \mu \tan \sigma_2}{1 + \mu \tan \sigma_1} . \quad . \quad . \quad . \quad . \ (11.40)$$

1, 2. See footnotes on next page.

Substituting $\sigma_2 = \theta - \sigma_1$,

$$\eta = \frac{1 - \mu \tan(\theta - \sigma_1)}{1 + \mu \tan \sigma_1}$$

For maximum efficiency, $\dfrac{d\eta}{d\sigma_1} = 0$, i.e.

$$\{1 + \mu \tan \sigma_1\} \times \mu \sec^2(\theta - \sigma_1) = \{1 - \mu \tan(\theta - \sigma_1)\} \times \mu \sec^2 \sigma_1$$

i.e. $\qquad \cos^2 \sigma_1 + \dfrac{\mu}{2} \sin 2\sigma_1 = \cos^2(\theta - \sigma_1) - \dfrac{\mu}{2} \sin 2(\theta - \sigma_1)$

i.e. $\qquad \cos^2(\theta - \sigma_1) - \cos^2 \sigma_1 = \dfrac{\mu}{2}\{\sin 2(\theta - \sigma_1) + \sin 2\sigma_1\}$

i.e. $\qquad 2 \sin \theta \sin(2\sigma_1 - \theta) = \mu \times 2 \sin \theta \cos(2\sigma_1 - \theta)$

i.e. $\qquad \tan(2\sigma_1 - \theta) = \mu = \tan \phi$

$$\therefore \ 2\sigma_1 - \theta = \phi$$

or $\qquad\qquad\qquad \sigma_1 - \sigma_2 = \phi \qquad . \qquad . \qquad . \qquad . \qquad . \qquad (11.41)$

Thus $\sigma_1 = \dfrac{\theta + \phi}{2}$ and $\sigma_2 = \dfrac{\theta - \phi}{2}$, so that, from equation (11.39),

$$\text{maximum efficiency} = \frac{\cos^2\left(\dfrac{\theta + \phi}{2}\right)}{\cos^2\left(\dfrac{\theta - \phi}{2}\right)}$$

$$= \frac{1 + \cos(\theta + \phi)}{1 + \cos(\theta - \phi)} \qquad . \qquad . \qquad (11.42)$$

[1] If wheel 2 is the driver, the reaction F is inclined at angle ϕ on the opposite side of the common normal to that shown in Fig. 11.12 and equation (11.40) becomes

$$\eta = \frac{1 - \mu \tan \sigma_1}{1 + \mu \tan \sigma_2}$$

[2] This is an approximation since ϕ is the angle between the *total* reaction (in the plane of the path of contact) and the normal to the tooth flank. The component of ϕ in the plane tangential to the pitch cylinders, ϕ', is given by

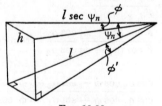

Fig. 11.13

$$h = l \tan \phi' = l \sec \psi_n \tan \phi, \text{ Fig. 11.13,}$$

or $\qquad\qquad\qquad \mu' = \mu \sec \psi_n \qquad . \qquad . \qquad . \qquad . \qquad . \qquad . \qquad (11.43)$

μ' is the virtual coefficient of friction.

11.17 Worm and wheel. Fig. 11.14 shows two views of a worm and wheel. The worm is basically a single- or multiple-start screw which meshes with a gear wheel whose teeth are inclined to the axis at the lead angle of the screw. In a cross-section of the wheel, the teeth are concave so as to give line contact with the worm instead of point contact, as would be obtained with straight teeth. The pitch of the worm (p) corresponds to the circular pitch of the wheel. The lead of the worm (L) is the axial advance of the thread per revolution and for a multiple-start worm having n threads. $L = np$.

The lead angle of the worm (σ) corresponds to the spiral angle of the wheel and is given by

$$\tan \sigma = \frac{L}{\pi d} \quad \text{where } d \text{ is the mean worm diameter . (11.44)}$$

If ω and Ω are the angular velocities of the worm and wheel respectively,

then
$$\frac{\omega}{\Omega} = \frac{\text{number of teeth on wheel}}{\text{number of threads on worm}} \quad . \quad . \quad . \quad (11.45)$$

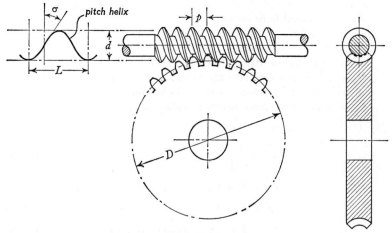

Fig. 11.14

$$\text{Efficiency of drive} = \frac{\tan \sigma}{\tan (\sigma + \phi)}^{*} \quad \text{where } \phi \text{ is the friction angle}$$

(11.46)

This is a maximum when

$$\sigma = \frac{\pi}{4} - \frac{\phi}{2} . \quad . \quad . \quad . \quad . \quad . \quad (11.47)$$

* See the authors' *Mechanics of Machines—Elementary Theory and Examples,* Art. 8.4.

11.18 Forces on worm and wheel. Fig. 11.15 shows the forces acting on the worm thread.

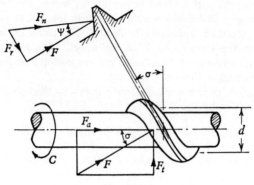

Fig. 11.15

F_t = tangential force on worm
  = axial force on wheel
F_a = axial force on worm
  = tangential or driving force on wheel
F_r = radial force on worm and wheel
F = resultant force tangential to pitch cylinder
F_n = total normal force between teeth

If C = torque on worm, then

$$F_t = \frac{C}{d/2} \qquad \ldots \qquad (11.48)$$

$$F_a = F_t \cot \sigma \qquad \ldots \qquad (11.49)$$

$$F = F_t \operatorname{cosec} \sigma \qquad \ldots \qquad (11.50)$$

$$F_r = F \tan \psi = F_t \operatorname{cosec} \sigma \tan \psi \qquad (11.51)$$

$$F_n = F \sec \psi = F_t \operatorname{cosec} \sigma \sec \psi \qquad (11.52)$$

Allowing for the effect of friction, the force between the teeth, F, is inclined at the friction angle ϕ to the normal to the tooth axis,* so that

$$F_a = F_t \cot (\sigma + \phi) \qquad \ldots \qquad (11.53)$$

and $$F = F_t \operatorname{cosec} (\sigma + \phi) \qquad \ldots \qquad (11.54)$$

---

* This is an approximation, as for spiral gears ; the virtual angle of friction is as shown in equation (11.43).

1. *Two mating gear wheels of module 6 mm have 19 and 47 teeth of 20°*
obliquity and with addenda of 6 mm. Find the number of pairs of teeth
in contact and the angle turned through by the larger wheel while one pair
of teeth is in contact.

What is the ratio of sliding to rolling motion at the point where the tip
of a tooth on the larger wheel is just leaving contact with its mating tooth?

Explain carefully the advantages obtained by increasing the addendum
height for the smaller wheel while maintaining the same working depth.

(U. Lond.)

P.C.D. of pinion $= 19 \times 6 = 114$ mm

$$\therefore R_1 = 57 \text{ mm} \quad \text{and} \quad \rho_1 = 63 \text{ mm}$$

P.C.D. of wheel $= 47 \times 6 = 282$ mm

$$\therefore R_2 = 141 \text{ mm} \quad \text{and} \quad \rho_2 = 147 \text{ mm}$$

Path of contact $= \sqrt{(\rho_1{}^2 - R_1{}^2 \cos^2 \psi)} + \sqrt{(\rho_2{}^2 - R_2{}^2 \cos^2 \psi)}$
$$- (R_1 + R_2) \sin \psi \quad \text{from equation (11.4)}$$

$$= \sqrt{(63^2 - 57^2 \cos^2 20°)} + \sqrt{(147^2 - 141^2 \cos^2 20°)}$$
$$- (57 + 141) \sin 20°$$

$$= 29{\cdot}2 \text{ mm}$$

$$\therefore \text{ arc of contact} = \frac{29{\cdot}2}{\cos 20°} = 31{\cdot}1 \text{ mm}$$

$$\therefore \text{ contact ratio} = \frac{\text{arc of contact}}{\text{circular pitch}} = \frac{31{\cdot}1}{\pi \times 6} = 1{\cdot}647$$

Therefore the number of pairs of teeth in contact varies between 1 and 2.

Angle turned through by wheel $= \dfrac{\text{arc of contact}}{\text{pitch circle radius}}$

$$= \frac{31{\cdot}1}{141} = 0{\cdot}22 \text{ rad} \quad \text{or} \quad 12{\cdot}62°$$

In Fig. 11.16, let C_1 and C_2 be the points on the pinion and wheel
teeth respectively at the last point of contact, C.

Then velocity of $C_1 = v_{c_1} = \omega_1 O_1 C$, at right angles to $O_1 C \ (= Ca)$

and velocity of $C_2 = v_{c_2} = \omega_2 O_2 C$, at right angles to $O_2 C \ (= Cb)$

The velocity of C in space is along the line of contact $(= Cc)$

The components of v_{c_1} and v_{c_2} along the common normal are equal and
hence bac is perpendicular to Cc.

The velocity of sliding is the velocity of one tooth relative to its
mating tooth along the common tangent at C; this is represented by ab.

The velocity of rolling with respect to a tooth is the velocity of the point of contact in space relative to that of the point of contact on the tooth. Thus, for the pinion, this velocity is represented by ca, and for the wheel, it is represented by cb.

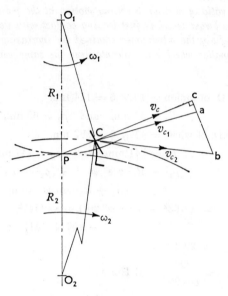

FIG. 11.16

Therefore, for the pinion,

$$\frac{\text{velocity of sliding}}{\text{velocity of rolling}} = \frac{ab}{ca}$$

$$= 4 \cdot 31$$

and for the wheel,

$$\frac{\text{velocity of sliding}}{\text{velocity of rolling}} = \frac{ab}{cb}$$

$$= 0 \cdot 81$$

This ratio is termed *specific sliding*.

Increasing the pinion addendum whilst decreasing the wheel addendum to maintain the same working depth is known as tooth correction and it is usually applied to prevent undercutting of the pinion teeth. Tooth correction also has the following advantages:

(a) Increase in pinion tooth strength. With equal tooth thickness at the pitch circle, the thickness of a pinion tooth at the base circle is less than that of a wheel tooth and hence the pinion tooth is weaker. Tooth correction increases the thickness of the pinion tooth and reduces that of the wheel tooth and hence makes their strengths more equal.

(b) Reduction of maximum sliding velocity. The velocity of sliding $= (\omega_1 + \omega_2) \times PC$ and hence, from Fig. 11.3, the maximum sliding velocity occurs either at the first or last points of contact A or B. For equal addenda, AP is larger than PB, but with tooth correction, AP and PB become more nearly equal, so that the maximum sliding velocity is reduced.

(c) Reduction of surface stress. For a constant driving torque, the force between the teeth remains constant and the compressive stress at the point of contact is then inversely proportional to the relative radius of curvature of the two surfaces at this point. This relative radius of curvature has its minimum value when contact occurs at the base circle of the pinion. Hence the maximum surface stress is reduced by reducing the wheel addendum so that the first point of contact is away from this base circle.

2. *A pair of spur gears have 14 and 42 teeth respectively. They are generated by a rack-cutter with involute teeth of module 6 mm and the pressure angle 20°. The working depth has the standard value of 2m. Using the minimum amount of correction necessary when generating, find the outside diameter of each wheel and each tooth thickness at the pitch circle.*

(I. Mech. E.)

$$\text{P.C.D. of wheel} = 42 \times 6 = 252 \text{ mm}$$
$$\text{P.C.D. of pinion} = 14 \times 6 = 84 \text{ mm}$$
$$\text{Standard rack addendum} = m = 6 \text{ mm}$$

If the standard addendum of the rack exceeds the maximum permissible pinion addendum, $R_1 \sin^2 \psi$, undercutting of the pinion teeth will occur. To correct this, the rack is moved away from the pinion centre a distance $c = m - R_1 \sin^2 \psi$, and to maintain the same depth of tooth, the outside diameter of the pinion blank is increased by $2c$. Since the pitch circle diameter is unaltered, the pinion addendum is thus increased by c.

The wheel teeth are also corrected by moving the rack *towards* the wheel centre a distance c and *decreasing* the outside diameter of the wheel blank by $2c$. This results in a *decrease* in the wheel addendum of c, so that the same working depth, $2m$, is maintained. The pressure angle also remains constant.

No interference between the rack and the wheel will occur since $R_2 \sin^2 \psi > m$.

For the pinion, minimum correction, $c = m - R_1 \sin^2 \psi$
$$= 6 - 42 \sin^2 20°$$
$$= 1 \cdot 085 \text{ mm}$$

$$\therefore \text{ corrected pinion addendum} = 6 + 1.085$$
$$= 7.085 \text{ mm}$$
$$\therefore \text{ outside diameter of pinion blank} = 84 + 2 \times 7.085$$
$$= \underline{98.17 \text{ mm}}$$
$$\text{Corrected wheel addendum} = 6 - 1.085$$
$$= 4.915 \text{ mm}$$
$$\therefore \text{ outside diameter of wheel blank} = 252 + 2 \times 4.915$$
$$= \underline{261.83 \text{ mm}}$$

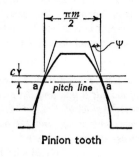

Pinion tooth

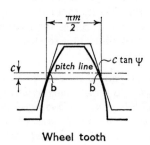

Wheel tooth

FIG. 11.17

When the rack is working at the standard centre distance, the thickness of the tooth is one-half of the circular pitch,

i.e. $$\text{standard tooth thickness} = \frac{\pi m}{2}$$

Therefore, when the cutter is moved outwards a distance c, Fig. 11.17, thickness of pinion tooth on pitch circle,

$$\text{aa} = \frac{\pi m}{2} + 2c \tan \psi$$
$$= \frac{\pi \times 6}{2} + 2 \times 1.085 \tan 20° = \underline{10.214 \text{ mm}}$$

When the cutter is moved inwards a distance c, thickness of wheel tooth on pitch circle,

$$\text{bb} = \frac{\pi m}{2} - 2c \tan \psi$$
$$= \frac{\pi \times 6}{2} - 2 \times 1.085 \tan 20° = \underline{8.636 \text{ mm}}$$

3. *A pinion has 21 involute teeth of module 5 mm, a pressure angle of 20° and an addendum of 6 mm. It drives an internally-toothed wheel having 80 teeth and an addendum of 5 mm. Calculate the arc of approach and arc of recess. What would be the maximum possible addendum for the internally-toothed wheel if interference is to be avoided ?* (U. Lond.)

$$R_1 = \frac{21 \times 5}{2} = 52 \cdot 5 \text{ mm} \qquad \therefore \rho_1 = 52 \cdot 5 + 6 = 58 \cdot 5 \text{ mm}$$

$$R_2 = \frac{80 \times 5}{2} = 200 \text{ mm} \qquad \therefore \rho_2 = 200 - 5 = 195 \text{ mm}$$

Path of approach

$$= R_2 \sin \psi - \sqrt{(\rho_2{}^2 - R_2{}^2 \cos^2 \psi)} \qquad . \qquad . \qquad \text{from equation (11.16)}$$
$$= 200 \sin 20° - \sqrt{(195^2 - 200^2 \cos^2 20°)} = 16 \cdot 4 \text{ mm}$$

$$\therefore \text{ arc of approach} = \frac{16 \cdot 4}{\cos 20°} = \underline{17 \cdot 45 \text{ mm}}$$

Path of recess

$$= \sqrt{(\rho_1{}^2 - R_1{}^2 \cos^2 \psi)} - R_1 \sin \psi \qquad . \qquad . \qquad \text{from equation (11.17)}$$
$$= \sqrt{(58 \cdot 5^2 - 52 \cdot 5^2 \cos^2 20°)} - 52 \cdot 5 \sin 20° = 13 \cdot 54 \text{ mm}$$

$$\therefore \text{ arc of recess} = \frac{13 \cdot 54}{\cos 20°} = \underline{14 \cdot 4 \text{ mm}}$$

Maximum addendum of annulus

$$= R_2 - \sqrt{(R_2{}^2 - 2R_1 R_2 \sin^2 \psi + R_1{}^2 \sin^2 \psi)} \quad . \quad \text{from equation (11.19)}$$
$$= 200 - \sqrt{(200^2 - 2 \times 52 \cdot 5 \times 200 \sin^2 20° + 52 \cdot 5^2 \sin^2 20°)}$$
$$= 200 - 194 \cdot 6 = \underline{5 \cdot 4 \text{ mm}}$$

4. *A pair of single helical gears is required to give a speed reduction of 4·2 : 1. The gears are to have a normal module of 3 mm, a pressure angle of 20° and a helix angle of 30°. If the shaft centre-lines are to be approximately 400 mm apart, determine the number of teeth on each wheel and the exact centre distance. (This should be given to the nearest 0·01 mm.)*

The pinion is supported in bearings equally spaced on either side of the centre line of the gear. If the speed of the pinion is 1000 rev/min and 75 kW is being transmitted, find the end-thrust on the pinion shaft and the load on each bearing. Assume that the end-thrust is carried by a separate thrust bearing. (U. Lond.)

Let suffixes 1 and 2 denote the pinion and wheel respectively.

Then $\qquad \dfrac{D_2}{D_1} = \dfrac{T_2}{T_1} = 4 \cdot 2 \quad$ and $\quad \dfrac{D_1 + D_2}{2} \backsimeq 400 \text{ mm}$

Hence $\qquad D_1 = 154$ mm $\quad$ and $\quad D_2 = 646$ mm

$$m = m_n \sec 30°$$
$$= 3 \times 1\cdot155 = 3\cdot465 \text{ mm}$$
$$\therefore T_1 = \frac{D_1}{m} = \frac{154}{3\cdot465} = \underline{44} \text{ (to nearest whole number)}$$

and $\qquad T_2 = 4\cdot2 \times 44 = \underline{185}$

$$\text{Exact centre distance} = \frac{44 + 185}{2} \times 3\cdot465 = \underline{396\cdot5 \text{ mm}}$$

$$\text{Torque on pinion} = \frac{75 \times 10^3 \times 60}{2\pi \times 1000} = 716 \text{ N m}$$

$$\therefore \text{ driving force on pinion, } F_d = \frac{716}{\left(\dfrac{44}{2} \times \dfrac{3\cdot465}{10^3}\right)} = 9400 \text{ N}$$

$$\therefore \text{ axial thrust, } F_a = F_d \tan \sigma$$
$$= 9400 \times \tan 30° = \underline{5425 \text{ N}}$$

Total force on tooth, normal to shaft axis $= F_d \sec \psi$

$$= 9400 \sec 20° = 10\,000 \text{ N}$$

$$\therefore \text{ load on each bearing} = \frac{10\,000}{2} = \underline{5000 \text{ N}}$$

If the *normal* pressure angle is 20°, then

$$\psi = \tan^{-1}(\sec 30° \tan 20°) = 22° \, 48'$$

and $\qquad\qquad$ load on each bearing $= \underline{5100 \text{ N}}$

No account can be taken of the effect of the moment due to the axial force F_a, since the distance between the bearings is unknown.

5. *Two shafts inclined at 70° are to be connected by spiral gears with a normal pitch of 12 mm and to have a 2 to 1 velocity ratio. Determine the pitch diameters of the wheels and the spiral angles if the distance apart of the shafts is fixed at 125 mm. Sliding of the teeth is to be a minimum as far as is practicable.*

If the pinion rotates at 240 rev/min, what is the speed of sliding between the teeth ? $\qquad$ (U. Lond.)

From equation (11.38),

$$v_s = \omega_1 R_1 \sin \sigma_1 + \omega_2 R_2 \sin \sigma_2$$

But $\qquad\qquad 2 = \dfrac{\omega_1}{\omega_2} = \dfrac{D_2 \cos \sigma_2}{D_1 \cos \sigma_1}$ $\qquad$. $\qquad$ from equation (11.35)

$$\therefore \omega_2 R_2 = \omega_1 R_1 \frac{\cos \sigma_1}{\cos \sigma_2}$$

$$\therefore v_s = \omega_1 R_1\left\{\sin \sigma_1 + \frac{\cos \sigma_1}{\cos \sigma_2}\sin \sigma_2\right\}$$

$$= \omega_1 R_1 \frac{\sin(\sigma_1 + \sigma_2)}{\cos \sigma_2} = \omega_1 R_1 \frac{\sin \theta}{\cos \sigma_2}$$

$$C = \tfrac{1}{2}(D_1 + D_2) = R_1 + \frac{\omega_1 R_1}{\omega_2}\frac{\cos \sigma_1}{\cos \sigma_2}$$

i.e. $$125 = R_1\left\{1 + \frac{2\cos \sigma_1}{\cos \sigma_2}\right\}$$

$$\therefore v_s = \omega_1 \times \frac{125}{\left\{1 + \dfrac{2\cos \sigma_1}{\cos \sigma_2}\right\}} \times \frac{\sin \theta}{\cos \sigma_2}$$

$$= \frac{125\omega_1 \sin \theta}{\cos \sigma_2 + 2\cos \sigma_1} \qquad . \qquad . \qquad . \qquad . \qquad (1)$$

Since ω_1 and θ are constants, the minimum value of v_s occurs when $\cos \sigma_2 + 2\cos \sigma_1$ is a maximum,

i.e. when $$\frac{d}{d\sigma_1}\{\cos(\theta - \sigma_1) + 2\cos \sigma_1\} = 0$$

or $$2\sin \sigma_1 = \sin(\theta - \sigma_1)$$

from which $$\cot \sigma_1 = \frac{2 + \cos \theta}{\sin \theta}$$

$$= \frac{2 + \cos 70°}{\sin 70°} = 2{\cdot}492$$

$$\therefore \sigma_1 = \underline{21° 52'} \quad \therefore \sigma_2 = 70° - 21° 52' = \underline{48° 8'}$$

$$\frac{D_2 \cos 48° 8'}{D_1 \cos 21° 52'} = 2$$

and $$D_1 + D_2 = 2 \times 125 = 250 \text{ mm}$$

from which $$D_1 = \underline{66{\cdot}1 \text{ mm}} \quad \text{and} \quad D_2 = \underline{183{\cdot}9 \text{ mm}}$$

$$v_s = \frac{125 \times \left(240 \times \dfrac{2\pi}{60}\right) \times \sin 70°}{\cos 48° 8' + 2 \times \cos 21° 52'} \quad \text{from equation (1)}$$

$$= \underline{1{\cdot}17 \text{ m/s}}$$

6. *A pair of screw wheels are used to drive a camshaft at half the speed of the crankshaft, the two shafts being at right angles. The teeth have a module of 6 mm in normal section. What should be the helix angles of both wheels and the pitch diameters if the camshaft wheel has 16 teeth and the pitch diameters are equal? If the driven wheel is right-handed, find the end-thrust on each shaft per N m torque of driver, neglecting the effect of friction.*

If such a pair of wheels were required to transmit power with a high efficiency, state what modifications would be necessary to the above design.

(U. Lond.)

From equation (11.35), $\dfrac{\cos \sigma_2}{\cos \sigma_1} = 2,$ since $D_1 = D_2$, Fig. 11.18.

Also $\sigma_1 + \sigma_2 = 90°$

$\therefore \tan \sigma_1 = 2$

$\therefore \sigma_1 = 63° \, 26'$ and $\sigma_2 = \underline{26\,°34'}$

$D_1 = D_2 = T_2 m_2 = T_2 m_n \sec \sigma_2$

$= 16 \times 6 \sec 26° \, 34' = \underline{\quad \cdot 4 \text{ mm}}$

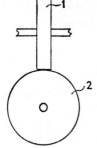

Driving force on wheel 1 for 1 N m torque,

$$F_d = \frac{1}{0 \cdot 1074/2}$$

$$= 18 \cdot 63 \text{ N}$$

$\therefore$ axial force on wheel 1,

$$F_{a_1} = 18 \cdot 63 \tan 63° \, 26'$$

$$= \underline{37 \cdot 2 \text{ N}}$$

Normal force between teeth,

$$F_t = 18 \cdot 63 \sec 63° \, 26'$$

$\therefore$ axial force on wheel 2,

$$F_{a_2} = 18 \cdot 63 \sec 63° \, 26' \sin 26° \, 34'$$

$$= \underline{18 \cdot 63 \text{ N}}$$

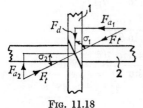

Fig. 11.18

Assuming the gear ratio and centre distance to be fixed, the maximum efficiency would be obtained if the diameter ratio were altered to give $\sigma_1 - \sigma_2 = \phi$ (see Art. 11.16).

7. *Two spiral gear wheels A and B have 45 and 15 teeth at spiral angles 20° and 50° respectively. Both wheels are of the same hand. A is 150 mm diameter. Find the distance between the shafts and the angle between the shafts.*

If the teeth are of 20° involute form and the coefficient of friction is 0·08, find the efficiency of the gears (a) if A is the driver, (b) if B is the driver.

(U. Lond.)

Since the teeth have the same hand, $\sigma_a + \sigma_b = \theta$,

i.e. shaft angle $= 20° + 50° = \underline{70°}$

$$\frac{T_a}{T_b} = \frac{D_a \cos \sigma_a}{D_b \cos \sigma_b}. \qquad \text{from equation (11.35)}$$

i.e.
$$\frac{45}{15} = \frac{150 \cos 20°}{D_b \cos 50°}$$

$$\therefore D_b = 73 \cdot 25 \text{ mm}$$

$$\therefore \text{ centre distance} = \tfrac{1}{2}(150 + 73 \cdot 25) = \underline{111 \cdot 625 \text{ mm}}$$

Virtual coefficient of friction in plane tangential to pitch cylinder

$$= \mu \sec \psi_n . \qquad . \qquad . \qquad . \qquad \text{from equation (11.43)}$$

$$= 0 \cdot 08 \times \sec 20° = 0 \cdot 08512$$

If A is the driver,

$$\eta = \frac{1 - \mu' \tan \sigma_b}{1 + \mu' \tan \sigma_a} . \qquad . \qquad . \qquad \text{from equation (11.40)}$$

$$= \frac{1 - 0 \cdot 0851 \times \tan 50°}{1 + 0 \cdot 0851 \times \tan 20°} = 0 \cdot 8715$$

If B is the driver,

$$\eta = \frac{1 - \mu' \tan \sigma_a}{1 + \mu' \tan \sigma_b}$$

$$= \frac{1 - 0 \cdot 0851 \times \tan 20°}{1 + 0 \cdot 0851 \times \tan 50°} = \underline{0 \cdot 8798}$$

8. *A pair of spiral gears, of equal pitch diameter with their axes at right angles, are to have a speed ratio of 4 to 1. The centre distance is to be between 100 mm and 105 mm.*

If these gears are to be cut with a 2·5 module milling cutter, determine: (a) the exact helical angles; (b) the exact centre distance; (c) the blank diameters; (d) the virtual number of teeth in each gear. (U. Lond.)

From equation (11.35), $\dfrac{\cos \sigma_2}{\cos \sigma_1} = 4$ since $D_1 = D_2$

Also
$$\sigma_1 + \sigma_2 = 90°$$

$$\therefore \tan \sigma_1 = 4$$

$$\therefore \sigma_1 = \underline{75° 58'} \quad \text{and} \quad \sigma_2 = \underline{14° 2'}$$

$$D_1 = T_1 m_1 \sec \sigma_1 = T_1 \times 2 \cdot 5 \sec 75° 58' = 10 \cdot 3 T_1$$

But D_1 is equal to the centre distance, which lies between 100 and 105 mm

$$\therefore T_1 = 10 \quad \text{and} \quad T_2 = 40$$

$$\therefore \text{ exact centre distance} = 10 \times 10 \cdot 3 = \underline{103 \text{ mm}}$$

$$\text{Blank diameter} = \text{P.C.D.} + 2 \times \text{addendum}$$
$$= D_1 + 2\,m_n$$
$$= 103 + 2 \times 2{\cdot}5 = \underline{108 \text{ mm}}$$

For driver, virtual number of teeth,

$$T_e = T_1 \sec^3 \sigma_1 \quad . \quad \text{from equation (11.34)}$$
$$= 10 \sec^3 75° 58' = \underline{71}$$

For follower, $\qquad T_e = T_2 \sec^3 \sigma_2$
$$= 40 \sec^3 14° 2' = \underline{44}$$

9. *A worm gear of speed ratio 5 connects two shafts at right angles. The worm has 4 teeth of normal module 20 mm. The pitch diameter of the worm is 40 mm. Calculate the tooth angles of the worm and wheel and the distance between the centres of the shafts.*

If the efficiency of the gear is 90%, the worm being the driver, find approximately the coefficient of friction between the tooth surfaces.

(U. Lond.)

$$\frac{\text{Speed of worm}}{\text{Speed of wheel}} = \frac{\text{teeth on wheel}}{\text{threads on worm}}$$

i.e. $\qquad 5 = \dfrac{T}{4}$

$$\therefore T = 20$$

$$\tan \sigma = \frac{L}{\pi d} \quad . \quad . \quad \text{from equation (11.44)}$$

$$= \frac{4p}{\pi d}$$

$$= \frac{4 \times (20 \sec \sigma)}{\pi \times 40} \quad \text{since } p = p_n \sec \sigma$$

$$\therefore \sin \sigma = \frac{2}{\pi}$$

$$\therefore \sigma = \underline{39° 31'}$$

This is the lead angle of the worm and the spiral angle of the wheel.

$$\text{P.C.D. of wheel} = T \times m = T \times \frac{p}{\pi}$$

$$= \frac{20 \times 20 \sec 39° 31'}{\pi}$$

$$= 165 \text{ mm}$$

$$\therefore \text{ centre distance} = \tfrac{1}{2}(165 + 40) = \underline{\underline{102{\cdot}5 \text{ mm}}}$$

From equation (11.46), $\eta = \dfrac{\tan \sigma}{\tan (\sigma + \phi)}$

$$\therefore \ 0 \cdot 9 = \frac{\tan 39° \ 31'}{\tan (39° \ 31' + \phi)}$$

$$\therefore \ \tan (39° \ 31' + \phi) = \frac{0 \cdot 8252}{0 \cdot 9} = 0 \cdot 9167$$

$$\therefore \ \phi = 3°$$

$$\therefore \ \mu = \tan 3° = \underline{0 \cdot 0524}$$

10. *A worm whose pitch cylinder diameter is 90 mm gears with a worm-wheel which has a pitch cylinder diameter of 510 mm, the centre distance between worm and worm-wheel being therefore 300 mm. The worm, which is below the worm-wheel, has a right-hand thread; the angle of lead being 25°, and the teeth normal to the thread have the usual angle of incidence of 20°. The bearings on the worm-wheel shaft are 225 mm apart, and the worm-wheel is placed centrally between them.*

Neglecting friction, find the loads of the worm-wheel shaft on its bearings in magnitude, sense and direction when the worm rotates clockwise at 1000 rev/min and 50 kW is being transmitted at the teeth. If friction were being taken into account, at what stage would it appear first in your calculations?

(U. Lond.)

Torque on worm, $C = \dfrac{50 \times 10^3 \times 60}{2\pi \times 1000} = 477 \ \text{N m}$

$$\therefore \ F_t = \frac{C}{d/2} = \frac{477}{0 \cdot 045}$$

$$= 10\ 600 \ \text{N} = \begin{array}{l} \text{tangential force on worm and} \\ \text{axial force on wheel} \end{array}$$

Referring to Fig. 11.15,

$$F = F_t \cosec 25°$$

and

$$F_r = F \tan 20°$$

$$\therefore \ F_r = F_t \cosec 25° \tan 20°$$

$$= 10\ 600 \times 2 \cdot 3662 \times 0 \cdot 3640$$

$$= 9130 \ \text{N} = \text{radial force on worm and wheel}$$

$$F_a = F_t \cot 25°$$

$$= 10\ 600 \times 2 \cdot 1445$$

$$= 22\ 700 \ \text{N} = \begin{array}{l} \text{axial force on worm and} \\ \text{tangential force on wheel} \end{array}$$

Fig. 11.19 shows the forces exerted *by* the worm *on* the wheel.

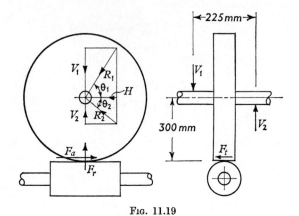

FIG. 11.19

$$V_1 = \frac{F_t \times 0.3}{0.225} + \frac{F_r}{2}$$

$$= \frac{10\,600 \times 0.3}{0.225} + \frac{9130}{2}$$

$$= 14\,140 + 4565 = 18\,705 \text{ N}$$

$$V_2 = \frac{F_t \times 0.3}{0.225} - \frac{F_r}{2}$$

$$= 14\,140 - 4565 = 9575 \text{ N}$$

$$H = \frac{F_a}{2} = \frac{22\,700}{2} = 11\,350 \text{ N}$$

$$\therefore\ R_1 = \sqrt{(11\,350^2 + 18\,705^2)} = \underline{21\,880 \text{ N}}$$

$$\therefore\ R_2 = \sqrt{(11\,350^2 + 9575^2)} = \underline{14\,850 \text{ N}}$$

$$\theta_1 = \tan^{-1}\frac{18\,705}{11\,350} = \underline{58°\ 45'}$$

$$\theta_2 = \tan^{-1}\frac{9575}{11\,350} = \underline{40°\ 10'}$$

Axial force on bearings = $\underline{10.6 \text{ kN}}$

If friction is taken into account, the force between the teeth, F, is inclined at the friction angle ϕ to the normal to the tooth axis so that the angle between F and F_a becomes $25° + \phi$.

11. Two mating gear wheels have 18 and 36 teeth of 10 mm module and 20° pressure angle. Determine the addendum height for each wheel if it is to be half the maximum possible for true involute action.

If the larger wheel rotates at 450 rev/min, find the velocity at the point of contact of the surface of each tooth at the moment when the tip of the tooth on the smaller wheel makes contact, and hence find the velocity of sliding at that instant. (*U. Lond.*)

(*Ans.:* 6·5 mm; 17·5 mm; 10·15 m/s; 8·325 m/s; 5·05 m/s)

12. Two gear wheels of diameters 75 mm and 250 mm have involute teeth of 5 mm module, 20° angle of obliquity. The addendum is the same for each wheel and is as large as possible while avoiding interference. Find (*a*) the length of the addendum, (*b*) the contact ratio, (*c*) the sliding velocity at the first point of contact when the smaller wheel is driving at 2000 rev/min. (*U. Lond.*)

(*Ans.:* 22·75 mm; 1·57; 3·5 m/s)

13. Two mating gear wheels of 10 mm module have 21 and 49 teeth of 20° obliquity, and addenda of 10 mm. The larger wheel rotates at 500 rev/min.

(*a*) Find the velocity of the point of contact on each of two mating teeth, perpendicular to the line of obliquity, at the instant when the tip of the tooth on the larger wheel makes contact.

(*b*) Show that correct working contact can still be made if the addendum height on the pinion is increased to 17·5 mm while the same working depth is maintained, and explain the advantage of this modification. (*U. Lond.*)

(*Ans.:* 5·75 m/s; 1·25 m/s)

14. A gear blank with overall diameter corresponding to a given number of standard involute teeth is cut with one less than this number. Show that the equivalent addendum correction coefficient is 0·5.

A standard cutter, module 6 mm and pressure angle 20°, is used to produce a pair of gears of 20 and 40 teeth respectively, each with this amount of correction. Calculate the blank diameters.

If the increase in centre distance is made equal to the combined correction, calculate its value and the corresponding pressure angle of engagement. Comment on the use of this value for centre distance. (*I. Mech. E.*)

(*Ans.:* 138 mm; 258 mm; centre distance = 186 mm; 24° 42′)

15. Explain briefly why corrections may be used for involute spur gears even though unnecessary to avoid undercutting. The B.S. pinion correction coefficient for a tooth sum exceeding 60 is $0·4\left(1 - \dfrac{t}{T}\right)$ or $0·02(30 - t)$, whichever is greater, the combined correction for pinion and wheel being zero. Using this amount of correction for a simple speed reduction gear of 25 and 50 teeth of 6 mm module, find the blank diameters. If the pressure angle is 20°, find the length of the path of face contact for the pinion and the velocity of sliding between the teeth at the end of contact for a pinion speed of 500 rev/min. (*I. Mech. E.*)

(*Ans.:* c = 1·2 mm; 164·4 mm; 309·6 mm; 16·46 mm; 1·29 m/s)

16. A pinion with 30 involute teeth, of 4 mm module, gears with a rack. If the pressure angle is 20°, and the addenda for pinion and rack are the same, determine: (*a*) the maximum addendum, if interference is to be avoided, (*b*) the length of the resulting path of contact. (*U. Lond.*)

(*Ans.:* 7·03 mm; 36·24 mm)

17. Explain carefully what you understand by involute interference; illustrate your answer with sketches.

A straight tooth pinion with teeth of 6 mm module and 20° pressure angle meshes with a rack. If the addendum of the pinion and the rack is 6 mm, find the minimum number of teeth on the pinion for interference to be avoided. If the pinion had three teeth less than this number, how much addendum correction would be required ? (*U. Lond.*)　　　　(*Ans.*: 18; 0·744 mm)

18. A pinion of 20 teeth is in mesh with an internally-toothed wheel with 80 teeth. The teeth have involute profiles with a pressure angle of 20° and the module is 5 mm. The addendum on the wheel is as large as possible while avoiding interference and the contact ratio is to be 2·0. Find the addendum length on pinion and wheel. (*U. Lond.*)　　　　(*Ans.*: 5·5 mm; 2·5 mm)

19. Show that gear teeth with involute profiles will give a constant velocity ratio between two gear wheels.

A pinion with 20 teeth is in mesh with an internally-toothed wheel with 90 teeth. The teeth have involute profiles with a pressure angle of 20° and a module of 5 mm. The addenda on pinion and wheel are 6·25 mm and 3·75 mm respectively. Find the length of the path of contact. (*U. Lond.*)

(*Ans.*: 24·95 mm)

20. A pair of helical gear wheels, on parallel axes, are to have involute teeth, of 20° normal pressure angle and 3 mm normal module ; the gear ratio is to be 2 : 1 and the helix or spiral angle as near to 12° as possible. The centre distance between these gears is to be exactly 165 mm. The pinion is to be supported in two bearings, 80 mm apart, these bearings being placed symmetrically with respect to the pinion.

Determine : (*a*) the number of teeth in each gear and the helix angle (to an accuracy of 1 minute) ; (*b*) the radial force on the more heavily loaded of the two pinion bearings when the pinion is transmitting 1 kW at 720 rev/min.

Neglect the effects of friction and the weight of the pinion and shaft. (*U. Lond.*)　　　　(*Ans.* : 72 ; 36 ; 10° 57′ ; 76·8 N)

21. A pair of screw gears of speed ratio 1·25 connects two shafts at 60° angle. The gears are to be cut with a standard milling cutter of 3 mm module. Making the pitch circle diameters of the wheels equal and each approximately 100 mm, find the requisite numbers of teeth on the wheels, the angles at which the teeth should be cut, and the exact pitch circle diameter. For what number of teeth should the milling cutter be chosen in cutting each wheel ? (*U. Lond.*)

(*Ans.*: 24 ; 30 ; 40° 54′ ; 19° 6′ ; 95·26 mm ; 55 ; 36)

22. A pair of spiral gears is to provide a speed ratio of four to one between two shafts. The axes of the shafts are at right angles, and the centre distance between them is to be between 100 mm and 106 mm. The slower wheel is to have a pitch circle diameter twice that of the faster wheel. The teeth are to be of 3 mm module.

Find (*a*) the spiral angles, (*b*) the numbers of teeth on each wheel, and (*c*) the exact centre distance. (*U. Lond.*)

(*Ans.*: 63° 26′ ; 26° 34′ ; 10 ; 40 ; 100·7 mm)

23. Two horizontal shafts are connected by a pair of spiral gear wheels A and B. The angle between the shafts is 60°. A is the driver and rotates 1½ times as fast as B. A has 40 teeth with a helix angle of 25°. The normal circular pitch is 12 mm. The driving torque applied to A is 30 N m.

Find (*a*) the pitch circle diameters ; (*b*) the end-thrust on each shaft, neglecting friction and the effect of the obliquity of involute tooth pressure. (*U. Lond.*)

(*Ans.*: 168·6 mm ; 279·6 mm ; 166 N ; 225 N)

24. Two shafts are to be connected by spiral gears with a velocity ratio of 3 to 1. The angle between the shafts is 45° and the least distance between the shaft axes is to be 225 mm. The normal module is to be 5 mm and the pinion is to have 20 teeth. Determine the pitch circle diameters and the spiral angles, which are to have the same hand.

If the pinion rotates at 240 rev/min, what will be the speed of rubbing between the teeth ? (*U. Lond.*)

(*Ans.*: 103·53 mm ; 346·47 mm ; 15° ; 30° ; 1·06 m/s)

25. A spiral wheel reduction gear, of ratio 3 to 2, is to be used on a machine, with the angle between the shafts 80°. The approximate centre distance between the shafts is 125 mm, the normal circular pitch of the teeth is 10 mm and the wheel diameters are equal. Find the number of teeth on each wheel, the pitch circle diameter and the spiral angles. Find the efficiency of the drive if the friction angle is 5°. (*U. Lond.*)

(*Ans.*: 24 ; 36 ; 128·2 mm ; 53° 24′ ; 26° 36′ ; 85·54%)

26. A pair of spiral gears is required to connect two shafts 210 mm apart, the shaft angle being 70°. The velocity ratio is to be 1½ to 1, the faster wheel having 80 teeth and a pitch circle diameter of 120 mm. Find the spiral angles for each wheel. If the torque on the faster wheel is 70 N m, find the axial thrust on each shaft, neglecting friction. (*U. Lond.*)

(*Ans.*: 54° 38′ ; 15° 22′ ; 321 N ; 988 N)

27. A screw gear has teeth with a helix angle α and the number of teeth on the gear is n. Show that the number of teeth n_1 for which the gear cutter should be chosen is given by $n_1 = n \sec^3 \alpha$.

Two horizontal shafts are connected by a pair of screw gears having a normal tooth section of 6 mm module. The driving gear has 16 teeth and a helix angle of 45°, right-handed. The driven wheel has 12 teeth and a helix angle of 15°, and also is right-handed.

Make a sketch of the arrangement, stating the angle between the shafts and the normal distance between them. (*U. Lond.*) (*Ans.*: 60° ; 105·14 mm)

28. A pair of spiral gears is to be designed to connect two shafts at right angles. The gears are to be of equal diameter, the centre distance is to lie between 200 mm and 195 mm, and the normal module of the teeth is to be 3 mm. The speed ratio between the shafts is to be 3 to 1.

Determine the number of teeth and the helix angle for each gear, and the exact centre distance. (*U. Lond.*)

(*Ans.*: 21 ; 63 ; 71° 34′ ; 18° 26′ ; 199·27 mm)

29. Two non-intersecting shafts are inclined at an angle of 72° and are to be connected by spiral gears. The gear reduction ratio is to be 2½ to 1 and the centre distance is to be within the limits of 100 ± 0·5 mm. If the minimum number of teeth is not to be less than 20, find, for a spiral angle of 40° on the driving wheel, a suitable *standard* normal module. Give the exact centre distance and estimate the efficiency of transmission when the friction angle is 5°. (*U. Lond.*)   (*Ans.*: 30 and 75 teeth ; 1·5 mm ; 95·67 mm ; 88·1%)

30. Show that, in a pair of spiral gears connecting inclined shafts, the efficiency is a maximum when the spiral angle of the driving wheel is half the sum of the shaft and friction angles.

Two shafts, inclined at an angle of 65°, and with a least distance between them of 170 mm, are to be connected by spiral gears of normal circular pitch 15 mm to give a reduction ratio of 3 : 1. Find suitable diameters and numbers of teeth, and evaluate the efficiency, if the spiral angles are determined by the condition of maximum efficiency. The friction angle is 7°. (*U. Lond.*)

(*Ans.* : 88·5 mm ; 246 mm ; 15 ; 45 ; 85·5%)

31. A pair of spiral gears is required to connect two shafts at right angles, the speed ratio between these gears being 3 : 1. The gears are to be at 144 mm centres as nearly as possible and are to be cut with a normal module of 3 mm. The gear having the smaller number of teeth is to be the driver and is to have a helical angle of exactly 70° (i.e. a lead angle of 20°).

(*a*) Determine the number of teeth in each gear, the pitch diameters, and the exact centre distance.

(*b*) If these gears are to be cut with a milling cutter, determine the virtual number of teeth in each gear so that the correct cutter may be selected. (*U. Lond.*)

(*Ans.* : 16 ; 48 ; 140·4 mm ; 153·2 mm ; 146·8 mm ; 114 ; 58)

32. A pair of spiral gears, of equal pitch diameter, is to be used to connect two shafts having their axes at right angles. The speed ratio is to be 2 : 1 and the centre distance as near to 125 mm as possible. The gears are to be cut with a 3 mm module milling cutter.

Determine : (*a*) the exact helical angles ; (*b*) the exact centre distance ; (*c*) the blank diameters, assuming an addendum length of 3 mm ; (*d*) the lead of the gear having the smaller number of teeth. (*U. Lond.*)

(*Ans.* : 63° 26′ ; 26° 34′ ; 120·29 mm ; 126·46 mm ; 125·4 mm ; 756·5 mm)

33. Two shafts at right angles are connected by a pair of spiral gears. The normal module is 4 mm and the wheels have 36 and 24 teeth respectively. Find the helix angle of the slower of the two wheels for the centre distance to be 180 mm. (The helix angle should be found with sufficient accuracy to give the centre distance within ±0·1 mm.)

Develop an expression for the helix angle which will give a minimum value for the centre distance and find this helix angle and centre distance. (*U. Lond.*)

(*Ans.* : 53° 9′ ; 41° 8′ ; 168·65 mm)

34. A pair of spiral gears having a speed ratio of 3½ to 1 has to connect two shafts whose axes are at an angle of 80°. The gear having the smaller number of teeth is to be the driver. The gears are to be cut with 2·5 mm module (normal) milling cutters, are to be of equal pitch diameters and are to be at 156 mm centre distance as nearly as possible.

(*a*) Determine the number of teeth and the helical angle in each gear, also the exact centre distance.

(*b*) Comment, with the aid of a sketch if necessary, on the possibility of the gear having the larger number of teeth being used as the driver. (*U. Lond.*)   (*Ans.* : 63 ; 18 ; 6° 30′ ; 73° 30′ ; 158·5 mm)

35. A pair of skew gears has to connect two shafts at right angles. The gears are to be of equal diameter, the centre distance is to be as near to 150 mm as possible, and the normal module is to be 2·5 mm. The speed ratio between the shafts is to be 2½ to 1.

Determine the number of teeth and the helix angle of each gear, also the exact centre distance. (*U. Lond.*)

(*Ans.* : 55 ; 22 ; 21° 48′ ; 68° 12′ ; 148·1 mm)

36. A pair of spiral gears is required to connect two shafts at right angles, the centre distance being 240 mm. If the velocity ratio is to be 2 to 1 and the faster wheel is to have 90 teeth at a pitch circle diameter of 135 mm, determine the spiral angles for each wheel. (*U. Lond*).   (*Ans.* : 38° 3′ ; 51° 57′)

37. Show that the maximum efficiency is obtained with a pair of spiral gears when the helix angle of the teeth on the driving wheel exceeds that of the teeth on the driven wheel by the angle of friction.

A pair of spiral gears is required to connect two shafts with an included angle of 65°. The normal module is 4 mm, the gear ratio 2·3 : 1, and the smaller of the two wheels has 20 teeth and is the driving wheel. If the coefficient of friction is 0·1228, find the helix angles to give maximum efficiency and determine the shaft centre distance. What is the torque on the output shaft if the input torque is 20 N m ? (*U. Lond.*) (*Ans.:* 36° ; 29° ; 154·6 mm ; 41·6 N m)

38. Derive from first principles an expression for the efficiency of a worm gear connecting two perpendicular shafts, stating the expression in terms of the helix angle α of the driving member of the gear and the effective angle of friction ϕ relating to the contact between the sliding surfaces.

In a gear of this kind, the worm (which is the driving member) has 4 threads of pitch diameter 75 mm. The worm-wheel has 22 teeth of 412·5 mm pitch diameter. The effective coefficient of friction between the sliding surfaces is 0·050. The driving torque applied to the worm is 50 N m. Calculate the end-thrust exerted on each shaft and the efficiency of the gear. (*U. Lond.*)

(*Ans.:* Wheel, 1333 N ; worm, 1208 N ; 90·6%)

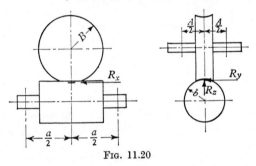

FIG. 11.20

39. In the worm gear shown in Fig. 11.20, the worm and worm-wheel shafts are supported in bearings symmetrically spaced ' a ' and ' A ' apart respectively. ' b ' and ' B ' are the mean radii from the axes of small surface areas ' s ' in contact on the central plane. The resolved components of the normal pressure between these contact areas are represented by R_x, R_y, and R_z ; R_x and R_y are parallel to the worm and worm-wheel axes respectively and R_z is perpendicular to them.

In a particular case, $R_x = 1\cdot03$ and $R_z = 0\cdot52 R_y$; $B = 132$ mm, $b = 26$ mm, $A/B = 2$ and $a/b = 4$.

If the power transmitted is 37 kW and the speeds of the worm and worm-wheel are 2000 and 500 rev/min respectively, calculate the axial thrusts and the loads on the supporting bearings. (*U. Lond.*)

(*Ans.:* Wheel — axial thrust, 5000 N ; bearing loads, 4590 N and 2835 N ; worm — axial thrust, 5150 N ; bearing loads, 3610 N and 1200 N)

CHAPTER 12

EPICYCLIC GEARS

12.1 Simple epicyclic trains. Referring to the epicyclic train shown in Fig. 12.1, S is the sun-wheel, A the annulus, having internal teeth, and P is a planet wheel which can rotate freely on a pin attached to the arm L. The arm L rotates freely about the axis of S.

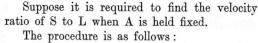
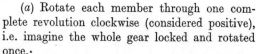
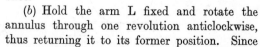

FIG. 12.1

Suppose it is required to find the velocity ratio of S to L when A is held fixed.

The procedure is as follows :

(*a*) Rotate each member through one complete revolution clockwise (considered positive), i.e. imagine the whole gear locked and rotated once.·

(*b*) Hold the arm L fixed and rotate the annulus through one revolution anticlockwise, thus returning it to its former position. Since the arm is fixed, wheels A, P and S form a simple train and the revolutions of P and S due to the rotation of A are $-T_a/T_p$ and $+T_a/T_s$ respectively.

(*c*) Add the corresponding rotations of each member in operations (*a*) and (*b*) to obtain the resultant motion.

These operations are set out in tabular form thus :

	L	A	P	S
(*a*) Turn whole gear clockwise 1 rev.	+1	+1	+1	+1
(*b*) Hold arm L and turn A anti-clockwise 1 rev . . .	0	−1	$-\dfrac{T_a}{T_p}$	$+\dfrac{T_a}{T_s}$
(*c*) Resulting motion [= (*a*) + (*b*)].	1	0	$1 - \dfrac{T_a}{T_p}$	$1 + \dfrac{T_a}{T_s}$

The last line of the table gives the relative motion of the arm, sun-wheel and planet when the annulus is fixed. It is always the fixed wheel which is given −1 rev in line (*b*).

When all the members are rotating, a modification of the method is necessary. The whole gear is given $+a$ rev in line (*a*). In line (*b*), the arm is held fixed and *any* wheel is given $+b$ rev. The motion of the other wheels is then found as before. The resulting motion found by the addition of lines (a) and (b) is in terms of the constants a and b, which are then evaluated from the known speeds of two of the members.

Thus, in tabular form:

	L	A	P	S
(a) Give whole gear $+a$ rev	$+a$	$+a$	$+a$	$+a$
(b) Hold arm L and give A $+b$ rev	0	$+b$	$+\dfrac{T_a}{T_p}.b$	$-\dfrac{T_a}{T_s}.b$
(c) Resulting motion	a	$a+b$	$a+\dfrac{T_a}{T_p}.b$	$a-\dfrac{T_a}{T_s}.b$

The above procedure applies only to members which form part of the epicyclic train. *Any wheel external to the epicyclic train must be considered separately from the table.*

12.2 Torques on gear trains. In the gear units shown in Fig. 12.2, let C_a, C_b and C_c be respectively the applied input torque, the resisting torque on the output shaft and the torque to hold the casing fixed. If there is to be no acceleration of the system, the net torque applied to the unit *about any one axis* must be zero,

i.e. $$C_a + C_b + C_c = 0 . \qquad . \qquad . \qquad . \quad (12.1)$$

Also, if there is no acceleration, the kinetic energy of the system remains constant so that the net work done per second is zero,

i.e. $$C_a\omega_a + C_b\omega_b = 0 . \qquad . \qquad . \qquad . \quad (12.2)$$

The appropriate signs must be given to both torques and speeds in equations (12.1) and (12.2). In Fig. 12.2(a), the direction of the fixing torque C_c will depend on the relative magnitudes of C_a and C_b.

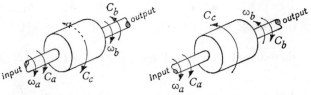

Input and output shafts rotating in the same direction
(a)

Input and output shafts rotating in opposite directions
(b)

FIG. 12.2

If, allowing for friction, the efficiency of the unit is η, then

$$\eta = \frac{\text{output power}}{\text{input power}}$$

so that equation (12.2) becomes

$$\eta C_a \omega_a + C_b \omega_b = 0 \qquad . \qquad . \qquad . \qquad . \qquad (12.3)$$

If the casing is not fixed, C_c represents either an input or output torque and equation (12.2) becomes

$$C_a \omega_a + C_b \omega_b + C_c \omega_c = 0 \qquad . \qquad . \qquad . \qquad . \qquad (12.4)$$

12.3 Compound epicyclic trains. A compound train consists of two or more co-axial simple trains with members forming part of two consecutive trains.

Consider the compound train shown in Fig. 12.3, where the annulus A_1 of the train $A_1 S_1 L$ also forms the arm of the train $A_2 S_2 A_1$. Let A_2 be fixed.

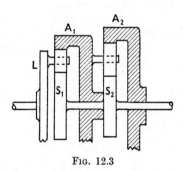

FIG. 12.3

First obtain the speed ratio of L to S_1 when A_1 is fixed, thus :

	L	A_1	S_1
Give whole train $+1$ rev. .	$+1$	$+1$	$+1$
Hold L and give A_1 -1 rev	0	-1	$+\dfrac{T_{a_1}}{T_{s_1}}$
Add	1	0	$1+\dfrac{T_{a_1}}{T_{s_1}}$

Thus $\quad \dfrac{N_l}{N_{s_1}} = \dfrac{1}{1 + T_{a_1}/T_{s_1}}$ when A_1 is fixed . $\qquad$. $\qquad$. $\qquad$ (1)

Then consider the whole train with A_1 as the arm, using the above result in the second line of the table when A_1 is fixed.

	L	A_1	S_1S_2	A_2
Give whole train $+1$ rev .	$+1$	$+1$	$+1$	$+1$
Hold A_1 and give A_2 -1. rev	$\dfrac{T_{a_2}/T_{s_2}*}{1+T_{a_1}/T_{s_1}}$	0	$+\dfrac{T_{a_2}}{T_{s_2}}$	-1
Add	$.1+\dfrac{T_{a_2}/T_{s_2}}{1+T_{a_1}/T_{s_1}}$	1	$1+\dfrac{T_{a_2}}{T_{s_2}}$	0

Thus the ratio $\dfrac{N_l}{N_{s_1}} = \dfrac{1+\dfrac{T_{a_2}/T_{s_2}}{1+T_{a_1}/T_{s_1}}}{1+T_{a_2}/T_{s_2}}$

* From equation (1).

The advantage of this type of gear over a simple train is that if A_2 is released and A_1 is fixed, a different ratio N_l/N_{s_1} is obtained.

12.4 Acceleration of gear trains. The acceleration of a gear train may be obtained by finding the equivalent inertia, I_e, of the train referred to the shaft to which the accelerating torque C is applied. The acceleration of this shaft is then given by

$$C = I_e \alpha$$

This equivalent inertia may be obtained from the general principle that the net energy supplied to the system per unit time is equal to the rate of change of its kinetic energy,

i.e. $\quad C\omega = \dfrac{d}{dt}$ (K.E. of system)

ω being the speed of the shaft to which the torque is applied.

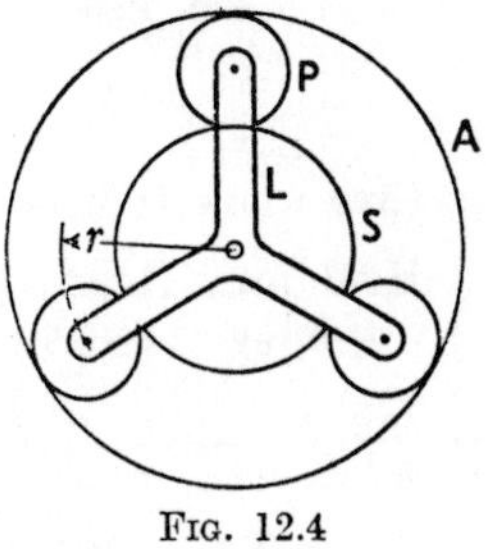

Consider a simple epicyclic gear train having three planets and a fixed annulus A, as shown in Fig. 12.4. Let S be the driving shaft and L the driven shaft.

Then total K.E. of system

FIG. 12.4

$$= \tfrac{1}{2}I_s\omega_s{}^2 + \tfrac{1}{2}I_l\omega_l{}^2 + 3(\tfrac{1}{2}I_p\omega_p{}^2 + \tfrac{1}{2}m_p v_p{}^2)$$

$$= \dfrac{\omega_s{}^2}{2}\left[I_s + I_l\left(\dfrac{\omega_l}{\omega_s}\right)^2 + 3\left\{I_p\left(\dfrac{\omega_p}{\omega_s}\right)^2 + m_p\left(\dfrac{\omega_l \times r}{\omega_s}\right)^2\right\}\right]$$

If C_s is the net accelerating torque applied to shaft S (after deducting any torque on S necessary to overcome a resisting torque on L), the energy supplied to the system in unit time $= C_s\omega_s$,

$$\therefore C_s\omega_s = \frac{d}{dt}\left(\frac{\omega_s^2}{2} \times I\right) \quad \text{where } I = I_s + I_l\left(\frac{\omega_l}{\omega_s}\right)^2 + \cdots$$

$$= I\omega_s\alpha_s$$

$$\therefore C_s = I\alpha_s$$

The quantity

$$\left[I_s + I_l\left(\frac{\omega_l}{\omega_s}\right)^2 + 3\left\{I_p\left(\frac{\omega_p}{\omega_s}\right)^2 + m_p\left(\frac{\omega_l \times r}{\omega_s}\right)^2\right\}\right] \qquad . \quad (12.5)$$

is evidently the equivalent inertia of the system referred to shaft S.

ω_s, ω_l and ω_p are the *absolute* angular velocities of S, L and P respectively and the speed ratios are obtained by the usual tabular method.

1. *In the epicyclic gear shown in Fig. 12.5, the shaft A is driven at 750 rev/min. Determine the speed of shaft D when the wheel E is held stationary.*

To enable the speed of shaft D to be varied while the speed of shaft A remains constant at 750 rev/min, the wheel E may be coupled to shaft D through a variable-speed gear which is not shown. Calculate the speed ratio required in the variable-speed gear if shaft D is to run at 118 rev/min in the same direction as shaft A.

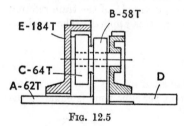

B-58T

E-184T

C-64T

A-62T

D

Fig. 12.5

Determine the torque transmitted from B to C when the input at A is 4 kW with shaft D at 118 rev/min.

(U. Glas.)

	D	A	E	B, C
Give whole train $+a$ rev .	$+a$	$+a$	$+a$	$+a$
Hold arm D and give A $+b$ rev	0	$+b$	$-\dfrac{62.64}{58.184}b$	$-\dfrac{62}{58}b$
Add.	a	$a+b$	$a-0\cdot3715b$	$a-1\cdot068b$

When $N_a = 750$ rev/min and $N_e = 0$,

$$a + b = 750$$

and

$$a - 0\cdot3715b = 0$$

Hence

$$N_d = a = \underline{203 \text{ rev/min}}$$

When $N_a = 750$ rev/min and $N_d = 118$ rev/min,

$$a + b = 750$$

and $$a = 118$$

$$\therefore b = 632$$

Hence $$N_e = a - 0.3715b$$

$$= 118 - 235$$

$$= -117 \text{ rev/min}$$

$$\therefore \text{ speed ratio, } \frac{N_e}{N_d} = -\frac{117}{118}$$

$$\text{Torque on A} = \frac{4 \times 1000 \times 60}{2\pi \times 750}$$

$$= 50.9 \text{ N m}$$

Assume that the teeth are of 2 mm module so that the radii, in **mm**, are equal to the numbers of teeth.

Then force at circumference of A

$$= \frac{50.9}{62 \times 10^{-3}}$$

$$= 821 \text{ N}$$

$$= \text{force at circumference of B}$$

$\therefore$ torque on B, which is transmitted to C

$$= 821 \times 58 \times 10^{-3}$$

$$= 47.6 \text{ N m}$$

2. *In the reduction gear shown in Fig. 12.6, there is an epicyclic train consisting of a sun-wheel A on the input shaft, a planet wheel B carried on an arm K on the output shaft, and an annular gear C. The latter can be held fixed, or can be connected directly to the input shaft through the spur gear drive DEFGH, as shown. Find the speed ratio from input to output for the two cases: (a) with EFG disengaged, and C held against rotation; (b) with EFG engaged. For case (b), determine also the torque transmitted between A and D, and between F and G, based on an input torque of 100 N m.*
                                                                    (U. Glas.)

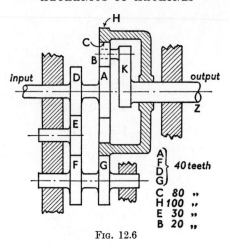

Fig. 12.6

The train EFG is not part of the epicyclic train and must therefore be treated separately.

	Z, K	A, D	C, H
Give whole train $+a$ rev .	$+a$	$+a$	$+a$
Hold K and give C $+b$ rev .	0	$-\dfrac{80}{40}.b$	$+b$
Add	a	$a-2b$	$a+b$

(a) When C is fixed,

$$a + b = 0$$
$$\therefore b = -a$$
$$\therefore \frac{N_d}{N_z} = \frac{a-2b}{a} = 3$$

(b) When train EFG is engaged,

$$N_h = -N_d \times \frac{T_f}{T_d} \times \frac{T_g}{T_h}$$
$$= -N_d \times \frac{40}{40} \times \frac{40}{100} = -\frac{N_d}{2\cdot5}$$
$$\therefore \frac{N_d}{N_h} = -2\cdot5 = \frac{a-2b}{a+b}$$

from which
$$b = -7a$$
$$\therefore \frac{N_d}{N_z} = \frac{a-2b}{a} = \underline{15}$$

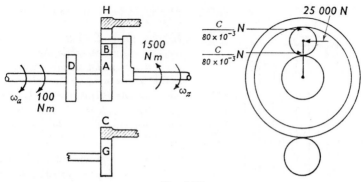

FIG. 12.7

Referring to Fig. 12.7, input torque on A and D

= 100 N m, in the same direction as the rotation,

∴ resisting torque on Z

= 100 × 15 = 1500 N m, in the opposite direction to the rotation.

Assume that the teeth are of 2 mm module, so that the radii, in mm, are equal to the numbers of teeth.

Then force on axis of B = $\dfrac{1500}{(40 + 20) \times 10^{-3}} = 25\,000$ N

Let the torque exerted on C by G be C.

Then force between teeth on C and B = $\dfrac{C}{80 \times 10^{-3}}$ N

Since there is no angular acceleration of B, the force between the teeth on A and B is also $\dfrac{C}{80 \times 10^{-3}}$ N, both these forces being opposite in direction to the 25 000 N force, and for equilibrium of B,

$$2 \times \frac{C}{80 \times 10^{-3}} = 25\,000 \qquad \therefore C = 1000 \text{ N m}$$

Hence   torque on A $= C \times \dfrac{40}{80} = 500$ N m

and   torque applied by G to C $= C \times \dfrac{40}{100} = 400$ N m

i.e.   torque transmitted between A and D = 500 N m

and   torque transmitted between F and G = 400 N m

3. *Fig. 12.8 shows a compound epicyclic gear train, gears S_1 and S_2 being rigidly attached to the shaft Q. If the shaft P rotates at 1000 rev/min clockwise, while the annulus A_2 is driven in the opposite direction at 500 rev/min, determine the speed and direction of rotation of the shaft Q. The numbers of teeth in the wheels are, S_1, 24; S_2, 40; A_1, 100; A_2, 120.*   (U. Lond.)

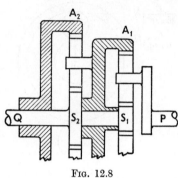

FIG. 12.8

Consider first the train PA_1S_1 in order to find the speed ratio of P to S_1 when annulus A_1 is fixed.

	P	A_1	S_1
Give whole train +1 rev . .	+1	+1	+1
Hold P and give A_1 −1 rev .	0	−1	$+\dfrac{100}{24}$
Add	+1	0	$+\dfrac{31}{6}$

$$\therefore N_p = \frac{6}{31} N_{s_1} \text{ when } A_1 \text{ is fixed}$$

The whole gear is now considered with A_1 as the arm and the speed ratio of P to S_1 already found is used to complete the table. Since no member is fixed, the whole train is given an initial rotation $+a$ rev.

	A_1	A_2	S_1, S_2, Q	P
Give whole train $+a$ rev . .	$+a$	$+a$	$+a$	$+a$
Hold A_1 and give A_2 $+b$ rev .	0	$+b$	$-\dfrac{120}{40}b$	$-\dfrac{6}{31}\cdot\dfrac{120}{40}b$
Add	a	$a+b$	$a-3b$	$a-\dfrac{18}{31}b$

Hence $\qquad N_p = a - \dfrac{18}{31}b = 1000$

and $\qquad N_{a_2} = a + b = -500$

so that $\qquad a = 450 \quad \text{and} \quad b = -950$

$$\therefore N_q = a - 3b = 450 + 3 \times 950$$

$$= \underline{3300 \text{ rev/min}} \text{ (in same direction as P)}$$

4. *Find the velocity ratio of the two co-axial shafts in the gear shown in Fig. 12.9, in which S_1 is the driver.*
The tooth numbers of the gears are $S_1, 40$; $A_1, 120$; $S_2, 30$; $A_2, 100$, and the sun-wheel S_2 is fixed.

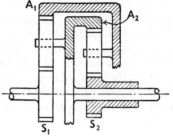

Determine also the magnitude and direction of the torque required to fix S_2, if a torque of 300 N m is applied in a clockwise direction to S_1. (U. Lond.)

Consider first the train $S_1A_1A_2$, for which A_2 is the arm, in order to find the speed ratio of S_1 to A_2 when annulus A_1 is fixed.

Fig. 12.9

	A_2	A_1	S_1
Give whole train $+1$ rev .	$+1$	$+1$	$+1$
Hold A_2 and give A_1 -1 rev .	0	-1	$+\dfrac{120}{40}$
Add	1	0	4

$$\therefore N_{s_1} = 4N_{a_2} \text{ when } A_1 \text{ is fixed}$$

The whole gear is now considered with A_1 as the arm and the speed ratio of S_1 to A_2 already found is used to complete the table.

	A_1	A_2	S_1	S_2
Give whole train $+1$ rev .	$+1$	$+1$	$+1$	$+1$
Hold A_1 and give S_2 -1 rev .	0	$+\dfrac{30}{100}$	$+4 \times \dfrac{30}{100}$	-1
Add	1	$1\cdot3$	$2\cdot2$	0

$$\therefore \frac{N_{s_1}}{N_{a_2}} = \underline{\frac{22}{13}}$$

Input torque on S_1,

$$C_{s_1} = 300 \text{ N m, in direction of rotation,}$$

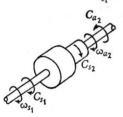

$\therefore$ resisting torque on A_2,

$$C_{a_2} = 300 \times \tfrac{22}{13} = 507 \cdot 7 \text{ N m, opposite to}$$
direction of rotation.

Therefore, referring to Fig. 12.10,

$$C_{s_1} + C_{s_2} = C_{a_2}$$
$$\therefore C_{s_2} = 507 \cdot 7 - 300$$
$$= \underline{207 \cdot 7 \text{ N m}} \text{ (clockwise)}$$

Fig. 12.10

5. *The compound epicyclic gear shown in Fig. 12.11 has a driving shaft D and a driven shaft E to which the arms F and G are fixed. The arms carry planet wheels which mesh with the annular wheels A and B, and the sun-wheels H and J. The sun-wheel H is part of B. Wheels J and K are fixed to the shaft D. K engages with a planet wheel carried on B and this planet wheel engages with the fixed annular wheel C. The numbers of teeth on the wheels are: A, 71; B, 65; C, 65; H, 29; J, 21; K, 21.*

Find the speeds of shaft E and wheel A when the driving shaft D makes 2000 rev/min. If the input at D is 20 kW, find the torque on the driven shaft E and the torque on the stationary member C, neglecting friction.

(U. Lond.)

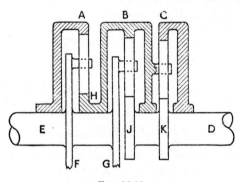

Fig. 12.11

Consider first the train A, HB, EFG, JKD for which EFG is the arm in order to find the speed ratios of A and E to J when B is fixed.

	EFG	JKD	HB	A
Give whole train +1 rev . .	+1	+1	+1	+1
Hold EFG and give B −1 rev .	0	$+\dfrac{65}{21}$	−1	$+\dfrac{29}{71}$
Add	+1	$+\dfrac{86}{21}$	0	$+\dfrac{100}{71}$
Multiply by $\dfrac{21}{86}$	0·244	1	0	0·343

The whole gear is now considered with B as the arm and the speed ratios of A and E to J already found are used to complete the table.

	HB	A	C	EFG	JKD
Give whole train +1 rev	+1	+1	+1	+1	+1
Hold B and give C −1 rev	0	$+\dfrac{65}{21}.0\cdot343$	−1	$+\dfrac{65}{21}.0\cdot244$	$+\dfrac{65}{21}$
Add	+1	+2·064	0	+1·756	+4·10
Multiply by $\dfrac{2000}{4\cdot10}$	—	1006	—	856	2000

i.e. the speeds of A and E are respectively 1006 and 856 rev/min, in the same direction as D.

Input torque on D,

$$C_d = \frac{20 \times 1000 \times 60}{2\pi \times 2000}$$

$$= 95\cdot5 \text{ N m, in direction of rotation,}$$

∴ resisting torque on E,

$$C_e = 95\cdot5 \times \frac{2000}{856}$$

$$= 223 \text{ N m, opposite to direction of rotation.}$$

Therefore, referring to Fig. 12.12,

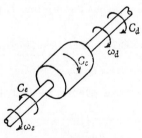

Fig. 12.12

$$C_d + C_c = C_e$$

$$\therefore C_c = 223 - 95\cdot5$$

$$= 127\cdot5 \text{ N m}$$

6. *In the epicyclic gear shown in Fig. 12.13, the wheel A, with 100 internal teeth, gears with the three planet wheels B, of 30 teeth, which are carried by the cage C. The planet wheels gear with D, of 40 teeth, and are on a circle of*

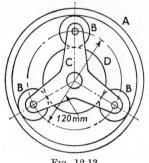

120 mm diameter. Each wheel B has a mass of 500 g and a moment of inertia about its axis of rotation of 0·15 g m². The moments of inertia of A and C about the axis of D are, respectively, 4·8 and 2·4 g m². The wheel D is fixed and A rotates at 700 rev/min. Determine the speed of the cage C and of each planet wheel about its own axis in rev/min, and calculate the torque required at A to give it an angular acceleration of 50 rad/s².

Fig. 12.13

(U. Lond.)

	A	B	C	D
Give whole train $+1$ rev .	$+1$	$+1$	$+1$	$+1$
Hold C and give D $\cdot$ -1 rev .	$+\dfrac{40}{100}$	$+\dfrac{40}{30}$	0	-1
Add	$\dfrac{7}{5}$	$\dfrac{7}{3}$	1	0
Multiply by 500 . . .	700	$1166{\cdot}7$	500	0

i.e. speed of C $= \underline{500 \text{ rev/min}}$, in same direction as A.

The speed of a planet about its own axis is its speed relative to the cage C,

i.e. $\qquad\qquad 1166{\cdot}7 - 500 = \underline{666{\cdot}7 \text{ rev/min}}$

Linear speed of planet centre $= \omega_c \times 0{\cdot}06$ m/s

Equivalent inertia of system referred to shaft A

$$= I_a + I_c\left(\frac{\omega_c}{\omega_a}\right)^2 + 3\left\{ I_b\left(\frac{\omega_b}{\omega_a}\right)^2 + m_b\left(\frac{\omega_c \times 0{\cdot}06}{\omega_a}\right)^2 \right\}$$

$$= 4{\cdot}8 + 2{\cdot}4 \times \left(\frac{5}{7}\right)^2 + 3\left\{ 0{\cdot}15 \times \left(\frac{5}{3}\right)^2 + 500 \times \left(\frac{5 \times 0{\cdot}06}{7}\right)^2 \right\}$$

$$= 10 \text{ g m}^2$$

$\therefore C_a = 10 \times 10^{-3} \times 50 = \underline{0{\cdot}5 \text{ N m}}$

7. *Fig. 12.14 is the diagram of an epicyclic train, in which possible movement of the annular wheel D is restrained by pistons acting against fluid under pressure in the cylinders G. The driving shaft is connected with the sun-wheel B which meshes with the three planet wheels C; these in turn mesh with the annular wheel D. Compound with wheels C, the planet wheels E mesh with F which is connected to the driven shaft. The driving shaft rotates at 5000 rev/min.*

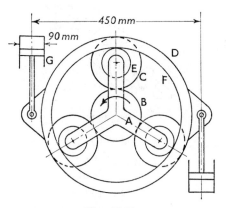

Fɪɢ. 12.14

Find (a) the power transmitted when the fluid pressure on the pistons is 1·75 MN/m²; (b) the initial acceleration of D when the torque exerted on B is instantaneously increased by 1·8 N m and the inertia of the mass on the driven shaft is very large.

Component of gear train	Teeth	Mass (kg)	Radius of gyration (mm)
B with shaft . . .	21	2	50
Compound planets (each) .	—	3	40
D with attached masses . .	65	12·5	360
F	61		
Spider carrying planets . .	—	5	140

(U. Lond.)

(a) The numbers of teeth on C and E are 22 and 18 respectively. Solving the gear train in the usual way, it is found that

$$\frac{N_f}{N_b} = 0.031$$

Restraining torque on D, $C_d = 1.75 \times 10^6 \times \dfrac{\pi}{4} \times 0.09^2 \times 0.45$

$$= 5000 \text{ N m}$$

$$N_f < N_b$$

$$\therefore \ C_b < C_f$$

Hence C_d acts in the same direction as C_b, Fig. 12.15, so that

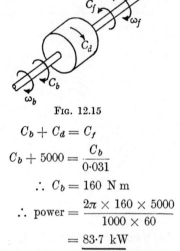

Fig. 12.15

$$C_b + C_d = C_f$$

i.e.
$$C_b + 5000 = \frac{C_b}{0.031}$$

$$\therefore \ C_b = 160 \text{ N m}$$

$$\therefore \ \text{power} = \frac{2\pi \times 160 \times 5000}{1000 \times 60}$$

$$= \underline{83.7 \text{ kW}}$$

(b) Since the inertia of F is very large, F may be regarded, for acceleration purposes, as fixed and D as free.

The relevant velocity ratios for this condition are:

$$\frac{N_a}{N_b} = 0.22, \quad \frac{N_c}{N_b} = 0.526, \quad \frac{N_d}{N_b} = -0.0321$$

Equivalent inertia of system referred to shaft B

$$= I_b + I_a \left(\frac{\omega_a}{\omega_b}\right)^2 + I_d \left(\frac{\omega_d}{\omega_b}\right)^2 + 3I_c \left(\frac{\omega_c}{\omega_b}\right)^2 {}^*$$

$$= 2 \times 0.05^2 + 5 \times 0.14^2 \times 0.22^2 + 12.5 \times 0.36^2 \times 0.0321^2$$
$$\qquad + 3 \times 3 \times 0.04^2 \times 0.526^2$$

$$= 0.0154 \text{ kg m}^2$$

$$\therefore \ 1.8 = 0.0154\alpha_b$$

$$\therefore \ \alpha_b = 117 \text{ rad/s}^2$$

$$\therefore \ \alpha_d = 117 \times 0.0321 = \underline{3.75 \text{ rad/s}^2}$$

* The radius of rotation of the planets is unknown.

8. *A simple epicyclic gear train consists of a spider, an annulus having 60 teeth and a planet having 20 teeth. There is no sun-wheel. The teeth have a module of 2·5 mm. The planet wheel is freely pinned to the spider and it meshes with the annulus, its polar axis being parallel to the common axis of rotation of the spider and the annulus.*

The moment of inertia of the spider (without the planet wheel) about its axis of rotation is 1·8 g m². The planet wheel has a mass of 1 kg and its polar radius of gyration is 20 mm. Initially the annulus is rotating clockwise at 150 rev/min and the spider is rotating clockwise at 100 rev/min. If a braking torque is applied to the annulus bringing it to rest, find the resulting speed of the spider. (U. Lond.)

Fig. 12.16(*a*) shows the arrangement of the train, the radius of the annulus being 75 mm and that of the planet 25 mm. The initial speeds of the annulus and spider are respectively 150 and 100 rev/min clockwise and it will be found that the speed of the planet is 250 rev/min, also clockwise.

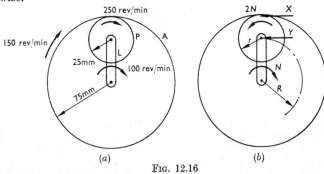

(*a*) (*b*)

Fig. 12.16

If the final speed of the arm is N rev/min clockwise and the annulus is stationary, the corresponding speed of the planet is $-2N$ rev/min.

Let the impulse applied by the annulus to the planet be X, Fig. 12.16(*b*), and let the impulse applied by the planet to the arm be Y.

Then, for the planet and arm

$$\text{angular impulse} = \text{change of angular momentum}$$

i.e.
$$X \times r = I_p(-2N - 250) \times \frac{2\pi}{60}$$

or
$$X = \frac{1 \times 0 \cdot 02^2(-2N - 250)}{0 \cdot 025} \times \frac{2\pi}{60}$$

$$= -0 \cdot 001\ 675(2N + 250) \qquad . \qquad . \qquad (1)$$

and
$$Y \times R = I_l(N - 100) \times \frac{2\pi}{60}$$

or
$$Y = \frac{1 \cdot 8 \times 10^{-3}}{0 \cdot 05} (N - 100) \times \frac{2\pi}{60}$$

$$= 0 \cdot 003\ 77(N - 100) \qquad . \qquad . \qquad . \qquad (2)$$

For the linear motion of the planet, the net impulse applied to it by the annulus and arm is $X - Y$.

Then   impulse = change of momentum

i.e. $$X - Y = m_p(N - 100) \times \frac{2\pi}{60} \times R$$

$$= 1(N - 100) \times \frac{2\pi}{60} \times 0 \cdot 05$$

$$= 0 \cdot 005\ 23(N - 100) \qquad . \qquad . \qquad . \qquad (3)$$

Subtracting (2) from (1) and equating to (3),

$$N = \underline{38 \cdot 9 \text{ rev/min}}$$

9. A gear box with co-axial input and output shafts is mounted on trunnions so that the torque required to prevent rotation of the gear box can be measured. In a given test the speeds of the input and output shafts are 3000 and 600 rev/min in the same direction ; the input power is 20 kW and the torque on the gear box is 235 N m applied in the same direction as the shafts rotate. Determine the efficiency of the gear box.

If the direction of rotation of the output shaft is reversed, the input power and speeds remaining as before, determine the torque required to prevent rotation of the gear box and state the direction in which it should be applied. Assume the efficiency of the gear box to be unchanged. (*I. Mech. E.*)

(*Ans.:* 93·89% ; 362 N m opposite to direction of rotation of input shaft)

10. An epicyclic speed reducing gear consists of a fixed outer annulus B, a sunwheel A attached to the input shaft and a spider carrying three planet wheels attached to the output shaft Z. Obtain an expression relating the speed of the shaft Z to that of A in terms of the numbers of teeth in wheels A and B. In an example the speed reduction is to be 5·5 to 1 and the number of teeth in the smallest wheel must not be less than 19. Also to enable the gear to be assembled the sum of the teeth in the co-axial wheels A and B must be divisible by the number of planets. Find the arrangement with the smallest numbers of teeth and also the arrangement with the next larger numbers that will satisfy these requirements. (*I. Mech. E.*)

(*Ans.:* A 24, B 108, planet 42 ; A 36, B 162, planet 163)

11. In the gear shown in Fig. 12.17 the input shaft X is directly connected to the sun-wheel A of the epicyclic gear and it rotates the annular wheel B through the geared side shaft, turning about a fixed axis, and the pinion D and external wheel B_0 on the casing of B. The teeth are all of the same pitch and the numbers are as follows : $E = E_1 = 75$, $D = 30$, $B_0 = 120$, $A = 40$, $C = 30$, $B = 100$. If the speed of the input shaft is 112 rev/min, find the speed of the output shaft Z. (*I. Mech. E.*) (*Ans.:* $+52$ rev/min)

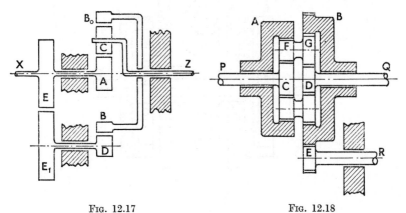

FIG. 12.17   FIG. 12.18

12. Fig. 12.18 shows an epicyclic gear train in which the internal teeth on the gears A and B have the same diametral pitch. Gear A has 88 teeth and gear B has 100 teeth both internally and externally. The sun-wheel C which is keyed to shaft P has 44 teeth and the pinion E keyed to shaft R has 25. There are floating compound planet wheels F, G. The power input to shaft R is 8 kW at 1440 rev/min.

If wheel A is fixed and the power output is shared equally between shafts P and Q, determine :

(*a*) the speeds of shafts P and Q and their directions of rotation relative to that of shaft R ;

(*b*) the torque required to hold wheel A fixed, if friction losses are neglected. (*U. Lond.*)   (*Ans.:* $+3000$ rev/min, $+5250$ rev/min ; $232 \cdot 5$ N m)

13. A form of epicyclic differential gear for a heavy motor vehicle is shown diagrammatically in Fig. 12.19. The two shafts C and D are connected to the driving wheels of the vehicle. The transmission shaft from the engine is connected to the worm A, of 4 threads, driving the worm-wheel B which has 25 teeth. The internally-toothed annulus E meshes with the planet wheels P carried on pins on the shaft C. The sun-wheel R is connected by a central shaft to the wheel S. The pinions Q, revolving on stationary axes, mesh with S and with annulus F. E and F have each 72 teeth, R has 24 teeth and S has 18 teeth.

Find an equation connecting the rotational speed of the worm A with the speeds of shaft C and D, the last two speeds being not necessarily equal. Show that the torques exerted on C and D are equal and in the same direction. (*U. Lond.*)   (*Ans.:* $N_a = 25(N_c + N_d)/3$)

14. Fig. 12.20 shows a variable-speed epicyclic reduction gear with friction drive. A is the motor shaft and B the driven shaft. The former carries the pinion C and the friction discs D and E. The latter carries an arm with a pin on which the pinion F can revolve freely. This pinion gears externally with C and internally with the ring wheel G. The external teeth on G gear with the pinion H on the lay shaft K. K is driven through the friction drive from D and E by means of the hardened rollers M and N which are in contact with D and E respectively, and also with the friction disc L secured to K. The

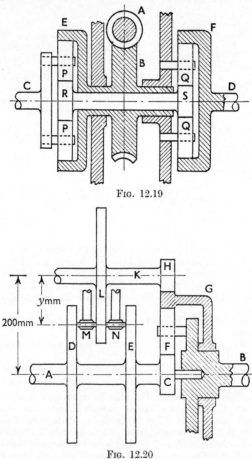

Fig. 12.19

Fig. 12.20

distance between the axes of A and K is 200 mm. The motor shaft A runs at 600 rev/min.

Find the axial travel of the rollers in order to change the speed of K from 200 to 1000 rev/min, assuming no slip.

If the limits of speed of B are to be ±100 rev/min and the number of teeth externally and internally on G are equal, show that the numbers of teeth on C and H must also be equal, and find the ratio of the numbers of teeth on C and G. (*U. Lond.*)   (*Ans.:* 75 mm; 1 : 3)

15. In a Wilson gear box, a gear is engaged by bringing the annulus of a simple sun and planet epicyclic gear to rest. The planet arm of the gear is integral with the annulus of a second epicyclic gear and the propeller shaft is driven by the planet arm of this second gear. The pinions of the two gears are integral with the driving shaft.

Show by means of a sketch the arrangement of the combined gears and calculate the gear ratio for the following proportions:

First gear :   Fixed annulus 80 teeth, driving pinion 24 teeth
Second gear :   Annulus 78 teeth, driving pinion 26 teeth

(*I. Mech. E.*)   (*Ans.*: 26 : 11)

16. In the epicyclic gear shown in Fig. 12.21, the wheels B and C are attached to the driving shaft X and gear with the annular wheels D and E through the pinions F and H. The pinion F is free to revolve on a pin carried on the arm A which is keyed to the driven shaft Y. The pinion H revolves on a pin carried by the annular wheel D. The numbers of teeth on certain wheels and pinions are as follows : B, 28 ; D, 78 ; C, 24 ; E, 80. The wheel E is fixed by means of a band brake. If the shaft X rotates at a speed of 2000 rev/min, find the speed of Y. (*U. Lond.*)   (*Ans.*: 868 rev/min in same direction as X)

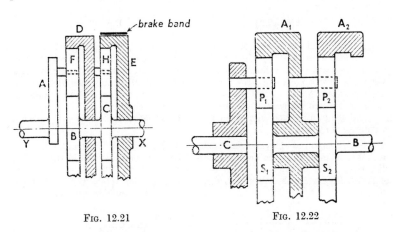

Fig. 12.21   Fig. 12.22

17. Part of a pre-selective gear box is shown in Fig. 12.22. The pinions S_1 and S_2 are driven directly from the engine by the shaft B and gear through pinions P_1 and P_2, with the annular wheels A_1 and A_2, which may be locked in turn to give two speeds to the driven shaft C ; the first speed is when A_1 is locked and the second speed when A_2 is locked. The pinion P_2 rotates freely on a pin carried by A_1 and P_1 rotates freely on a pin carried by the arm which is keyed to the driven shaft C. Assuming the tooth numbers on A_1 and A_2 to be equal and the tooth numbers on S_1 and S_2 to be equal, find the ratio of the number of teeth on A_1 to the number of teeth on S_1 when the second speed is 1·75 times the first speed and in the same direction.

Evaluate the first and second speeds for this ratio when the engine speed is 3000 rev/min. (*U. Lond*).   (*Ans.*: 3 ; 750 rev/min ; 1312·5 rev/min)

18. A compound epicyclic gear is shown in Fig. 12.23. If the shaft R rotates at 1000 rev/min while the annulus A_1 rotates at 1000 rev/min in the opposite direction, determine the speed and direction of rotation on the shaft Q. The numbers of teeth on the wheels are : sun-wheel $S_1 = 35$, sun-wheel $S_2 = 20$, annulus $A_1 = 105$, annulus $A_2 = 60$. (*U. Lond.*)

(*Ans.:* 11 000/13 rev/min in the opposite direction to R)

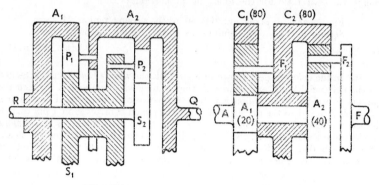

Fɪɢ. 12.23   Fɪɢ. 12.24

19. Fig. 12.24 shows part of an epicyclic gear box. Wheels A_1 and A_2 are all in one piece with shaft A, which is the driving shaft. The annular wheel C_1 is held stationary. The rotating frame F_2, which carries the first pair of planet wheels, is in one piece with the driven shaft F. The annular wheel C_2 rotates freely on shaft A and carries the second pair of planet wheels. The numbers of teeth on the wheels concerned are given in the diagram.

Find the number and sign of the revolutions shaft F makes for every positive revolution of shaft A. (*U. Lond.*)   (*Ans.:* +7/15)

20. A compound epicyclic gear is shown in Fig. 12.25. The shaft P is driven at 3000 rev/min while the annulus A_2 is driven at 1000 rev/min in the opposite direction. The numbers of teeth in the gears are S_1, 16 ; S_2, 24 ; A_1, 60 ; A_2, 90. Determine the speed and direction of rotation of shaft Q. (*U. Lond.*)

(*Ans.:* 1308 rev/min opposite to P)

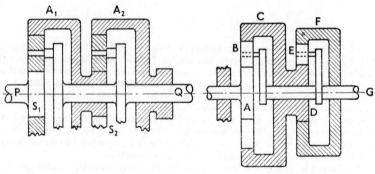

Fɪɢ. 12.25   Fɪɢ. 12.26

21. A compound epicyclic gear is shown in Fig. 12.26. C and D form a compound wheel which rotates freely on shaft G. The planet wheels B and E rotate on pins fixed in arms attached to shaft G. C and F have internal teeth; the others have external teeth with the following numbers: A, 40 ; B, 30 ; D, 50 ; E, 20. If A rotates at 500 rev/min and wheel F is fixed, find the speed of shaft G. (*U. Lond.*)   (*Ans.:* 1000/7 rev/min in opposite direction to A).

22. Fig. 12.27 shows a compound epicyclic gear in which the casing C contains an epicyclic train and this casing is inside the larger casing D. Determine the velocity ratio of the output shaft B to the input shaft A when the casing D is held stationary.   The number of teeth on the wheels are as follows :

Wheel on A . . . 80	Small pinion on F . 20	
Annular wheel on B . 160	Large pinion on F . 66	
Annular wheel on C . 100		
Annular wheel on D . 120	(*U. Lond.*)	(*Ans.:* −5·54)

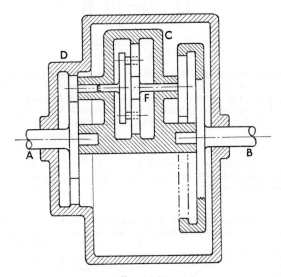

Fig. 12.27

23. A compound epicyclic gear train, Fig. 12.28, is composed of a main train and an auxiliary train. The main train consists of the shaft A carrying the arms on which the planet wheels D are mounted which mesh with the annular wheel C and the sun-wheel E which is secured to the shaft B. The auxiliary train consists of planet wheels G mounted on arms carried by C which mesh with the fixed annular wheel F and with the sun-wheel H which is also secured to the shaft B. Find the velocity ratio of B to A.

If the train transmits 8 kW when A rotates at 600 rev/min, determine the torque transmitted by the pinions E and H respectively. All the wheels have the same diametral pitch and the tooth numbers are as follows : D, 25 ; E, 25 ; G, 30 ; H, 30. (*U. Lond.*)   (*Ans.:* 16 : 7 ; 31·8 N m ; 23·8 N m)

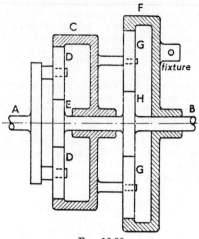

Fig. 12.28

24. In an epicyclic gear with a fixed outer annulus the sun-wheel is to rotate at 1000 rev/min and the spider, which carries three planet wheels, is to be driven at 200 rev/min. The teeth of all wheels are to be of 5 mm module and the pitch circle diameter of the annulus is to be as near 375 mm as possible.

Determine the total kinetic energy of the gear, the following values being given :

	Mass (kg)	Radius of gyration (mm)
Sun-wheel with shaft and coupling .	4·5	40
Spider with shaft and coupling .	10	75
Planet wheel (each) . . .	2	50

(*U. Lond.*)   (*Ans.:* 78·4 J)

25. In the compound epicyclic gear train shown in Fig. 12.29, of the three independent wheels concentric about O, wheel A is keyed to the driving shaft,

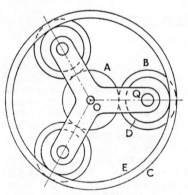

E to the driven shaft and wheel C is fixed ; the planet wheels B and D rotate together about the axis Q on the three-armed planet carrier. Wheels A, B and D have 20, 30 and 25 teeth respectively, and all the wheels have a module pitch of 6 mm. The moments of inertia of the wheels A and E, the carrier, and each planet pair are 0·004, 2, 0·08 and 0·04 kg m² respectively ; the total mass of the three sets of planet wheels is 15 kg.

Find the velocity ratio between the driving and driven shafts, and the torque required to give the driven shaft an acceleration of 1 rad/s². (*U. Lond.*)

Fig. 12.29

(*Ans.:* 45/1 ; 1·575 N m)

26. Fig. 12.30 shows an epicyclic gear train, S and R being sun-wheels connected to shafts X and Y respectively. There are three *floating* compound planet wheels PQ. A is a fixed casing which carries an internal gear having 88 teeth. B has internal and external teeth and is driven by pinion C, which is mounted on shaft Z carried in bearings in the casing A. The particulars of the wheels are :

Member	Number of teeth	P.C.D. (mm)	Mass (kg)	Radius of gyration (mm)
B	100 int., 100 ext.		20	240
S	44	240	6	90
R	32		5	60
P, Q			3·5	60
C	20		2·5	105

Calculate : (*a*) the kinetic energy of the gear train when shaft Z is driven at 1000 rev/min ; (*b*) the torque which must be applied to shaft Z to overcome the inertia of the gear if shaft X is to have an angular acceleration of 25 rad/s² (*U. Lond.*)   (*Ans.*: 2·82 kJ ; 7·69 N m)

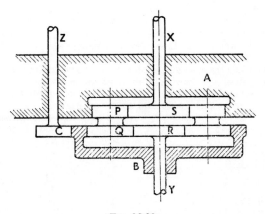

Fig. 12.30

27. In an epicyclic gear drive, a sun-wheel A, with 30 teeth, meshes with three planet wheels, B, each with 20 teeth ; these rotate in bearings in a spider C, which itself rotates on the same axis as A. The planet wheels mesh with a fixed outer annulus with internal teeth. The moments of inertia of A and C are 4·5 and 30 g m² respectively. The radius from the axis of A to the centre of each planet wheel B is 125 mm ; *each* wheel B has a mass of 1·5 kg and a moment of inertia of 2 g m².

Find the equivalent inertia at the spider to replace the system, and find the angular acceleration at the spider when a torque of 1 N m is applied to wheel A. (*U. Lond.*)   (*Ans.* : 0·1878 kg m² ; 17·65 rad/s²)

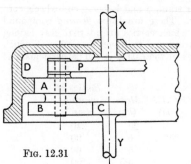

Fig. 12.31

28. In the epicyclic gear shown in Fig. 12.31, a disc P attached to shaft X carries three pins on which three sets of compound wheels, A and B, revolve freely. Wheels A mesh with an annular wheel D fixed to the casing, and wheels B mesh with wheel C attached to shaft Y. Find (a) the number of teeth on D, (b) the velocity ratio between X and Y, (c) the constant torque applied to X to raise the speed of Y to 2500 rev/min in 5 s.

Wheel	Number of teeth	Module (mm)	Mass (kg)	Radius of gyration (mm)
Disc P with shaft X	—	—	2·7	70
A	16	8⎫	0·6 total	36
B	32	5⎬		
C	16	5	1·4	40
with shaft Y				

(*U. Lond.*)   (*Ans.*: 46; +4/27; 0·982 N m)

29. An epicyclic gear train consists of a sun-wheel S, two annular wheels A_1 and A_2 concentric with S, and three pairs of compound wheels P_1 and P_2 mounted on a carrier and symmetrically spaced round the axis of S; the wheels P_1 mesh with S and A_1, while the wheels P_2 mesh with A_2. The driving shaft is connected to the annular wheel A_2, the driven shaft to the sun-wheel, and the annular wheel A_1 is fixed. Attached to the driven shaft is a rotor whose moment of inertia is 75 g m²; other particulars are given below:

Wheel	Number of teeth	Mass (kg)	Radius of gyration (mm)
A_1	53	—	—
A_2, with shaft	52	30	250
P_1	18	5⎫	100
P_2	17	4⎬	
S, with shaft	17	9	50
Carrier with 3 pairs of pinions .	—	40	200

Determine the time taken by a torque of 35 N m applied to wheel A_2 to bring the speed of S up to 2000 rev/min from rest. (*U. Lond.*)   (*Ans.*: 166 s)

30. An epicyclic gear train has a fixed outer annulus, a sun-wheel with 30 teeth and three planets, each having 30 teeth, carried on a spider. The gears are of 4 mm module. Determine the torque which must be applied to the sun-wheel to accelerate it at a rate of 20 rad/s², using the following data:

	Mass (kg)	Radius of gyration (mm)
Sun-wheel	3	50
Planet (each) . . .	2	50
Spider	5	100

(*U. Lond.*)   (*Ans.*: 0·3955 N m)

CHAPTER 13

FREE VIBRATIONS

13.1 Free vibrations. Free, or natural, vibrations occur in an elastic system when a body is acted upon only by the internal restoring forces of the system. Since these forces are proportional to the displacement of the body from the equilibrium position, the acceleration of the body is also proportional to the displacement and is always directed towards the equilibrium position, so that the body moves with S.H.M.

The number of degrees of freedom of a system is the number of different modes of vibration which the system may possess.

13.2 Simple harmonic motion. A body moves with simple harmonic motion if its acceleration is proportional to its displacement from a fixed point and is always directed towards that point. Let the line OP, of length a, rotate about a fixed point O, with constant angular velocity ω rad/s, Fig. 13.1. Then, if time is measured from the position OB, the angle turned through by OP in time t,

$$\psi = \omega t$$

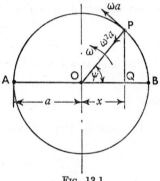

Fig. 13.1

If Q is the projection of P on the diameter AB, the displacement of Q from its mid-position,

$$x = a \cos \psi \qquad . \qquad . \qquad . \qquad . \qquad (13.1)$$

The maximum displacement, a, is termed the *amplitude* of the motion. The velocity of Q is the component of the velocity of P parallel to AB,

i.e. $$v = \omega a \sin \psi = \omega \sqrt{(a^2 - x^2)} \qquad . \qquad . \qquad (13.2)$$

and $$v_{max} = \omega a, \quad \text{when } x = 0 \qquad . \qquad . \qquad . \qquad (13.3)$$

The acceleration of Q is the component of the acceleration of P parallel to AB,

i.e. $$f = \omega^2 a \cos \psi = \omega^2 x \qquad . \qquad . \qquad . \qquad (13.4)$$

and $$f_{max} = \omega^2 a \quad \text{when } x = a \qquad . \qquad . \qquad (13.5)$$

307

Thus the acceleration of Q is proportional to its displacement from the fixed point O and is always directed towards O, so that the motion of Q is simple harmonic.

These formulae give merely the numerical relationships between displacement, velocity and acceleration without regard to direction.

The *periodic time* is the time taken for one complete revolution of P,

i.e.
$$t_p = \frac{2\pi}{\omega} \text{ s}$$

but, from equation (13.4), $\omega^2 = \dfrac{f}{x}$

$$\therefore t_p = 2\pi\sqrt{\frac{x}{f}} = 2\pi\sqrt{\frac{\text{displacement}}{\text{acceleration}}} \text{ s} \quad . \qquad . \quad (13.6)$$

The *frequency*, $n = \dfrac{1}{t_p}$ vibrations per second (Hz)   .   .   (13.7)

Similar equations apply in the case of angular simple harmonic motion. Thus if the amplitude of the motion is ϕ, the angular velocity of the body at any angular displacement, θ, is given by

$$\Omega = \omega\sqrt{(\phi^2 - \theta^2)} \quad . \qquad . \qquad . \qquad . \quad (13.8)$$

and
$$\Omega_{\max} = \omega\phi \quad . \qquad . \qquad . \qquad . \quad (13.9)$$

The angular acceleration is given by

$$\alpha = \omega^2\theta \quad . \qquad . \qquad . \qquad . \quad (13.10)$$

and
$$\alpha_{\max} = \omega^2\phi \quad . \qquad . \qquad . \qquad . \quad (13.11)$$

The periodic time, $t_p = 2\pi\sqrt{\dfrac{\text{angular displacement}}{\text{angular acceleration}}}$

$$= 2\pi\sqrt{\frac{\theta}{\alpha}} \quad . \qquad . \qquad . \qquad . \quad (13.12)$$

and
$$n = \frac{1}{2\pi}\sqrt{\frac{\alpha}{\theta}} \text{ Hz} \quad . \qquad . \qquad . \qquad . \quad (13.13)$$

13.3 Linear motion of an elastic system. If a body of mass m, controlled by an elastic system, is given a displacement x, then restoring force $= Sx$, where S is the stiffness of the system (i.e. the restoring force per unit displacement).

When released, the acceleration of the body is given by

$$\text{force} = \text{mass} \times \text{acceleration}$$

i.e.
$$Sx = mf$$

i.e.
$$\frac{x}{f} = \frac{m}{S}$$

i.e.
$$t_p = 2\pi \sqrt{\frac{m}{S}} \text{ s} \quad . \quad . \quad \text{from equation (13.6)}$$

$$= 2\pi \sqrt{\frac{\delta}{g}} \text{ s}$$

where δ is the static deflection under the mass m $\quad . \quad . \quad .$ (13.14)

or
$$n = \frac{1}{2\pi} \sqrt{\frac{g}{\delta}} \text{ Hz}$$

$$\simeq \frac{1}{2\sqrt{\delta}} \text{ Hz} \quad \text{where } \delta \text{ is in metres} \quad . \quad (13.15)$$

13.4 Angular motion of an elastic system. If a body of moment of inertia I, controlled by an elastic system, is given an angular displacement θ, then restoring torque $= q\theta$, where q is the torsional stiffness of the system (i.e. the restoring torque per unit angular displacement). When released, the angular acceleration of the body is given by

$$\text{torque} = \text{moment of inertia} \times \text{angular acceleration}$$

i.e.
$$q\theta = I\alpha$$

i.e.
$$\frac{\theta}{\alpha} = \frac{I}{q}$$

$$\therefore t_p = 2\pi \sqrt{\frac{I}{q}} \text{ s}$$

or
$$n = \frac{1}{2\pi} \sqrt{\frac{q}{I}} \text{ Hz} \quad . \quad . \quad . \quad . \quad (13.16)$$

13.5 Differential equation of motion. If a body of mass m is acted upon by a restoring force S per unit displacement from the equilibrium position, the equation of motion is

$$m\frac{d^2x}{dt^2} = -Sx$$

The negative sign arises because the restoring force Sx is opposite in direction to the displacement x.

This may be written

$$\frac{d^2x}{dt^2} + \omega^2 x = 0 \quad \text{where } \omega^2 = \frac{S}{m} \quad . \quad . \quad . \quad (13.17)$$

or
$$(D^2 + \omega^2)x = 0 \quad \text{where } D \text{ represents } \frac{d}{dt}$$

so that $$D = \pm i\omega$$

Hence
$$x = C_1 e^{iwt} + C_2 e^{-iwt}$$
$$= C_1\{\cos \omega t + i \sin \omega t\} + C_2\{\cos \omega t - i \sin \omega t\}$$
$$= A \cos \omega t + B \sin \omega t \qquad . \qquad . \qquad . \qquad . \qquad . \qquad (13.18)$$

where $\quad A = C_1 + C_2 \quad$ and $\quad B = i(C_1 - C_2)$

This equation represents an oscillatory motion of periodic time $t_p = \dfrac{2\pi}{\omega}$. The constants A and B are determined by the initial conditions of the motion.

Thus, if $\qquad\qquad x = a \quad$ when $\quad t = 0,$

and $\qquad\qquad\qquad \dfrac{dx}{dt} = 0 \quad$ when $\quad t = 0,$

then $\qquad\qquad A = a$ and $B = 0, \quad$ so that $x = a \cos \omega t \quad . \qquad . \quad (13.19)$

In Fig. 13.1 these conditions correspond to the motion commencing when P coincides with B, so that $\psi = \omega t$.

Differentiating equation (13.19) twice,

$$v = -\omega a \sin \omega t = \omega a \cos\left(\omega t + \frac{\pi}{2}\right) \qquad . \qquad . \qquad (13.20)$$

and $\qquad\qquad f = -\omega^2 a \cos \omega t = \omega^2 a \cos(\omega t + \pi) \qquad . \qquad . \qquad (13.21)$

Equations (13.20) and (13.21) correspond to equations (13.2) and (13.4) except that the correct signs are automatically obtained by differentiation.

Equations (13.20) and (13.21) show that the velocity leads the displacement by 90° and the acceleration leads the displacement by 180°.

The corresponding equations for angular motion are

$$I\frac{d^2\theta}{dt^2} = -q\theta$$

i.e. $\qquad\qquad \dfrac{d^2\theta}{dt^2} + \omega^2\theta = 0 \quad$ where $\quad \omega^2 = \dfrac{q}{I} \qquad . \qquad . \qquad . \qquad (13.22)$

$$\therefore \quad \theta = A \cos \omega t + B \sin \omega t \qquad . \qquad (13.23)$$

If $\qquad\qquad\qquad \theta = \phi \quad$ when $\quad t = 0,$

and $\qquad\qquad\qquad \dfrac{d\theta}{dt} = 0 \quad$ when $\quad t = 0,$

then $\qquad\qquad\qquad \theta = \phi \cos \omega t \qquad . \qquad . \qquad . \qquad (13.24)$

$$\Omega = -\omega\phi \sin \omega t \qquad . \qquad . \qquad . \qquad (13.25)$$

and $\qquad\qquad\qquad \alpha = -\omega^2\phi \cos \omega t \qquad . \qquad . \qquad . \qquad (13.26)$

13.6   Simple and compound pendula.   If a simple pendulum is given a *small* angular displacement, θ, Fig. 13.2,

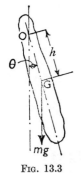

restoring moment about $O = mgl \sin \theta$

$$\backsimeq mgl\theta, \quad \text{since } \theta \text{ is small}$$

$$\therefore \ mgl\theta = I_0\alpha = ml^2\alpha$$

$$\therefore \ \frac{\theta}{\alpha} = \frac{l}{g}$$

$$\therefore \ t_p = 2\pi \sqrt{\frac{l}{g}} \ \text{s} \qquad . \qquad . \qquad . \ (13.27)$$

FIG. 13.2

If the mass of the pendulum is not concentrated at a point, let the radius of gyration about the centre of gravity, G, be k and the distance of the point of suspension from G be h, Fig. 13.3.

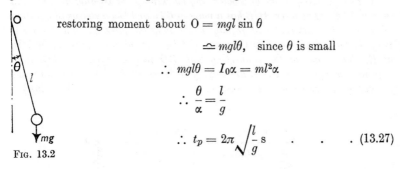

Then, for a *small* angular displacement, θ

restoring moment about $O = mgh \sin \theta$

$$\backsimeq mgh\theta$$

$$\therefore \ mgh\theta = I_0\alpha = m(k^2 + h^2)\alpha$$

$$\therefore \ \frac{\theta}{\alpha} = \frac{k^2 + h^2}{gh}$$

$$\therefore \ t_p = 2\pi \sqrt{\frac{k^2 + h^2}{gh}} \ \text{s} \ . \qquad . \ (13.28)$$

FIG. 13.3

13.7   Two degrees of freedom.   A system possesses two degrees of freedom if two distinct modes of vibration are possible.   If each part of the system is given a displacement from its equilibrium position, the restoring force or moment on each is a function not only of its own displacement but also of the displacement of the other part or parts.   The equations of motion obtained must be solved as simultaneous differential equations, but it will always be found that the motions of each are *simple harmonic and of the same frequency.*   The equations of motion may therefore be greatly simplified by assuming that $x_1 = a_1 \cos \omega t$ and $x_2 = a_2 \cos \omega t$ (or $\theta_1 = \phi_1 \cos \omega t$ and $\theta_2 = \phi_2 \cos \omega t$ in the case of angular motion), which leads to a quadratic equation in ω^2 from which the two natural frequencies of vibration may be obtained.

(a) Single degree of freedom

1. *The swinging frame of a balancing machine is shown diagrammatically in Fig. 13.4. The lever ACB has a mass of 6 kg and has a radius of gyration of 75 mm about its centre of gravity, which is at the pivot C. The lever FD is of uniform section and has a mass of 1 kg. It carries a mass of 0·75 kg at E. The mass of the connecting link BD may be ignored. The compression springs at A and B are each of stiffness 5·6 kN/m. Determine the frequency of vibration of the system.* (U. Glas.)

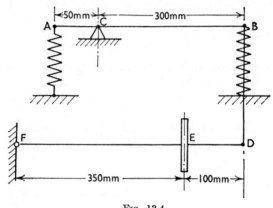

FIG. 13.4

$$I_{AB} \text{ about } C = 6 \times 0·075^2 = 0·033\ 75 \text{ kg m}^2$$

$$I_{FD} \text{ about } F = 1 \times \frac{0·45^2}{3} + 0·75 \times 0·35^2 = 0·1594 \text{ kg m}^2$$

Let ACB be displaced through a small angle θ and let the force in BD be P.

Then   restoring force at A $= 5·6 \times 10^3 \times 0·05\theta$ N

and   restoring force at B $= (5·6 \times 10^3 \times 0·3\theta - P)$ N

Therefore the equation of motion for AB is

$$5·6 \times 10^3 \times 0·05\theta \times 0·05 + (5·6 \times 10^3 \times 0·3\theta - P) \times 0·3 = 0·033\ 75\alpha$$

i.e. $518·5\theta - 0·3P = 0·033\ 75\alpha$ . . . (1)

and the equation of motion for FD is

$$P \times 0·45 = 0·1594.\alpha' \quad . \quad . \quad . \quad (2)$$

but $0·3\alpha = 0·45\alpha'$ since the *linear* accelerations of B and D are equal,

i.e. $\alpha' = \tfrac{2}{3}\alpha$ so that equation (2) becomes $P = 0·236\alpha$

Substituting in equation (1),

$$518.5\theta - 0.0708\alpha = 0.033\ 75\alpha$$

$$\therefore \frac{\alpha}{\theta} = 4955$$

$$\therefore n = \frac{1}{2\pi}\sqrt{4955} = \underline{11.2\ \text{Hz}}$$

2. *Two uniform rods AB and CD each having a mass of 4·5 kg per metre length are suspended at their upper ends and connected by a spring at their lower ends, as shown in Fig. 13·5. When hanging freely they are vertical and there is no force in the spring. The spring has a stiffness of 2·8 kN/m.*

The spring is now compressed slightly and released. Neglecting the effect of gravity, find the frequency of the resulting vibrations.

If AB moves through 1° on either side of the vertical, find the corresponding angular movement of CD and the maximum force in the spring. (U. Lond.)

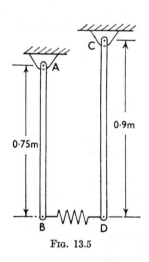

Fig. 13.5

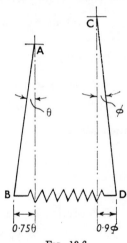

Fig. 13.6

Let the rods AB and CD be displaced through angles θ and ϕ respectively from their equilibrium positions, Fig. 13.6.

Then   extension of spring $= (0.75\theta + 0.9\phi)$ m

$$\therefore \text{force in spring} = 2.8 \times 10^3 \times (0.75\theta + 0.9\phi)\ \text{N} \qquad (1)$$

$$I_{\text{AB}}\ \text{about A} = 0.75 \times 4.5 \times \frac{0.75^2}{3} = 0.633\ \text{kg m}^2$$

$$I_{\text{CD}}\ \text{about C} = 0.9 \times 4.5 \times \frac{0.9^2}{3} = 1.094\ \text{kg m}^2$$

L

Therefore the equation of motion of AB is

$$0{\cdot}633\frac{d^2\theta}{dt^2} = -2{\cdot}8 \times 10^3(0{\cdot}75\theta + 0{\cdot}9\phi) \times 0{\cdot}75 \qquad . \qquad (2)$$

and the equation of motion of CD is

$$1{\cdot}094\frac{d^2\phi}{dt^2} = -2{\cdot}8 \times 10^3(0{\cdot}75\theta + 0{\cdot}9\phi) \times 0{\cdot}9 \, . \qquad . \qquad (3)$$

Assuming that $\theta = \alpha \cos \omega t$ and $\phi = \beta \cos \omega t$, then equation (2) becomes

$$-0{\cdot}633\alpha\omega^2 \cos \omega t = -2{\cdot}8 \times 10^3(0{\cdot}75\alpha \cos \omega t + 0{\cdot}9\beta \cos \omega t) \times 0{\cdot}75$$

i.e.

$$(0{\cdot}000\ 301\ 5\omega^2 - 0{\cdot}75)\alpha = 0{\cdot}9\beta \qquad . \qquad . \qquad . \qquad . \qquad . \qquad (4)$$

and equation (3) becomes

$$-1{\cdot}094\beta\omega^2 \cos \omega t = -2{\cdot}8 \times 10^3(0{\cdot}75\alpha \cos \omega t + 0{\cdot}9\beta \cos \omega t) \times 0{\cdot}9$$

i.e.

$$(0{\cdot}000\ 434\omega^2 - 0{\cdot}9)\beta = 0{\cdot}75\alpha \qquad . \qquad . \qquad . \qquad . \qquad (5)$$

Eliminating α and β between (4) and (5),

$$(0{\cdot}000\ 301\ 5\omega^2 - 0{\cdot}75)(0{\cdot}000\ 434\omega^2 - 0{\cdot}9) = 0{\cdot}675$$

from which

$$\omega^2 = 4560$$

$$\therefore \ \omega = 67{\cdot}5 \ \text{rad/s}$$

$$\therefore \ n = \frac{1}{2\pi} \times 67{\cdot}5 = \underline{10{\cdot}75 \ \text{Hz}}$$

Substituting in equation (4),

$$(0{\cdot}000\ 301\ 5 \times 4560 - 0{\cdot}75)\alpha = 0{\cdot}9\beta$$

$$\therefore \ \beta = 0{\cdot}694\alpha$$

but $\alpha = 1°$

$$\therefore \ \beta = 0{\cdot}694°$$

i.e.   the amplitude of CD is 0·694°.

Substituting $\theta = 1°$ and $\phi = 0{\cdot}694°$ in equation (1), the maximum force in the spring is

$$2{\cdot}8 \times 10^3\left[\left(0{\cdot}75 \times 1 \times \frac{\pi}{180}\right) + \left(0{\cdot}9 \times 0{\cdot}694 \times \frac{\pi}{180}\right)\right] = \underline{67{\cdot}2 \ \text{N}}$$

3. *The apparatus in Fig. 13·7 is intended for the determination of the coefficient of sliding friction between dry surfaces. A metal block is supported with its lower plane face AB horizontal and in contact with two cylindrical rollers which revolve as shown in opposite directions about fixed axes C and D. When initially disturbed so that the c.g. of the block, which is at a height h above AB, is displaced from the central position between the rollers it performs simple harmonic oscillations.*

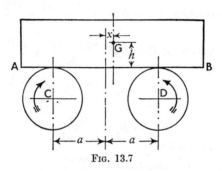

FIG. 13.7

FIG. 13.8

If the coefficient of friction μ between the block and the rollers is the same for both rollers, show that its magnitude is

$$\mu = \frac{4\pi^2 a}{gt^2 + 4\pi^2 h}$$

where t is the period of oscillation.

If the coefficients of friction are unequal, show that the centre of gravity of the block will oscillate about a mean position given by

$$x_0 = a\frac{\mu_1 - \mu_2}{\mu_1 + \mu_2}$$

and that the values of the two coefficients are

$$\left.\begin{matrix} \mu_1 \\ \mu_2 \end{matrix}\right\} = \frac{4\pi^2(a \pm x_0)}{gt^2 + 4\pi^2 h}$$

(U. Lond.)

When G is displaced a distance x to the right, let the reactions at C and D be R_1 and R_2 respectively, Fig. 13.8.

Then $\qquad$ restoring force $= \mu(R_2 - R_1)$

Moment of this force about G $= \mu(R_2 - R_1)h$

Reactions at C and D to balance this moment

$$= \frac{\mu(R_2 - R_1)h}{2a}$$

$\therefore$ total reaction at C $= R_1 = \dfrac{mg\,(a-x)}{2a} - \dfrac{\mu(R_2 - R_1)h}{2a}$

and $\qquad$ total reaction at D $= R_2 = \dfrac{mg(a+x)}{2a} + \dfrac{\mu(R_2 - R_1)h}{2a}$

$$\therefore R_2 - R_1 = \frac{mgx}{a} + \frac{\mu(R_2 - R_1)h}{a}$$

from which $\qquad R_2 - R_1 = \dfrac{mgx}{a - \mu h}$

$$\therefore \text{restoring force} = \frac{\mu mgx}{a - \mu h}$$

$$\therefore \frac{\mu mgx}{a - \mu h} = mf$$

$$\therefore \frac{x}{f} = \frac{a - \mu h}{\mu g}$$

$$\therefore t = 2\pi\sqrt{\frac{a - \mu h}{\mu g}} \quad \text{from which} \quad \mu = \frac{4\pi^2 a}{gt^2 + 4\pi^2 h}$$

When μ is different for each roller, restoring force $= \mu_2 R_2 - \mu_1 R_1$. Modifying the above equations,

$$R_2 - R_1 = \frac{mgx}{a} + \frac{(\mu_2 R_2 - \mu_1 R_1)h}{a}$$

i.e. $\qquad R_2\left(1 - \dfrac{\mu_2 h}{a}\right) - R_1\left(1 - \dfrac{\mu_1 h}{a}\right) = \dfrac{mgx}{a}$ . . . . (1)

also $\qquad\qquad\qquad R_2 + R_1 = mg$ . . . . (2)

Therefore, from equations (1) and (2),

$$R_1 = \frac{mg\left\{-\dfrac{x}{a} + \left(1 - \dfrac{\mu_2 h}{a}\right)\right\}}{\left\{2 - \dfrac{h}{a}(\mu_1 + \mu_2)\right\}} \quad \text{and} \quad R_2 = \frac{mg\left\{\dfrac{x}{a} + \left(1 - \dfrac{\mu_1 h}{a}\right)\right\}}{\left\{2 - \dfrac{h}{a}(\mu_1 + \mu_2)\right\}}$$

∴ restoring force

$$= \frac{\mu_2 mg\left\{\dfrac{x}{a} + \left(1 - \dfrac{\mu_1 h}{a}\right)\right\} - \mu_1 mg\left\{-\dfrac{x}{a} + \left(1 - \dfrac{\mu_2 h}{a}\right)\right\}}{\left\{2 - \dfrac{h}{a}(\mu_1 + \mu_2)\right\}} \ . \tag{3}$$

In the mean position, this force is zero and hence, denoting this displacement of G by x_0,

$$\mu_2\left\{\frac{x_0}{a} + \left(1 - \frac{\mu_1 h}{a}\right)\right\} = \mu_1\left\{-\frac{x_0}{a} + \left(1 - \frac{\mu_2 h}{a}\right)\right\}$$

from which

$$x_0 = a\left(\frac{\mu_1 - \mu_2}{\mu_1 + \mu_2}\right) \qquad . \qquad . \qquad . \tag{4}$$

From equation (3),

$$\frac{\mu_2 mg\left\{\dfrac{x}{a} + \left(1 - \dfrac{\mu_1 h}{a}\right)\right\} - \mu_1 mg\left\{-\dfrac{x}{a} + \left(1 - \dfrac{\mu_2 h}{a}\right)\right\}}{\left\{2 - \dfrac{h}{a}(\mu_1 + \mu_2)\right\}} = mf$$

i.e.

$$(\mu_1 + \mu_2)\frac{x}{a} - (\mu_1 - \mu_2) = \frac{f}{g}\left\{2 - \frac{h}{a}(\mu_1 + \mu_2)\right\}$$

i.e.

$$x - x_0 = \frac{f}{g}\left(\frac{2a}{\mu_1 + \mu_2} - h\right)$$

Since $f \propto (x - x_0)$, the block will perform simple harmonic oscillations about the position $x = x_0$

$$\therefore t = 2\pi\sqrt{\frac{x - x_0}{f}} = 2\pi\sqrt{\left[\frac{1}{g}\left(\frac{2a}{\mu_1 + \mu_2} - h\right)\right]}$$

$$\therefore gt^2 = 4\pi^2\left(\frac{2a}{\mu_1 + \mu_2} - h\right)$$

$$\therefore \mu_1 + \mu_2 = \frac{8\pi^2 a}{gt^2 + 4\pi^2 h}$$

But, from equation (4),

$$\mu_2 = \mu_1\left(\frac{a - x_0}{a + x_0}\right)$$

$$\therefore \mu_1\left(1 + \frac{a - x_0}{a + x_0}\right) = \frac{8\pi^2 a}{gt^2 + 4\pi^2 h}$$

$$\therefore \mu_1 = \frac{4\pi^2(a + x_0)}{gt^2 + 4\pi^2 h}$$

and

$$\mu_2 = \frac{4\pi^2(a + x_0)}{gt^2 + 4\pi^2 h} \times \left(\frac{a - x_0}{a + x_0}\right) = \frac{4\pi^2(a - x_0)}{gt^2 + 4\pi^2 h}$$

4. *A pendulum of mass 27 kg is suspended from a pivot, such that the distance of the pivot from the centre of gravity of the pendulum is 380 mm. When the amplitude of oscillation is small, the periodic time of oscillation is 1·60 s.*

If the pendulum is now made to swing with amplitude 45° on each side of the vertical, find the force exerted on the pivot, (a) at the extremity of the swing, and (b) at the centre of the swing. (U. Lond.)

$$t_p = 2\pi \sqrt{\left[\frac{k^2 + h^2}{gh}\right]} = 2\pi \sqrt{\frac{I_0}{mgh}} \quad \text{from equation (13.28)}$$

i.e. $$1·60 = 2\pi \sqrt{\frac{I_0}{27 \times 9·81 \times 0·38}}$$

$$\therefore I_0 = 6·52 \text{ kg m}^2$$

(a) At the extremity of the swing, the forces acting at G are (i) its weight, mg, and (ii) the inertia force, F_1, due to its acceleration, Fig. 13.9. The angular acceleration is given by

restoring torque $= I_0 \alpha$

i.e. $27 \times 9·81 \times 0·38 \sin 45° = 6·52\alpha$

$$\therefore \alpha = 10·92 \text{ rad/s}^2$$

$\therefore$ linear acceleration of G $= 10·92 \times 0·38 \text{ m/s}^2$

$$\therefore F_1 = mf = 27 \times 10·92 \times 0·38 = 112 \text{ N}$$

$$\therefore R = \sqrt{\{(27 \times 9·81)^2 + 112^2 - 2 \times 27 \times 9·81 \times 112 \times \cos 45°\}}$$

$$= \underline{202 \text{ N}}$$

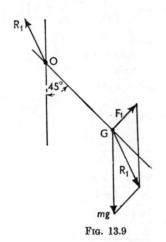

FIG. 13.9   FIG. 13.10

(b) At the centre of the swing, the forces acting at G are (i) its weight, mg, and (ii) the centrifugal force, F_2, due to its angular velocity, Ω, about O, Fig. 13.10. Ω is obtained by equating the loss of P.E. in the lowest position to the gain in K.E.,

i.e. $27 \times 9 \cdot 81(0 \cdot 38 - 0 \cdot 38 \cos 45°) = \dfrac{6 \cdot 52}{2}\Omega^2$

$$\therefore \ \Omega^2 = 9 \cdot 05 \ (\text{rad/s})^2$$

$$\therefore \ F_2 = m\Omega^2 r = 27 \times 9 \cdot 05 \times 0 \cdot 38$$

$$= 92 \cdot 8 \ \text{N}$$

$$\therefore \ R_2 = 27 \times 9 \cdot 81 + 92 \cdot 8 = 357 \cdot 6 \ \text{N}$$

NOTE: In any intermediate position the forces acting at G are (i) its weight, mg, (ii) the inertia force, F_1, at right angles to the axis, and (iii) the centrifugal force, F_2, along the axis.

5. *The moment of inertia of a turbine rotor of mass M, and of which the centre of gravity lies in the shaft axis C, is to be determined by the method*

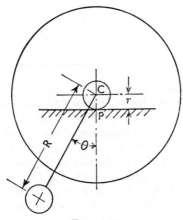

FIG. 13.11

shown in Fig. 13.11. A known mass m is attached at a distance R from the axis and the time of oscillation t is found by rolling the journal, of radius r, on hardened horizontal surfaces, through small distances on either side of the equilibrium position.

Deduce an expression for the moment of inertia of the rotor about the axis C in terms of the quantities given and show that small variations of r will have least effect when

$$R = \left(\frac{M}{m} + 1\right)r \qquad \text{(U. Lond.)}$$

The rotor may be considered as oscillating about the line of contact at P between the journal and the bearing surfaces for small angles of oscillation.

$$I_P = I_C + Mr^2 + m(R - r)^2$$

When displaced through a *small* angle θ, restoring torque $\simeq mgR\theta$

$$\therefore \; mgR\theta = I_P \cdot \alpha \qquad \therefore \; \frac{\theta}{\alpha} = \frac{I_P}{mgR}$$

$$\therefore \; t = 2\pi \sqrt{\frac{I_P}{mgR}}$$

$$\therefore \; I_P = \frac{mgRt^2}{4\pi^2} = I_C + Mr^2 + m(R - r)^2$$

$$\therefore \; I_C = \frac{mgRt^2}{4\pi^2} - Mr^2 - m(R - r)^2$$

For least effect upon I_C of small variations in r, $\dfrac{dI_C}{dr} = 0$

i.e. $$-2Mr + 2m(R - r) = 0$$

from which $$R = \left(\frac{M}{m} + 1\right)r$$

6. *The pendulum shown in Fig. 13.12 is suspended from a fixed pivot at O. The pendulum consists of a bar B, of mass 1 kg, and a block C of mass 6 kg. The centres of gravity G_1 and G_2 of B and C are at distances 150 mm and 375 mm from O. The radii of gyration of B and C, each about its own centre of gravity, are respectively 100 mm and 25 mm. A light spring is attached to the pendulum at a point P, 200 mm from O, and is anchored at a fixed point Q. When the pendulum is in equilibrium the line OG_1PG_2 is at 45° from the vertical and the angle OPQ is 90°. The spring has a stiffness of 700 N/m.*

Calculate the natural frequency of the pendulum for small oscillations about the equilibrium position. (U. Lond.)

The moment of inertia of the pendulum about O

$$= 1(0.1^2 + 0.15^2) + 6(0.025^2 + 0.375^2)$$

$$= 0.8834 \text{ kg m}^2$$

Let the pendulum be displaced through a *small* angle θ, Fig. 13.13.

Then additional pull in spring, $P = 0.2\theta \times 700 = 140\theta$ N

$\therefore$ restoring moment due to spring $= 140\theta \times 0.2 = 28\theta$ N m

Reduction in moment of pendulum about O

$$= 1 \times 9 \cdot 81 \times \left(0 \cdot 150\theta \times \frac{1}{\sqrt{2}}\right) + 6 \times 9 \cdot 81 \times \left(0 \cdot 3750\theta \times \frac{1}{\sqrt{2}}\right)$$

$$= 16 \cdot 650 \text{ N m}$$

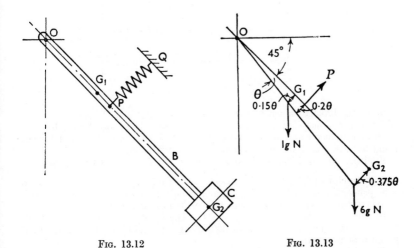

FIG. 13.12   FIG. 13.13

∴ total restoring moment

$$= (28 + 16 \cdot 65)\theta$$

$$= 44 \cdot 650 \text{ N m}$$

$$\therefore 44 \cdot 650 = 0 \cdot 8834\alpha$$

$$\therefore \frac{\alpha}{\theta} = \frac{44 \cdot 65}{0 \cdot 8834} = 50 \cdot 6$$

$$\therefore n = \frac{1}{2\pi}\sqrt{50 \cdot 6} = \underline{1 \cdot 134 \text{ Hz}}$$

7. A uniform bar AB, 2·5 m long and of mass 100 kg, is supported on a hinge at one end A and on a spring support at the other end B, so that it can vibrate in a vertical plane. The stiffness of the spring is 20 kN/m deflection. When in static equilibrium the bar is horizontal. The bar may be assumed to be flexurally rigid.

The end B of the bar is depressed 10 mm, then released. Calculate (a) the frequency of the resulting vibrations, and (b) the maximum bending moment at the mid-point of the bar. (*U. Lond.*)   (*Ans.:* 3·9 Hz; 400 N m)

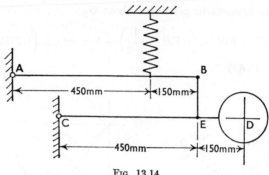

FIG. 13.14

8. In the linkage shown in Fig. 13.14 the bars AB and CD are uniform in section and each has a mass of 1·2 kg, while the disc at D, carried by the bar CD, is 150 mm diameter and has a mass of 1 kg. The mass of link BE is 1 kg, the spring has a mass of 1·5 kg and a stiffness of 400 N/m.

Find the natural time period of the system. (*U. Lond.*)   (*Ans.:* 0·856 s)

9. A tractor and trailer, with masses m_1 and m_2 respectively, are connected by a spring of stiffness k. Neglecting any damping forces in the system, derive an expression for the natural frequency of longitudinal oscillations.

In such a system $m_1 = 2$ t, $m_2 = 1$ t. The spring consists of two sections in series; one having a stiffness of 360 kN/m and the other 90 kN/m. Find the natural frequency.

The node of the vibration is in the weaker spring, which is 250 mm long and is adjacent to the 1 t mass. Find the distance of the node from this mass.

(*U. Lond.*) $\left(Ans: n = \dfrac{1}{2\pi}\sqrt{\dfrac{k(m_1 + m_2)}{m_1 m_2}} \text{ Hz}; \ 1\cdot655 \text{ Hz}; \ 208 \text{ mm}\right)$

10. Fig. 13.15 shows a block of mass 18 kg whose lower face CD is horizontal

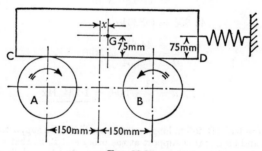

FIG. 13.15

and rests on two rollers A and B. The centre of gravity of the block is at G and a horizontal spring of stiffness 900 N/m is attached to the blocks as shown and may act in either tension or compression. The two rollers are revolved in opposite directions at such a speed as to ensure that slipping always takes place between the block and the rollers. The coefficient of friction is equal to 0·2 for each contact line. When G is midway between A and B the spring is unstressed. The block is given a horizontal displacement x from this position.

(a) Show that the block will perform simple harmonic oscillations.

(b) Calculate the periodic time of the oscillations. (*U. Lond.*)

(*Ans.:* 0·782 s)

11. A connecting rod of mass 450 kg is suspended on a knife-edge at a point 0·9 m from the c.g. It is initially set swinging over a range of 35° on either side of the vertical. When the oscillations become small the period of the double swing is timed as 2·75 s. Determine the moment of inertia of the rod about an axis through its c.g. perpendicular to the plane of oscillation and calculate the pressure on the knife-edge at the extreme swing of 35°. (*U. Lond.*)

(*Ans.:* 397 kg m²; 3·85 kN)

12. A compound pendulum swings through an angle of 60° on either side of the vertical. The distance between the centre of gravity and the centre of suspension is equal to the radius of gyration about the centre of gravity. Find the angle of the pendulum to the vertical at which the reaction at the hinge is equal to the weight of the pendulum. (*Ans.:* 44° 2′)

13. A 1·25 t flywheel rolls on parallel ways on its 150 mm diameter shaft. It is arranged to oscillate, rolling through a small angle, by adding a mass of 160 kg at 400 mm radius from the shaft centre and the oscillations occur at the rate of 100 per 499 s. Calculate the radius of gyration of the wheel alone about its own centre, and estimate the period of oscillation, with the same off-balance loading but with the wheel rolling on its own 625-mm radius rim instead of on the 150-mm diameter shaft. (*U. Glas.*) (*Ans.:* 0·545 m; 7·39 s)

14. A uniform rod of length 400 mm and diameter 25 mm is used as a pendulum, being fitted with a light knife-edge at a distance of 100 mm from the upper end of the rod. The knife-edge rests upon a fixed horizontal surface, forming the pivot of the pendulum. Find the periodic time of the pendulum for small oscillations.

If the knife-edge, instead of being made sharp, were rounded to a radius of 7 mm, what would be the percentage change in the periodic time of the pendulum? Assume that the point of contact with the horizontal surface is 100 mm from the upper end of the rod, as before. (*U. Lond.*)

(*Ans.:* 0·97 s; 3·34%)

15. An armature of mass 100 kg is mounted on an 80-mm diameter shaft. In order to find the eccentricity of the centre of gravity of the armature, the shaft is placed horizontally on horizontal knife-edges and set into small rolling oscillations. The periodic time of small oscillations is found to be 5·5 s. The radius of gyration about an axis through the centre of gravity is 150 mm. Calculate the centrifugal force which would be set up by the armature when rotating at 1500 rev/min. (*U. Lond.*) (*Ans.:* Eccentricity = 3·17 mm; 7·82 kN)

16. An unbalanced turbine wheel fixed to a shaft of 120 mm diameter, the total mass being 180 kg, is arranged for balancing with the shaft resting on parallel ways. When disturbed slightly from rest it oscillates with a period of 12 s. It may be assumed that the unbalance is in the relatively heavy rim and at a mean radius of 450 mm. After balancing, by the removal of the necessary material from the rim, the rotor is mounted in ball bearings and when a mass of 2 kg is clamped at a radius of 450 mm the period of oscillation is 10·5 s. Determine the mass of material removed to effect balance and the moment of inertia of the balanced rotor. Any formula used should be established.

(*U. Lond.*) (*Ans.:* 1·546 kg; 24·24 kg m²)

17. The bar shown in Fig. 13.16 has a mass of 5 kg and k^2 about the c.g. is 0·055 m². Find the natural period of oscillation. (*U. Lond.*) (*Ans.:* 0·967 s)

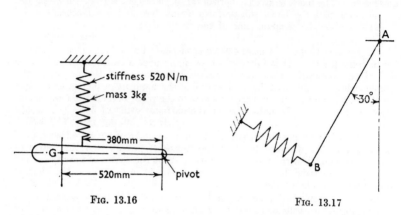

FIG. 13.16   FIG. 13.17

18. In Fig. 13.17, AB is a uniform rod 400 mm long and of mass 4 kg which is pivoted on a fixed pin at A. A tension spring is attached at the end B of the rod. In the position of equilibrium the rod is inclined at 30° to the vertical and the axis of the spring is perpendicular to the rod. The stiffness of the spring is 45 N/m. Neglecting friction, calculate the natural frequency of vibration of the system for oscillations of small amplitude. (*U. Lond., modified.*)

(*Ans.:* 1·29 Hz)

(b) Two degrees of freedom

19. *Fig. 13.18 shows a system consisting of two bodies and three springs that may be set into free vertical vibration.*

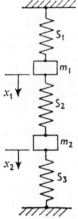

Determine the natural frequencies of this system and compare the amplitudes of the movements of m_1 and m_2 for each frequency. Work from first principles.

$$m_1 = 70 \ kg, \quad m_2 = 140 \ kg, \quad S_1 = 60 \ kN/m,$$
$$S_2 = 45 \ kN/m, \quad S_3 = 30 \ kN/m$$

(U. Lond.)

Let the instantaneous displacements of m_1 and m_2 be x_1 and x_2 respectively.

Then the equation of motion of m_1 is

$$m_1 \frac{d^2x_1}{dt^2} = -S_1x_1 + S_2(x_2 - x_1) \qquad . \qquad (1)$$

and the equation of motion of m_2 is

$$m_2 \frac{d^2x_2}{dt^2} = -S_2(x_2 - x_1) - S_3x_2 \qquad . \qquad (2)$$

FIG. 13.18

Assuming that $x_1 = a_1 \cos \omega t$ and $x_2 = a_2 \cos \omega t$, equation (1) becomes

$$-m_1 a_1 \omega^2 \cos \omega t = -S_1 a_1 \cos \omega t + S_2 (a_2 \cos \omega t - a_1 \cos \omega t)$$

i.e. $\quad \{m_1 \omega^2 - (S_1 + S_2)\} a_1 = -S_2 a_2 \quad . \quad . \quad . \quad . \quad . \quad (3)$

and equation (2) becomes

$$-m_2 a_2 \omega^2 \cos \omega t = -S_2 (a_2 \cos \omega t - a_1 \cos \omega t) - S_3 a_2 \cos \omega t$$

i.e. $\quad \{m_2 \omega^2 - (S_2 + S_3)\} a_2 = -S_2 a_1 \quad . \quad . \quad . \quad . \quad . \quad (4)$

Eliminating a_1 and a_2 between equations (3) and (4),

$$\frac{m_1 \omega^2 - (S_1 + S_2)}{S_2} = \frac{S_2}{m_2 \omega^2 - (S_2 + S_3)}$$

from which

$$\omega^4 - \left\{ \frac{S_1 + S_2}{m_1} + \frac{S_2 + S_3}{m_2} \right\} \omega^2 + \frac{S_1 S_2 + S_2 S_3 + S_3 S_1}{m_1 m_2} = 0$$

i.e. $\qquad \omega^4 - \left\{ \frac{60 + 45}{70} + \frac{45 + 30}{140} \right\} \times 10^3 \omega^2$

$$+ \left\{ \frac{60 \times 45 + 45 \times 30 + 30 \times 60}{70 \times 140} \right\} \times 10^6 = 0$$

which reduces to $\qquad \omega^4 - 2035 \omega^2 + 597\,000 = 0$

from which $\qquad \omega = 18 \cdot 87$ and $40 \cdot 98$ rad/s

$$\therefore \; n = \underline{3 \cdot 0 \text{ and } 6 \cdot 52 \text{ Hz}}$$

From equation (3), $\dfrac{a_1}{a_2} = \dfrac{-S_2}{m_1 \omega^2 - (S_1 + S_2)} = \dfrac{-45 \times 10^3}{70 \omega^2 - 105 \times 10^3}$

When $\qquad \omega = 18 \cdot 87$ rad/s, $\dfrac{a_1}{a_2} = \underline{0 \cdot 562}$

and when $\quad \omega = 40 \cdot 98$ rad/s, $\dfrac{a_1}{a_2} = \underline{-3 \cdot 59}$

These ratios show that the masses are in phase at the lower (fundamental) frequency and 180° out-of-phase at the higher frequency.

20. *A simple pendulum of length l_1 has a bob of mass m_1. Another simple pendulum of length l_2 and having a bob of mass m_2 is suspended from m_1. Show that the two natural frequencies of oscillation are given by the equation*

$$\frac{\omega^4}{g^2} - \left(\frac{l_1 + l_2}{l_1 l_2}\right)\left(\frac{m_1 + m_2}{m_1}\right)\frac{\omega^2}{g} + \frac{m_1 + m_2}{m_1 l_1 l_2} = 0$$

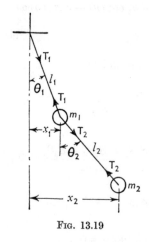

FIG. 13.19

Let the *small* angular displacements of the upper and lower pendula be θ_1 and θ_2 respectively and let the corresponding linear displacements of m_1 and m_2 be x_1 and x_2 respectively, Fig. 13.19. Then, if T_1 and T_2 are the tensions in the two strings, the equation of motion of m_1 is

$$m_1 \frac{d^2 x_1}{dt^2} = -T_1 \theta_1 + T_2 \theta_2 \quad . \qquad . \quad (1)$$

and the equation of motion of m_2 is

$$m_2 \frac{d^2 x_2}{dt^2} = -T_2 \theta_2 \qquad . \qquad . \quad (2)$$

but $\qquad\qquad x_1 \simeq l_1 \theta_1 \quad \text{and} \quad x_2 - x_1 \simeq l_2 \theta_2$

also $\qquad\qquad T_2 \simeq m_2 g \quad \text{and} \quad T_1 \simeq m_1 g + T_2 \simeq (m_1 + m_2)g$

$$\therefore \; m_1 \frac{d^2 x_1}{dt^2} = -\frac{(m_1 + m_2)g}{l_1}x_1 + \frac{m_2 g}{l_2}(x_2 - x_1) \quad . \qquad . \quad (3)$$

and $\qquad m_2 \frac{d^2 x_2}{dt^2} = -\frac{m_2 g}{l_2}(x_2 - x_1) \quad . \qquad . \qquad . \qquad . \qquad . \quad (4)$

Assuming that $x_1 = a_1 \cos \omega t$ and $x_2 = a_2 \cos \omega t$, equation (3) becomes

$$-m_1 a_1 \omega^2 \cos \omega t = -\frac{(m_1 + m_2)g}{l_1}a_1 \cos \omega t + \frac{m_2 g}{l_2}(a_2 \cos \omega t - a_1 \cos \omega t)$$

i.e. $\qquad m_1 \omega^2 \frac{a_1}{a_2} = \frac{(m_1 + m_2)g}{l_1}\frac{a_1}{a_2} - \frac{m_2 g}{l_2}\left(1 - \frac{a_1}{a_2}\right) \quad . \qquad . \quad (5)$

and equation (4) becomes

$$-m_2 a_2 \omega^2 \cos \omega t = -\frac{m_2 g}{l_2}(a_2 \cos \omega t - a_1 \cos \omega t)$$

i.e.
$$\frac{\omega^2}{g}l_2 = 1 - \frac{a_1}{a_2} \qquad . \qquad . \qquad . \qquad . \qquad . \qquad (6)$$

Substituting for $\dfrac{a_1}{a_2}$ in equation (5),

$$m_1\omega^2\left(1 - \frac{\omega^2}{g}l_2\right) = \frac{(m_1 + m_2)g}{l_1}\left(1 - \frac{\omega^2}{g}l_2\right) - m_2\omega^2$$

i.e.
$$-m_1 l_2 \frac{\omega^4}{g^2} + \left\{(m_1 + m_2)\left(1 + \frac{l_2}{l_1}\right)\right\}\frac{\omega^2}{g} - \frac{m_1 + m_2}{l_1} = 0$$

i.e.
$$\frac{\omega^4}{g^2} - \left(\frac{l_1 + l_2}{l_1 l_2}\right)\left(\frac{m_1 + m_2}{m_1}\right)\frac{\omega^2}{g} + \frac{m_1 + m_2}{m_1 l_1 l_2} = 0$$

NOTE: The ratio of the amplitudes may be obtained from equation (6)

21. *A uniform beam of length 2a and mass m is supported horizontally on two equal springs, each of stiffness S, at points distant b and c from the middle. Derive the equation of the two frequencies of small oscillations in the vertical plane. Show that these frequencies are equal only if*

$$b = c = a/\sqrt{3}. \qquad \text{(U. Lond.)}$$

Let the linear displacement of the beam be x and the angular displacement of the beam be θ, Fig. 13.20.

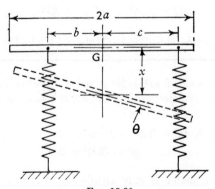

FIG. 13.20

The restoring force $= S(x - b\theta) + S(x + c\theta)$

$$= 2Sx - S\theta(b - c)$$

$$\therefore \; m\frac{d^2x}{dt^2} = -2Sx + S\theta(b - c) \qquad . \qquad (1)$$

The restoring moment about G $= -S(x - b\theta)b + S(x + c\theta)c$

$$= -Sx(b - c) + S\theta(b^2 + c^2)$$

$$\therefore \; \frac{m(2a)^2}{12}\frac{d^2\theta}{dt^2} = Sx(b - c) - S\theta(b^2 + c^2) . \qquad (2)$$

Assuming that $x = k \cos \omega t$ and $\theta = \phi \cos \omega t$, equation (1) becomes

$$-mk\omega^2 \cos \omega t = -2Sk \cos \omega t + S\phi(b - c) \cos \omega t$$

i.e. $\qquad \{-m\omega^2 + 2S\}k = S\phi(b - c) \qquad . \qquad . \qquad . \qquad . \qquad (3)$

and equation (2) becomes

$$-\frac{ma^2}{3}\phi\omega^2 \cos \omega t = Sk(b - c) \cos \omega t - S\phi(b^2 + c^2) \cos \omega t$$

i.e. $\qquad \left\{-\frac{ma^2}{3}\omega^2 + S(b^2 + c^2)\right\}\phi = Sk(b - c) \qquad . \qquad . \qquad (4)$

Eliminating k and ϕ between equations (3) and (4):

$$\frac{-m\omega^2 + 2S}{S(b - c)} = \frac{S(b - c)}{-\dfrac{ma^2}{3}\omega^2 + S(b^2 + c^2)}$$

i.e. $\dfrac{m^2a^2}{3}\omega^4 - \left\{\dfrac{ma^2}{3}.2S + mS(b^2 + c^2)\right\}\omega^2 + 2S^2(b^2 + c^2) = S^2(b - c)^2$

i.e. $\qquad m^2a^2\omega^4 - mS\{2a^2 + 3b^2 + 3c^2\}\omega^2 + 3S^2(b + c)^2 = 0$

$$\therefore \; \omega^2 = \frac{S}{2ma^2}\{(2a^2+3b^2+3c^2)\pm\sqrt{[4a^4+9b^4+9c^4+18b^2c^2-24a^2bc]}\}$$

The roots are equal if $4a^4 + 9b^4 + 9c^4 + 18b^2c^2 - 24a^2bc = 0$.

This equation is symmetrical in b and c and hence can only be satisfied when $b = c$.

Replacing c by b, $4a^4 + 36b^4 - 24a^2b^2 = 0$

i.e. $\qquad\qquad (a^2 - 3b^2)^2 = 0 \qquad\qquad \therefore \; b = \frac{a}{\sqrt{3}}$

Therefore the equation is only satisfied by $b = c = \dfrac{a}{\sqrt{3}}$.

22. *In the system shown in Fig. 13.21, the mass m is carried by a spring of stiffness S_2. This spring is attached to a belt, passing over a pulley of total moment of inertia I_1 and of effective radius R_1. The belt is attached at the other end to another spring, of stiffness S_1 which, in turn, is fastened to a firm anchorage. It may be assumed that there is never any slip between the pulley and the belt. Concentric with the pulley and attached firmly to it, is a pinion with 20 teeth. This pinion engages with a wheel having 70 teeth, of moment of inertia I_2.*

In the instance considered, S_1 and S_2 are respectively 15 kN/m and 9 kN/m; I_1 and I_2 are respectively 12 kg m² and 24·5 kg m²; R_1 and m are respectively 400 mm and 100 kg.

Derive expressions for the two fundamental frequencies of vibration of this system, and evaluate these for the particular values given. (U. Lond.)

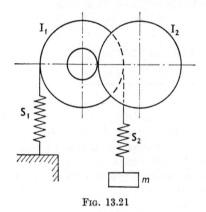

FIG. 13.21

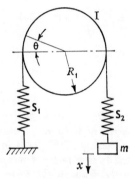

FIG. 13.22

Equivalent moment of inertia of system at pulley,

$$I = I_1 + \left(\frac{20}{70}\right)^2 \times I_2$$

$$= 12 + \frac{4}{49} \times 24·5 = 14 \text{ kg m}^2$$

Let the linear displacement of the mass be x and the angular displacement of the pulley be θ, Fig. 13.22.

Then the equation of motion of the mass is

$$m\frac{d^2x}{dt^2} = -(x - R_1\theta)S_2 \qquad . \qquad . \qquad . \qquad . \qquad (1)$$

and the equation of motion of the pulley is

$$I\frac{d^2\theta}{dt^2} = \{(x - R_1\theta)S_2 - R_1\theta S_1\}R_1 \qquad . \qquad . \qquad (2)$$

Assuming that $x = a \cos \omega t$ and $\theta = \phi \cos \omega t$, equation (1) becomes

$$-ma\omega^2 = (a - R_1\phi)S_2$$

i.e.
$$(m\omega^2 - S_2)a = -R_1S_2\phi \quad . \quad . \quad . \quad . \quad (3)$$

and equation (2) becomes

$$-I\phi\omega^2 = \{(a - R_1\phi)S_2 - R_1\phi S_1\}R_1$$

i.e.
$$\{I\omega^2 - R_1^2(S_1 + S_2)\}\phi = -R_1S_2a \quad . \quad . \quad . \quad (4)$$

Eliminating a and ϕ between equations (3) and (4),

$$(m\omega^2 - S_2)\{I\omega^2 - R_1^2(S_1 + S_2)\} = R_1^2S_2^2$$

or
$$Im\omega^4 - \{IS_2 + mR_1^2(S_1 + S_2)\}\omega^2 + R_1^2S_1S_2 = 0$$

i.e. $14 \times 100\omega^4 - \{14 \times 9 \times 10^3 + 100 \times 0{\cdot}4^2(15 + 9) \times 10^3\}\omega^2$
$$+ 0{\cdot}4^2 \times 15 \times 9 \times 10^6 = 0$$

i.e.
$$\omega^4 - 364\omega^2 + 15\,430 = 0$$

$$\therefore \ \omega = 6{\cdot}98 \text{ and } 17{\cdot}75 \text{ rad/s}$$

$$\therefore \ n = \underline{1{\cdot}11 \text{ Hz and } 2{\cdot}825 \text{ Hz}}$$

---

23. *A load having a mass of 500 kg is suspended by a light inextensible rope from a drum having an effective diameter of 1·2 m and a moment of inertia of 85 kg m². In order to reduce shock loading, a light spring having a stiffness of 35 kN/m is inserted between the rope and the load. Initially the system is at rest. A motor then applies a constant torque of 4 kN m directly to the drum in the sense to raise the load.*

Derive an expression for the tension in the rope after a time t. What will be the maximum tension induced in the rope? (U. Lond.)

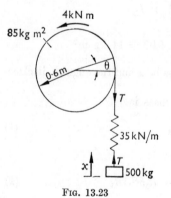

FIG. 13.23

Let T be the tension in the spring and x and θ be the displacements of the mass and drum respectively from the initial position, Fig. 13.23.

Then the equation of motion of the drum is

$$85\frac{d^2\theta}{dt^2} = 4 \times 10^3 - 0{\cdot}6T \quad . \quad . \quad (1)$$

and the equation of motion of the mass is

$$500\frac{d^2x}{dt^2} = T - 500g \quad . \quad . \quad . \quad (2)$$

Also $T = 500g + 35 \times 10^3(0{\cdot}6\theta - x)$ (3)

Substituting for T in equations (1) and (2),

$$\frac{85d^2\theta}{dt^2} = 4 \times 10^3 - 0.6\{500g + 35 \times 10^3(0.6\theta - x)\}$$

or
$$\frac{d^2\theta}{dt^2} = 12.43 - 247(0.6\theta - x) \qquad . \qquad . \qquad . \qquad . \qquad (4)$$

and
$$500\frac{d^2x}{dt^2} = \{500g + 35 \times 10^3(0.6\theta - x)\} - 500g$$

or
$$\frac{d^2x}{dt^2} = 70(0.6\theta - x) \qquad . \qquad . \qquad . \qquad . \qquad . \qquad (5)$$

Multiplying equation (4) by 0.6 and subtracting equation (5),

$$0.6\frac{d^2\theta}{dt^2} - \frac{d^2x}{dt^2} = 7.46 - 218.2(0.6\theta - x)$$

or
$$\frac{d^2u}{dt^2} + 218.2u = 7.46 \quad \text{where } u = 0.6\theta - x$$

$$\therefore \quad u = A \cos 14.77t + B \sin 14.77t + 0.0341$$

When $t = 0$, $\dfrac{du}{dt} = 0$, $\quad \therefore B = 0$

When $t = 0$, $u = 0$, $\quad \therefore A = -0.0341$

$$\therefore u = 0.0341(1 - \cos 14.77t)$$

Therefore, from equation (3),

$$T = 500 \times 9.81 + 35 \times 10^3 \times 0.0341(1 - \cos 14.77t)$$
$$= \underline{6099 - 1194 \cos 14.77t \text{ N}}$$

The maximum value of T occurs when $\cos 14.77t = -1$,

i.e. $\qquad T_{max} = 6099 + 1194 = \underline{7293 \text{ N}}$

24. *An electric motor is direct-coupled to a machine by a solid uniform shaft of 25 mm diameter and length 300 mm. The revolving parts of the motor and the driven machine have masses 250 kg and 200 kg respectively and the corresponding radii of gyration are 225 mm and 350 mm.*

The assembly is rotating freely at 240 rev/min, no current being supplied to the motor and no power being absorbed by the machine, when the current is suddenly switched on and off again. During this very short period of time the motor receives an accelerating torsional impulse of 10 N m s.

Working from the equations of motion, determine

(a) the maximum angle of twist induced in the shaft,

(b) the angular velocity of the motor and that of the machine at a time t after the impulse,

(c) the smallest value of t for which the speed of the motor is again 240 rev/min.

Assume that bearing friction is negligible and that the modulus of rigidity of the shaft is 80 GN/m².   (U. Lond.)

Referring to Fig. 13.24, let I_1 represent the motor and I_2 the generator.

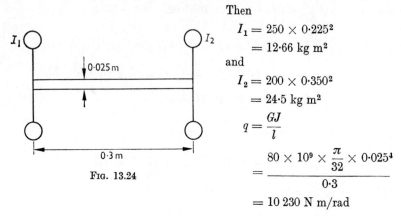

FIG. 13.24

Then

$$I_1 = 250 \times 0.225^2$$
$$= 12.66 \text{ kg m}^2$$

and

$$I_2 = 200 \times 0.350^2$$
$$= 24.5 \text{ kg m}^2$$

$$q = \frac{GJ}{l}$$

$$= \frac{80 \times 10^9 \times \dfrac{\pi}{32} \times 0.025^4}{0.3}$$

$$= 10\ 230 \text{ N m/rad}$$

When the current is switched off, the rotors oscillate freely and their equations of motion are

$$I_1 \frac{d^2\theta_1}{dt^2} = -q(\theta_1 - \theta_2)$$

and

$$I_2 \frac{d^2\theta_2}{dt^2} = q(\theta_1 - \theta_2)$$

Let ϕ = angle of twist, $\theta_1 - \theta_2$

Then $\dfrac{d^2\theta_1}{dt^2} = -\dfrac{q}{I_1}\phi$ and $\dfrac{d^2\theta_2}{dt^2} = \dfrac{q}{I_2}\phi$

$$\therefore \frac{d^2\phi}{dt^2} = -q\left(\frac{1}{I_1} + \frac{1}{I_2}\right)\phi = -\omega^2\phi \quad \text{where } \omega^2 = q\left(\frac{1}{I_1} + \frac{1}{I_2}\right)$$

$$\therefore \frac{d^2\phi}{dt^2} + \omega^2\phi = 0$$

$$\therefore \phi = A \cos \omega t + B \sin \omega t$$

$$\omega^2 = 10\ 230\left(\frac{1}{12.66} + \frac{1}{24.5}\right) = 1225$$

$$\therefore \omega = 35 \text{ rad/s}$$

When $t = 0$, $\phi = 0$ $\therefore A = 0$

Since the impulse is applied to rotor I_1, that rotor acquires an instantaneous increase in velocity of $\dfrac{10}{12\cdot66}$ rad/s relative to rotor I_2,

i.e. when $t = 0$, $\quad \dfrac{d\phi}{dt} = \dfrac{10}{12\cdot66} = 35B$

$$\therefore B = 0\cdot022\ 55$$

$$\therefore \phi = 0\cdot022\ 55 \sin 35t$$

$\therefore$ amplitude of $\phi = 0\cdot022\ 55$ rad $= \underline{1\cdot294°}$

$$\frac{d^2\theta_1}{dt^2} = -\frac{q}{I_1}\phi = -\frac{10\ 230}{12\cdot66} \times 0\cdot022\ 55 \sin 35t$$

$$= -18\cdot25 \sin 35t$$

$$\therefore \frac{d\theta_1}{dt} = \frac{18\cdot25}{35} \cos 35t + C_1 = 0\cdot521 \cos 35t + C_1$$

When $t = 0$, $\dfrac{d\theta_1}{dt} = 240 \times \dfrac{2\pi}{60} + \dfrac{10}{12\cdot66} = 25\cdot924$ rad/s

$$\therefore 25\cdot924 = 0\cdot521 + C_1$$

$$\therefore C_1 = 25\cdot403$$

$$\therefore \frac{d\theta_1}{dt} = \underline{25\cdot403 + 0\cdot521 \cos 35t}$$

$$\frac{d^2\theta_2}{dt^2} = \frac{q}{I_2}\phi = \frac{10\ 230}{24\cdot5} \times 0\cdot022\ 55 \sin 35t$$

$$= 9\cdot42 \sin 35t$$

$$\therefore \frac{d\theta_2}{dt} = -\frac{9\cdot42}{35} \cos 35t + C_2 = -0\cdot269 \cos 35t + C_2$$

When $t = 0$, $\dfrac{d\theta_2}{dt} = 240 \times \dfrac{2\pi}{60} = 25\cdot136$ rad/s

$$\therefore 25\cdot136 = -0\cdot269 + C_2$$

$$\therefore C_2 = 25\cdot406$$

$$\therefore \frac{d\theta_2}{dt} = \underline{25\cdot406 - 0\cdot269 \cos 35t}$$

When $\dfrac{d\theta_1}{dt} \doteqdot 25\cdot136$ rad/s again,

$$25\cdot136 = 25\cdot406 + 0\cdot521 \cos 35t$$

from which $\quad t = \underline{0\cdot0602}$ s

25. A mass m_1 of 3 kg is suspended from a spring of stiffness 600 N/m and to the underside of the mass m_1 another spring of stiffness 500 N/m is attached. This second spring carries a mass m_2 of 2 kg hanging vertically below m_1.

Assuming that the movements of the two masses x_1 and x_2 are of the form $x_1 = a_1 \sin \omega t$ and $x_2 = a_2 \sin \omega t$ respectively, show that one mode of vertical vibration corresponds to a pacer speed of 13·33 rad/s. (*U. Lond.*)

26. A mass m_1 is suspended from a fixed point by means of a spring of stiffness S_1. Attached to m_1 by means of a second spring of stiffness S_2 is another mass m_2. When the system is set in free vibration, if the inertia of the springs be neglected, show that

$$m_1 m_2 \omega^4 - \{m_1 S_2 + m_2(S_1 + S_2)\}\omega^2 + S_1 S_2 = 0$$

where ω is the phase velocity.

If m_1 and m_2 are each 225 kg, and S_1 and S_2 are 240 and 120 kN/m respectively, find the frequencies of oscillation and the ratio of the amplitudes of m_1 and m_2. (*U. Lond.*)   (*Ans.:* 2·812 and 6·79 Hz; 0·349 and −2·79)

27. An electric hoist A is carried at the free end of a horizontal cantilever beam as shown in Fig. 13.25, and supports a load B by means of a single rope. The effective length of the cantilever, measured to the centre of the hoist and rope, is 1·5 m. The hoist A has mass of 250 kg, the load B has mass of 1000 kg, and the mass of the beam can be neglected.

The second moment of area of the beam section about the horizontal axis is 15×10^{-6} m⁴ and the modulus of elasticity of the material 200 GN/m². The rope is elastic and has a stiffness of $3·5 \times 10^6$ N/m.

(*a*) Working from first principles, find the two frequencies corresponding to the two modes of vibration of the hoist and load.

(*b*) If the system is vibrating in the mode with the higher frequency and the amplitude of motion of the load B is $\pm0·25$ mm, calculate the amplitude of motion of the hoist A and the maximum tension in the rope. (*U. Lond.*)

(*Ans.:* 5·94 Hz and 26·08 Hz; 1·664 mm; 16·51 kN)

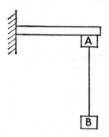

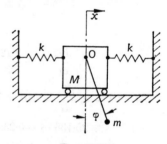

FIG. 13.25   FIG. 13.26

28. Fig. 13.26 shows a trolley, of mass M, which runs on a frictionless horizontal plane. At O the trolley carries a simple pendulum of length l with a body of mass m at its ends.

Two equal springs, each of stiffness k, are attached to the trolley and to the fixed walls. By using the independent co-ordinates x and ϕ as shown in the figure, deduce, for *small* free oscillations, the equations of motion:

$$(M + m)\ddot{x} + 2kx + ml\ddot{\phi} = 0$$
$$\ddot{x} + l\ddot{\phi} + g\phi = 0$$

Hence determine the natural frequencies of the system when $M = 10$ kg, $m = 0.5$ kg, $l = 0.3$ m and the stiffness of each spring is 720 N/m. (*U. Lond.*)

(*Ans.:* 1·925 Hz; 0·904 Hz)

29. A pendulum consists of a light string 0·6 m long and a bob of mass 1·5 kg, attached to the bob is a second string also of length 0·6 m with a bob of mass 0·5 kg at the lower end. Find the frequency of small oscillations and the ratios of the amplitudes of the motion of the bobs. (*U. Lond.*)

(*Ans.:* 0·525 Hz and 0·911 Hz; $+\frac{1}{3}$ or -1)

30. Two light rods AB and CD, Fig. 13.27, each of length 0·6 m, are hinged at A and C and carry concentrated masses of 1 kg each at B and D. The rods are coupled at their mid-points by a spring of stiffness 500 N/m. Calculate the two natural frequencies of small oscillations and the ratio of the amplitudes for the two modes of vibration. (*Ans.:* 0·644 Hz; 2·59 Hz; 1; -1)

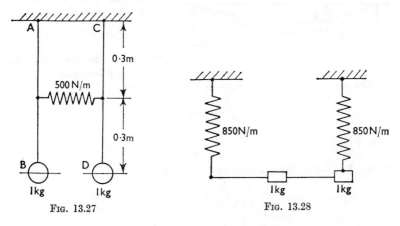

FIG. 13.27   FIG. 13.28

31. A light rod is supported at each end by two similar springs, each of stiffness 850 N/m, and carries two concentrated masses, each of mass 1 kg, one at one end and one in the centre, as shown in Fig. 13.28.

Determine the two natural frequencies of vibration of the system.

(*Ans.:* 4·06 Hz; 10·62 Hz)

32. A vehicle has a mass of 1 t and has a wheelbase of 2·4 m. The c.g. is 1 m in front of the back axle and the radius of gyration of the vehicle about the c.g. is 1 m. If each of the four springs has a stiffness of 12·5 kN/m, find the two natural frequencies of vibration. (*Ans.:* 1·082 Hz; 1·405 Hz)

33. A straight link AB, 0·9 m long, is supported horizontally by springs at A and B, of stiffness k_1 and k_2 respectively. The mass is 10 kg and the mass centre G is distant a from A and b from B. The moment of inertia about G is 0·6 kg m².

Find the relationship between k_1, k_2, a and b in order that one mode of free vibration shall be linear motion only and the other mode rotation only. If $a = 0.3$ m, and $k_1 = 14$ kN/m find the spring strength at B to give these two modes of vibration and calculate the two natural frequencies. (*U. Lond.*)

$$\left(Ans.:\ \frac{k_1}{k_2} = \frac{b}{a};\ 7\ \text{kN/m};\ 7\cdot32\ \text{Hz};\ 12\cdot64\ \text{Hz}\right)$$

CHAPTER 14

TRANSVERSE VIBRATIONS OF BEAMS

14.1 Single concentrated load. If the mass m, Fig. 14.1, is deflected

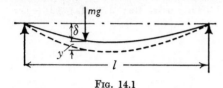

FIG. 14.1

a vertical distance y from its equilibrium position and released, the restoring force is $\dfrac{kEI.y}{l^3}$, where k depends on the position of the load and the end-fixing conditions.

The acceleration of the mass is then given by

$$\frac{kEIy}{l^3} = mf$$

from which

$$\frac{f}{y} = \frac{kEI}{ml^3} = \frac{g}{\delta}$$

where δ = static deflection under the load.

$$\therefore \ n = \frac{1}{2\pi}\sqrt{\frac{g}{\delta}} \ \text{Hz}$$

$$\simeq \frac{1}{2\sqrt{\delta}} \ \text{Hz} \quad \text{where } \delta \text{ is in metres} \quad . \qquad . \quad (14.1)$$

14.2 Effect of the mass of the beam. Assuming the shape of the vibrating beam to be similar to the static deflection curve for the concentrated load, the kinetic energy of the beam can be found in terms of the instantaneous velocity of the load. From this can be obtained the equivalent load to be added to the concentrated load to give the same total kinetic energy.

The principal cases are summarized below:

Case	Equivalent concentrated load
Simply supported beam, central load	$\frac{17}{35} \times$ mass of beam
Built-in beam, central load	$\frac{13}{35} \times$ mass of beam
Cantilever, end load	$\frac{33}{140} \times$ mass of beam

This approximation is only justified when the mass of the beam is small in comparison with that of the concentrated load.

336

14.3 Uniformly distributed load. (*a*) **Exact solution.** If y is the instantaneous displacement of a point on the beam (being a function of both x and t), then the inertia force per unit length is $m\dfrac{\partial^2 y}{\partial t^2}$, m being the mass per unit length. The restoring force in the beam per unit length is $EI\dfrac{\partial^4 y}{\partial x^4}$ so that

$$EI\frac{\partial^4 y}{\partial x^4} = -m\frac{\partial^2 y}{\partial t^2}$$

The negative sign arises because the inertia force $m\dfrac{\partial^2 y}{\partial t^2}$ is opposite in direction to the acceleration $\dfrac{\partial^2 y}{\partial t^2}$.

Since the beam is vibrating with S.H.M., $y = a \cos \omega t$ where a is the amplitude of the vibration.

$$\therefore \ \frac{\partial^4 y}{\partial x^4} = \frac{d^4 a}{dx^4} \cos \omega t$$

and

$$\frac{\partial^2 y}{\partial t^2} = -\omega^2 a \cos \omega t$$

$$\therefore \ EI\frac{d^4 a}{dx^4} \cos \omega t = -m(-\omega^2 a \cos \omega t)$$

i.e.

$$EI\frac{d^4 a}{dx^4} = m\omega^2 a$$

or

$$\frac{d^4 a}{dx^4} - \gamma^4 a = 0, \quad \text{where } \gamma^4 = \frac{m\omega^2}{EI}$$

This may be written

$$(D^4 - \gamma^4)a = 0 \quad \text{where } D \text{ represents } \frac{d}{dx}$$

or

$$(D^2 + \gamma^2)(D^2 - \gamma^2)a = 0$$

so that

$$D = \pm i\gamma \quad \text{and} \quad \pm\gamma$$

Hence

$$a = C_1 e^{i\gamma x} + C_2 e^{-i\gamma x} + C_3 e^{\gamma x} + C_4 e^{-\gamma x}$$

which may be expressed in the form

$$a = A \cos \gamma x + B \sin \gamma x + C \cosh \gamma x + D \sinh \gamma x$$

The constants are then solved from the appropriate end conditions and hence γ is obtained.

Then

$$\omega = \gamma^2 \sqrt{\frac{EI}{m}} = (\gamma l)^2 \sqrt{\frac{g}{k\Delta}}$$

where Δ is the maximum static deflection of the beam.

$$\therefore n = \frac{1}{2\pi}\omega \simeq \frac{1}{2\sqrt{\Delta}} \times \frac{(\gamma l)^2}{\sqrt{k}} \text{ Hz} \quad \text{where } \Delta \text{ is in metres.}$$

The principal cases are as follows:

Case	γl	k	n (Hz)
Simply supported beam	π	$\dfrac{384}{5}$	$\dfrac{0 \cdot 563}{\sqrt{\Delta}}$
Built-in beam	$\simeq \dfrac{3\pi}{2}$	384	$\dfrac{0 \cdot 571}{\sqrt{\Delta}}$
Cantilever	$1 \cdot 875$	8	$\dfrac{0 \cdot 621}{\sqrt{\Delta}}$

(b) **Approximate solution (Energy method).** Let the static deflection of an element of beam dx, distant x from one end, be y and the amplitude of vibration be a, Fig. 14.2. Then assuming that the shape of

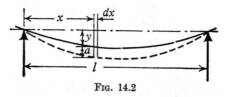

Fig. 14.2

the vibrating beam is similar to the static deflection curve, $a = cy$ where c is a constant.

If m is the mass of the beam per unit length, the additional load required to deflect the element a further distance a is $mg \, dx \, a/y$.

$$\text{Work done by this load moving through distance } a = \tfrac{1}{2}\left(mg \, dx \cdot \frac{a}{y}\right)a$$

$$= \tfrac{1}{2}mgc^2 y \, dx$$

$$\therefore \text{ total work done} = \tfrac{1}{2}mgc^2 \int_0^l y \, dx \, *$$

$$= \text{energy stored in beam at maximum displacement}$$

If ω is the angular speed of the point generating the S.H.M., the velocity of the element as it passes through its mid-position $= \omega a$.

$$\therefore \text{ kinetic energy of element} = \frac{m \, dx}{2}(\omega a)^2 = \frac{m}{2}\omega^2 c^2 y^2 \, dx$$

$$\therefore \text{ total kinetic energy of beam} = \frac{m}{2}\omega^2 c^2 \int_0^l y^2 \, dx$$

* The work done due to the change in potential energy is accounted for by a change in strain energy in the beam and does not affect the vibration.

Equating the kinetic energy of the beam in the mid-position to the strain energy at the maximum displacement, since the total energy of the system is constant,

$$\frac{m}{2}\omega^2 c^2 \int_0^l y^2 \, dx = \frac{mg}{2}c^2 \int_0^l y \, dx$$

i.e.

$$\omega = \sqrt{\left(g \int_0^l y \, dx \Big/ \int_0^l y^2 \, dx \right)}$$

i.e.

$$n = \frac{1}{2\pi} \sqrt{\left(g \int_0^l y \, dx \Big/ \int_0^l y^2 \, dx \right)} \text{ Hz}$$

$$\simeq \tfrac{1}{2} \sqrt{\left(\int_0^l y \, dx \Big/ \int_0^l y^2 \, dx \right)} \text{ Hz}$$

where y is in metres   (14.2)

The appropriate expression for y must then be substituted in this equation. The principal cases are as follows:

Case	y	n (Hz)
Simply supported beam .	$\dfrac{mg}{24EI}(x^4 - 2lx^3 + l^3x)$	$\dfrac{9\cdot87}{2}\sqrt{\dfrac{EI}{mgl^4}} = \dfrac{0\cdot564}{\sqrt{\Delta}}$
Built-in beam . . .	$\dfrac{mg}{24EI}x^2(l - x)^2$	$\dfrac{22\cdot4}{2}\sqrt{\dfrac{EI}{mgl^4}} = \dfrac{0\cdot572}{\sqrt{\Delta}}$
Cantilever . . .	$\dfrac{mg}{24EI}(6l^2x^2 - 4lx^3 + x^4)$	$\dfrac{3\cdot51}{2}\sqrt{\dfrac{EI}{mgl^4}} = \dfrac{0\cdot624}{\sqrt{\Delta}}$

NOTE. The origin for x is taken at either end for a simply supported or built-in beam and at the fixed end for a cantilever.

14.4 Several concentrated loads. (a) **Energy method.** Assuming that the shape of the vibrating beam is similar to the static deflection curve, equation (14.2) becomes, in the case of concentrated loads,

$$n = \tfrac{1}{2} \sqrt{\frac{\sum my}{\sum my^2}} \text{ Hz} \quad . \quad . \quad . \quad . \quad (14.3)$$

The value of y under each load is calculated with all the loads *acting together*.

The mass of the beam may be taken into account by adding to each concentrated load the mass of the beam between the centres of the sections into which the loads divide the beam, or by utilizing the formulae in paragraph 14.2.

The value of n from equations (14.2) and (14.3) will not be seriously affected if the shape of the vibrating beam is assumed to be any other reasonable curve, such as a sine curve or parabola. This is known as *Rayleigh's Principle*.

(b) **Dunkerley's empirical method.** If n_1, n_2, etc., are the frequencies of vibration of the concentrated loads *when each load acts on the beam alone* and n_b is the frequency of vibration of the beam under its own weight,

then $n_1 = \dfrac{1}{2\sqrt{\delta_1}}$, $n_2 = \dfrac{1}{2\sqrt{\delta_2}}$, etc., and $n_b = \dfrac{0 \cdot 563}{\sqrt{\Delta}}$ (if simply supported),

and n the frequency of vibration of the whole system, is given by

$$\frac{1}{n^2} = \frac{1}{n_1{}^2} + \frac{1}{n_2{}^2} + \ldots + \frac{1}{n_b{}^2}$$

i.e.
$$n = \frac{1}{2\sqrt{\left(\delta_1 + \delta_2 + \ldots + \dfrac{\Delta}{1 \cdot 27}\right)}} \text{ Hz} \qquad . \qquad . \quad (14.4)$$

where δ_1, δ_2, etc., are the static deflections at each load when acting alone and Δ is the maximum static deflection of the beam under its own weight.

1. *A beam of I section has a span of 3 m and is supported at the ends. The mass of the beam is 200 kg/m and the second moment of area of the section is 16×10^{-6} m^4. Two equal loads of one tonne are carried at points 1 m from each support.*

Find the natural frequency of transverse vibration of the system, if $E = 200$ GN/m^2.   (I. Mech. E.)

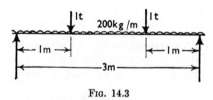

Fig. 14.3

Let $\delta =$ static deflection under each concentrated load (considered separately), Fig. 14.3,

and $\Delta =$ maximum static deflection of beam due to its own weight (considered separately),

then $\delta = \dfrac{Wa^2b^2}{3EIl}$ where W divides the span l into two parts a and b

$$= \frac{1 \times 10^3 \times 9 \cdot 81 \times 1^2 \times 2^2}{3 \times 200 \times 10^9 \times 16 \times 10^{-6} \times 3} = 0 \cdot 001\ 362 \text{ m}$$

and $\quad \Delta = \dfrac{5}{384}\dfrac{wl^4}{EI}$

$$= \frac{5 \times 200 \times 9 \cdot 81 \times 3^4}{384 \times 200 \times 10^9 \times 16 \times 10^{-6}} = 0 \cdot 000\ 646\ \text{m}$$

Using Dunkerley's formula,

$$n = \frac{1}{2\sqrt{\left(2\delta + \dfrac{\Delta}{1 \cdot 27}\right)}} \qquad . \qquad . \qquad . \qquad \text{from equation (14.4)}$$

$$= \frac{1}{2\sqrt{\left(2 \times 0 \cdot 001\ 362 + \dfrac{0 \cdot 000\ 646}{1 \cdot 27}\right)}} = 8 \cdot 8\ \text{Hz}$$

2. *A steel beam is simply supported over a span of 3·6 m and carries loads of 5 t at the centre and 3 t at 0·9 m from each end. Calculate the frequency of transverse vibration by the energy method, assuming that the vibrating form is (a) similar to the static deflection form and (b) sinusoidal. State how it is possible to tell which of these results is the more accurate.*

The static deflection at the 5 t load is 1·20 mm and at the 3 t load is 0·84 mm. (U. Glas.)

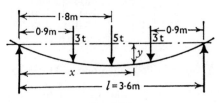

Fig. 14.4

(*a*) By the energy method, $n = \dfrac{1}{2}\sqrt{\dfrac{\Sigma\,my}{\Sigma my^2}}$ Hz from equation (14.3)

m (t)	y (mm)	my	my^2
3	0·84	2·52	2·12
5	1·20	6·00	7·20
3	0·84	2·52	2·12

$$\therefore \sum my = 11\cdot04 \text{ tonne mm} = 11\cdot04 \text{ kg m}$$

and $$\sum my^2 = 11\cdot44 \text{ tonne mm}^2 = 0\cdot011\ 44 \text{ kg m}^2$$

hence $$n = \frac{1}{2}\sqrt{\frac{11\cdot04}{0\cdot011\ 44}} = \underline{15\cdot52 \text{ Hz}}$$

(b) If the deflection curve is sinusoidal,

then $$y = y_0 \sin\frac{\pi}{l}x, \quad \text{where } y_0 \text{ is the maximum deflection.}$$

In this case, $y_0 = 1\cdot20$ mm

$$\therefore y = 1\cdot20 \sin\frac{\pi}{3\cdot6}x$$

$$\therefore \text{ deflection under 3 t load} = 1\cdot20 \sin\frac{\pi}{3\cdot6} \times 0\cdot9$$

$$= 0\cdot848 \text{ mm}$$

m (t)	y (mm)	my	my^2
3	0·848	2·544	2·16
5	1·20	6·00	7·20
3	0·848	2·544	2·16

$$\therefore \sum my = 11\cdot088 \text{ tonne mm} = 11\cdot088 \text{ kg m}$$

and $$\sum my^2 = 11\cdot52 \text{ tonne mm}^2 = 0\cdot011\ 52 \text{ kg m}^2$$

hence $$n = \frac{1}{2}\sqrt{\frac{11\cdot088}{0\cdot011\ 52}} = \underline{15\cdot51 \text{ Hz}}$$

If a shape is assumed for the vibrating beam which differs from the true shape, additional constraints would be necessary to cause the beam to take up this assumed shape. These additional constraints would stiffen the beam and hence increase the frequency. Thus the more unlikely the beam shape chosen, the higher will be the value of the frequency obtained. The smallest approximate result by the energy method will therefore be the most accurate.

For proof, see *Mechanical Vibrations*, J. P. Den Hartog, p. 200.

3. *A 50-mm diameter steel shaft AB, 2·5 m long, is supported in two short bearings 1·8 m apart, one being at the end A of the shaft. The shaft carries three concentrated loads as indicated:*

| Load in kg | . | . | . | 80 | 160 | 40 |
| Distance from A in m | | . | | 0·6 | 1·2 | 2·4 |

Calculate the deflection at each load and hence obtain a first approximation to the fundamental frequency of transverse vibration of the loaded shaft. Neglect the mass of the shaft. Take $E = 200$ GN/m². (U. Lond.)

At the fundamental frequency, the form of the vibrating beam will be as shown in Fig. 14.5 so that the inertia force on the 40-kg load will be

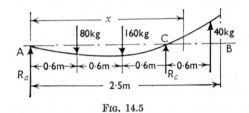

Fig. 14.5

in the opposite direction to those on the other two loads. In calculating the static deflection, therefore, the direction of the 40-kg load should be reversed to give the correct shape of the vibrating beam at the fundamental frequency.

$$EI = 200 \times 10^9 \times \frac{\pi}{64} \times 0·05^4 = 61\ 350 \text{ N m}^2$$

$$R_c = \frac{(80 \times 0·6 + 160 \times 1·2 - 40 \times 2·4) \times 9·81}{1·8} = 80 \times 9·81 \text{ N}$$

$$R_a = (80 + 160 - 40 - 80) \times 9·81 = 120 \times 9·81 \text{ N}$$

By Macaulay's method,

$$EI\frac{d^2y}{dx^2} = \{-120x + 80[x - 0·6] + 160[x - 1·2] - 80[x - 1·8]\} \times 9·81$$

Integrating twice and inserting the conditions that $y = 0$ when $x = 0$ and $y = 0$ when $x = 1·8$,

$$y = \frac{20 \times 9·81}{61\ 350}\{-x^3 + \tfrac{2}{3}[x - 0·6]^3 + \tfrac{4}{3}[x - 1·2]^3 - \tfrac{2}{3}[x - 1·8]^3 + 2·44x\} \text{ m}$$

When $\qquad x = 0.6$ m, $\quad y = \quad 0.004$ m

$\qquad\qquad\qquad x = 1.2$ m, $\quad y = \quad 0.0043$ m

$\qquad\qquad\qquad x = 2.4$ m, $\quad y = -0.00615$ m

m (kg)	y (m)	my	my^2
80	0.0040	0.320	0.001 28
160	0.0043	0.688	0.002 96
40	0.006 15	0.246	0.001 51

$$\therefore \; \sum my = 1.254 \quad \text{and} \quad \sum my^2 = 0.005\ 75$$

$$\therefore \text{ from equation (14.3) } n = \frac{1}{2}\sqrt{\frac{1.254}{0.005\ 75}} = \underline{7.37 \text{ Hz}}$$

NOTE. No account is taken of the negative deflection under the 40-kg load. In the numerator, both the load and the deflection are negative, giving a positive product. and in the denominator, the negative signs have no significance since the term my^2 represents the kinetic energy of the mass as it passes through its mid-position.

4. *In order to examine the effect of lateral vibration on a certain instrument, a vibrating table was made as shown in Fig. 14.6, which gives* (a) *the front elevation and* (b) *the side elevation. The four supporting strips have a cross-section of 50 mm by 10 mm and an effective length of 300 mm and their ends may be assumed to be rigidly fixed in direction.*

Estimate the natural frequency of vibration if the total effective mass of the platform is 28 kg. Take $E = 200$ GN/m².   (U. Lond.)

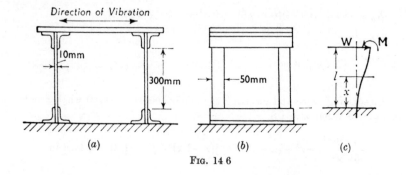

FIG. 14 6

The frequency of vibration is given by $n = \dfrac{1}{2\sqrt{\delta}}$, where δ is the static deflection under a *transverse* load of $28g$ N.

Considering a cantilever, Fig. 14.6(c), length l and concentrated end load W with a moment M applied to hold the free end vertical,

$$EI\frac{d^2y}{dx^2} = W(l-x) - M$$

$$\therefore EI\frac{dy}{dx} = W\left(lx - \frac{x^2}{2}\right) - Mx + A$$

when $\qquad x = 0,\ \dfrac{dy}{dx} = 0 \quad \therefore A = 0$

when $\qquad x = l,\ \dfrac{dy}{dx} = 0 \quad \therefore M = \dfrac{Wl}{2}$

$$\therefore EIy = W\left(\frac{lx^2}{2} - \frac{x^3}{6}\right) - \frac{Wl}{4}x^2 + B$$

when $\qquad x = 0,\ y = 0 \quad \therefore B = 0$

when $\qquad x = l,\ y = \dfrac{Wl^3}{12EI}$

In this case, $\qquad \delta = \dfrac{28 \times 9\cdot81 \times 0\cdot3^3 \times 12}{12 \times 200 \times 10^9 \times 4 \times 0\cdot05 \times 0\cdot01^3}$

$\qquad\qquad\qquad = 0\cdot000\ 1855$ m, taking the least value of I

$$\therefore n = \frac{1}{2\sqrt{0\cdot000\ 1855}} \qquad . \qquad . \qquad \text{from equation (14.1)}$$

$$= \underline{36\cdot7 \text{ Hz}}$$

5. *A steel cantilever shaft, 450 mm long, is 50 mm diameter for a length of 300 mm from the fixed end and 40 mm diameter for the remaining 150 mm. Find the natural frequency of transverse vibration using an arc of a cosine curve as an approximation to the vibrating form. Density of steel = 7·8 Mg/m³. E = 200 GN/m².*   (U. Glas.)

The static deflection of the free end will be found to be 9·65 μm, Fig. 14.7.

Hence, if the vibration form is assumed to be a cosine curve, its equation is

$$y = 9\cdot65 \times 10^{-6}\left(1 - \cos\frac{\pi}{0\cdot9}x\right)$$

M

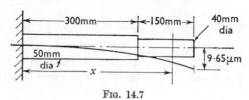

Fig. 14.7

$$\int_0^l my\,dx = \int_0^{0\cdot3}\left(\frac{\pi}{4}\times0\cdot05^2\times1\times7\cdot8\times10^3\right)\times9\cdot65\times10^{-6}$$

$$\times\left(1-\cos\frac{\pi}{0\cdot9}x\right)dx+\int_{0\cdot3}^{0\cdot45}\left(\frac{\pi}{4}\times0\cdot04^2\times1\times7\cdot8\times10^3\right)$$

$$\times9\cdot65\times10^{-6}\left(1-\cos\frac{\pi}{0\cdot9}x\right)dx$$

$$=9\cdot65\times10^{-6}\left\{15\cdot32\left[x-\frac{0\cdot9}{\pi}\sin\frac{\pi}{0\cdot9}x\right]_0^{0\cdot3}\right.$$

$$\left.+9\cdot81\left[x-\frac{0\cdot9}{\pi}\sin\frac{\pi}{0\cdot9}x\right]_{0\cdot3}^{0\cdot45}\right\}$$

$$=18\cdot07\times10^{-6}\text{ kg m}$$

$$\int_0^l my^2\,dx = \int_0^{0\cdot3}\left(\frac{\pi}{4}\times0\cdot05^2\times1\times7\cdot8\times10^3\right)\times(9\cdot65\times10^{-6})^2$$

$$\times\left(1-\cos\frac{\pi}{0\cdot9}x\right)^2dx+\int_{0\cdot3}^{0\cdot45}\left(\frac{\pi}{4}\times0\cdot05^2\times1\times7\cdot8\times10^3\right)$$

$$\times(9\cdot65\times10^{-6})^2\left(1-\cos\frac{\pi}{0\cdot9}x\right)^2dx$$

$$=(9\cdot65\times10^{-6})^2\left\{15\cdot32\left[\frac{3}{2}x-\frac{1\cdot8}{\pi}\sin\frac{\pi}{0\cdot9}x+\frac{0\cdot9}{4\pi}\sin\frac{\pi}{0\cdot45}x\right]_0^{0\cdot3}\right.$$

$$\left.+9\cdot81\left[\frac{3}{2}x-\frac{1\cdot8}{\pi}\sin\frac{\pi}{0\cdot9}x+\frac{0\cdot9}{4\pi}\sin\frac{\pi}{0\cdot45}x\right]_{0\cdot3}^{0\cdot45}\right\}$$

$$=0\cdot1007\times10^{-9}\text{ kg m}^2$$

From equation (14.2),

$$n=\frac{1}{2}\sqrt{\frac{\displaystyle\int_0^l my\,dx}{\displaystyle\int_0^l my^2\,dx}}=\frac{1}{2}\sqrt{\frac{18\cdot07\times10^{-6}}{0\cdot1007\times10^{-9}}}=\underline{\underline{211\cdot8\text{ Hz}}}$$

6. *A thin uniform disc 200 mm diameter is mounted at the end of a hori-zontal shaft 20 mm diameter and 250 mm long, the other end of which is firmly fixed. The disc mass, m, is 6 kg, and the mass of the shaft can be neglected. When set into transverse vibration, the shaft is assumed to take up the same curve as when deflected statically by the end load, m, the end slope and deflection being ½mgl²/EI and ⅓mgl³/EI respectively. Using an energy method which includes the kinetic energy of rotation of the disc about a dia-meter, determine the frequency.* $E = 200 \ GN/m^2$. (I. Mech. E.)

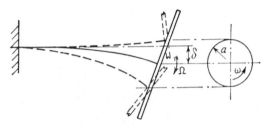

FIG. 14.8

Let δ = static deflection under the load, Fig. 14.8

a = amplitude of vibration under the load

ω = angular speed of the point generating the S.H.M.

θ = slope of disc in the equilibrium position

Ω = angular speed of the disc about its diameter in the equili-brium position.

Then $\qquad \theta = \dfrac{mgl^2}{2EI} \quad$ and $\quad \delta = \dfrac{mgl^3}{3EI} \quad \therefore \ \theta = \dfrac{3}{2l} \times \delta$

$\therefore \ \Omega$ = rate of change of the slope of the disc as it passes through the equilibrium position

$\qquad = \dfrac{3}{2l} \times$ rate of change of the displacement of the disc as it passes through the equilibrium position

$\qquad = \dfrac{3}{2l} \times \omega a$

$\therefore$ angular K.E. of disc in equilibrium position $= \tfrac{1}{2}I\Omega^2 = \dfrac{mk^2}{2}\left(\dfrac{3\omega a}{2l}\right)^2$

and linear K.E. of disc in equilibrium position $= \tfrac{1}{2}mv^2 = \dfrac{m}{2}(\omega a)^2$

$\therefore$ total K.E. of disc in equilibrium position $= \dfrac{m}{2}\omega^2 a^2\left(1 + \dfrac{9}{4}\dfrac{k^2}{l^2}\right)$ (1)

The additional load required to deflect the free end through a further distance $a = mg \times \dfrac{a}{\delta}$. Therefore the strain energy in the shaft is this additional load multiplied by the average distance moved, i.e. strain energy in shaft at maximum displacement $= \frac{1}{2}\left(\dfrac{mga}{\delta}\right)a$

$$= \frac{mga^2}{2\delta} \qquad . \qquad . \qquad . \qquad (2)$$

Equating (1) and (2), $\quad \dfrac{\omega^2}{g}\left(1 + \dfrac{9}{4}\dfrac{k^2}{l^2}\right) = \dfrac{1}{\delta}$

i.e.

$$\omega = \sqrt{\frac{g}{\delta\left(1 + \dfrac{9}{4}\dfrac{k^2}{l^2}\right)}}$$

but $\qquad \delta = \dfrac{mgl^3}{3EI} = \dfrac{6 \times 9{\cdot}81 \times 0{\cdot}25^3}{3 \times 200 \times 10^9 \times \dfrac{\pi}{64} \times 0{\cdot}02^4} = 0{\cdot}000\,195$ m

and $\qquad k^2 = \dfrac{r^2}{4} = \dfrac{0{\cdot}1^2}{4} = 0{\cdot}0025$ m^2

$$\therefore \; n = \frac{1}{2\pi}\sqrt{\frac{9{\cdot}81}{0{\cdot}000\,195\left(1 + \dfrac{9}{4} \times \dfrac{0{\cdot}0025}{0{\cdot}25^2}\right)}} = \underline{34{\cdot}1\ \text{Hz}}$$

NOTE: If the angular K.E. of the disc is neglected, $n = \underline{35{\cdot}7\ \text{Hz}}$

7. *Show that the transverse vibrations of a beam are governed by the equation*

$$\frac{\partial^4 y}{\partial x^4} + \mu^4 \frac{\partial^2 y}{\partial t^2} = 0 \quad where \; \mu^4 = \frac{\rho}{gEk^2}$$

k being the radius of gyration of the cross-section of the beam and ρ the density of the metal.

Hence show that for a cantilever of length l, the frequency of vibrations is given by

$$\omega = \frac{r_0^2}{l^2}\sqrt{\frac{gEk^2}{\rho}} \quad where \; r_0 \; is \; a \; root \; of \quad \cos r \cosh r = -1$$

(U. Lond.)

The fundamental equation for transverse vibrations of a beam is

$$EI\frac{\partial^4 y}{\partial x^4} = -m\frac{\partial^2 y}{\partial t^2}, \quad where \; m = \text{mass per unit length}$$

i.e.
$$\frac{\partial^4 y}{\partial x^4} = -\frac{\rho}{gEk^2}\frac{\partial^2 y}{\partial t^2}$$

i.e.
$$\frac{\partial^4 y}{\partial x^4} + \mu^4 \frac{\partial^2 y}{\partial t^2} = 0 \quad \text{where } \mu^4 = \frac{\rho}{gEk^2}.$$

Let $y = a \cos \omega t$ where a is the amplitude of the vibration at any point distant x from the fixed end.

Then
$$\frac{\partial^4 y}{\partial x^4} = \frac{d^4 a}{dx^4}\cos \omega t \quad \text{and} \quad \frac{\partial^2 y}{\partial t^2} = -\omega^2\, a \cos \omega t$$

$$\therefore \frac{d^4 a}{dx^4}\cos \omega t - \mu^4 \omega^2 a \cos \omega t = 0$$

i.e.
$$\frac{d^4 a}{dx^4} - \gamma^4 a = 0 \quad \text{where } \gamma^4 = \mu^4 \omega^2$$

$$\therefore a = A \cos \gamma x + B \sin \gamma x + C \cosh \gamma x + D \sinh \gamma x$$

when $\quad x = 0, \quad a = 0 \quad \therefore \ 0 = A + C$ . . . . (1)

when $\quad x = 0, \dfrac{da}{dx} = 0 \quad \therefore \ 0 = B + D$ . . . . (2)

when $\quad x = l, \dfrac{d^2 a}{dx^2} = 0$

$$\therefore 0 = -A \cos \gamma l - B \sin \gamma l + C \cosh \gamma l + D \sinh \gamma l \qquad (3)$$

when $\quad x = l, \dfrac{d^3 a}{dx^3} = 0$

$$\therefore 0 = A \sin \gamma l - B \cos \gamma l + C \sinh \gamma l + D \cosh \gamma l \quad . \qquad (4)$$

From (1), $C = -A$ and from (2), $D = -B$

$\therefore$ (3) becomes $\qquad -A(\cos \gamma l + \cosh \gamma l) = B(\sin \gamma l + \sinh \gamma l)$

and (4) becomes $\qquad A(\sin \gamma l - \sinh \gamma l) = B(\cos \gamma l + \cosh \gamma l)$

Hence, by division, $\qquad \dfrac{-\cos \gamma l - \cosh \gamma l}{\sin \gamma l - \sinh \gamma l} = \dfrac{\sin \gamma l + \sinh \gamma l}{\cos \gamma l + \cosh \gamma l}$

i.e. $\quad -\cos^2 \gamma l - 2 \cos \gamma l \cosh \gamma l - \cosh^2 \gamma l = \sin^2 \gamma l - \sinh^2 \gamma l$

i.e. $\qquad\qquad\qquad -2 \cos \gamma l \cosh \gamma l = 1 + 1$

or $\qquad\qquad\qquad\qquad \cos r \cosh r = -1 \quad \text{where } r = \gamma l$

If r_0 is the root of this equation, $r_0 = \gamma l = \mu \sqrt{(\omega l)}$

i.e.
$$\omega = \frac{r_0^2}{\mu^2 l^2} = \frac{r_0^2}{l^2}\sqrt{\frac{gEk^2}{\rho}}$$

(Unless otherwise stated, take $E = 200$ GN/m²)

8. A horizontal cantilever of length l is clamped at one end and carries a mass m at the other. Derive an expression for the time period of vibration of the cantilever when the load is given a small vertical displacement. The equation

$$t = 2\pi \sqrt{\frac{\text{displacement}}{\text{acceleration}}}$$

may be assumed, and the mass of the cantilever itself is to be neglected.

A horizontal flat steel strip 12 mm wide by 6 mm thick is clamped at one end with the 12 mm side horizontal, and carries a mass of 0·5 kg at the free end. Find the distance of the mass from the fixed end if the frequency of the natural vibrations is 50 Hz. Neglect the mass of the strip. (*U. Lond.*)

(Ans.: 139·6 mm)

9. A strip of steel 10 mm wide and 0·8 mm thick rests on knife-edged supports 200 mm apart and a body of mass 0·15 kg is fixed to the strip at mid-span. Find the natural time of vibration of the strip. If the greatest bending stress in the strip during a vibration is 100 MN/m², what is the amplitude of movement of the body and also the least pressure on each support ? Neglect the mass of the strip. (*U. Lond.*)

(Ans.: 0·1073 s ; 1·29 mm ; 0·404 N)

10. Obtain an expression for the natural frequency of transverse vibrations for a bar of length l carrying a uniformly distributed load m kg/m, and simply supported at the ends. Assume that the vibration deflection curve is similar to the static deflection curve.

Determine the natural frequency of a beam of I-section 6 m long, simply supported at the ends, which carries a uniformly distributed load of 750 kg/m. The second moment of area of the cross-section is 140×10^{-6} m⁴. (*U. Lond.*)

(Ans.: 8·46 Hz)

11. A beam carrying masses m_1, m_2, m_3, . . . at points along its length is deflected statically through distances y_1, y_2, y_3, . . . at these points. If, when in transverse vibration, it is assumed that the form is similar, prove that the frequency is

$$n = \frac{1}{2\pi} \sqrt{\frac{g \, \Sigma \, my}{\Sigma \, my^2}}$$

A shaft 50 mm diameter resting freely on supports $l = 0·8$ m apart carries three loads each of 36 kg, one at the centre and one 0·2 m from each end. Calculate the frequency of transverse vibration. It may be assumed for the loaded shaft that the deflection under the central load is $\dfrac{19}{384} \dfrac{mgl^3}{EI}$ and under each end load $\dfrac{9}{256} \dfrac{mgl^3}{EI}$. (*I. Mech. E.*)

(Ans.: 45·5 Hz)

12. A beam 6 m long, simply supported at each end, carries a load of 6 t/m run, extending from a point 1·2 m from one support to another point 4·2 m from the same support. EI for the beam is 110 MN m². The mass of the beam itself may be neglected.

Determine the frequency of transverse vibrations of the beam in a vertical plane,

(*a*) very roughly, assuming the whole load to be concentrated at its centre of gravity,

(*b*) more accurately, by treating the load as three equal concentrated loads applied at suitable points. (Dunkerley's method is suggested.) (*U. Lond.*)

(Ans.: 5·94 Hz ; 6·4 Hz)

13. Obtain from first principles an expression for the fundamental natural frequency of transverse vibration of a cantilever of length l m and mass m per m, it being assumed that the vibration deflection curve is of the same form as the static deflection curve. Hence find the frequency of transverse vibration of a steel turbine blade of uniform section, 125 mm long, having a mass of 2 kg/m length and least moment of inertia 2700 mm⁴. Ignore the effect of centrifugal loading. (*U. Lond.*)

$$(Ans.: \ 1\cdot76 \ \sqrt{\frac{EI}{mgl^4}} \ \text{Hz, } EI \text{ in m units ; 590 Hz})$$

14. A shaft is simply supported on bearings 3 m apart and carries five equal concentrated loads equally spaced with the end loads 0·3 m from each bearing. If the maximum static deflection is 2·5 mm, estimate the frequency of transverse vibration of the shaft when the static deflection curve is assumed to be (*i*) a sine wave and (*ii*) a parabola. (*U. Lond.*)   (*Ans.:* 11·37 Hz ; 11·30 Hz)

15. A steel joist has a span of 6 m and the ends are simply supported. The joist has a mass of 52 kg/m and the second moment of area of the cross-section is 120×10^{-6} m⁴. A load of 3 t is carried at a point 2·4 m from one support. Find from first principles the frequency of the natural transverse vibrations of the joist, if the effect of its own mass is neglected. Then find the approximate frequency when the mass of the joist is taken into account. (*I. Mech. E.*)
(*Ans.:* 7·01 Hz ; 6·82 Hz [$\delta = 5\cdot09$ mm ; $\Delta = 0\cdot3585$ mm])

16. A beam with second moment of area of cross-section 8×10^{-6} m⁴ and of mass 18 kg/m is simply supported at the end of a 3·6 m span and it carries a body of mass 250 kg at the centre. Find the frequency of transverse vibrations. (*I. Mech. E.*)   (*Ans.:* 12·2 Hz [$\delta = 1\cdot49$ mm ; $\Delta = 0\cdot241$ mm])

17. Two parallel steel beams span an opening of 6 m and jointly support a central load of 3 t. Each beam has a mass of 45 kg/m run and the moment of inertia of the section is 86×10^{-6} m⁴. Assuming the beams are simply supported at the ends, calculate the natural frequency of the transverse vibrations. (*U. Lond.*)   (*Ans.:* 7·72 Hz [$\delta = 3\cdot85$ mm ; $\Delta = 0\cdot433$ mm])

18. A small, imperfectly balanced machine is mounted on a rigid horizontal plate, supported on four vertical legs each 28 mm outside diameter, 25 mm inside diameter and 0·9 m long, rigidly welded to the plate but having their other extremities always position fixed but direction fixed or not at will. If the effective mass of the assembly is 45 kg, find the five lowest machine speeds which would give resonance corresponding to the five different conditions of the legs. (*U. Lond.*)   (*Ans.* 9·06, 8·16, 7·17, 5·99, 4·53 Hz)

19. A uniform beam, of mass 31 kg/m run, is simply supported on a span of 3·6 m. Taking EI for the beam as 7 MN m², calculate the frequency of transverse vibrations.
This frequency is to be reduced by 40% by fixing three equal masses to the beam, at the mid-point and the quarter points. Calculate how much these masses should be. (*U. Lond.*)   (*Ans.:* 57·75 Hz ; 45·8 kg)

20. A beam, 3 m long, is simply supported at the ends and carries a distributed load of 3 t/m with a concentrated load of 0·5 t at the middle. The second moment of area of the section about the neutral axis is 9×10^{-6} m⁴. Estimate the value of the lowest natural frequency of transverse vibration. (*U. Lond.*)   (*Ans.:* 4·07 Hz [$\delta = 1\cdot533$ mm ; $\Delta = 17\cdot25$ mm])

21. A uniform beam AB is simply supported on a span of 3·6 m. The beam itself has mass 18 kg/m run, and the relevant value of EI is 1·75 MN m². The beam carries an additional uniformly distributed mass of 180 kg/m run, between two points C and D which are respectively 0·9 m and 3·3 m from A, also two concentrated masses of 115 kg and 215 kg at E and F, which are respectively 1·2 m and 2·1 m from A.

Using such approximations as you deem justifiable estimate the lowest frequency of transverse vibrations of the beam. (The use of Dunkerley's method is suggested.) (*U. Lond.*) (*Ans.:* 8·4 Hz)

22. Assuming that the curve of deflection of a simply supported uniform beam when vibrating transversely is a parabola, find an expression for the frequency.

A steel shaft 150 mm diameter and 3 m long carries a load of 1·8 Mg, 1 m from one end. Find approximately the time of transverse vibration if the shaft runs in spherical bearings at the ends.

Steel has a density of 7·8 Mg/m³. (*U. Glas.*)

$$\left(Ans.:\ \frac{0·559}{\sqrt{\delta_{max}}};\ 0·0848\ s\right)$$

23. A light horizontal cantilever ABC of length $5a$ is made from two lengths of tube, AB = $2a$ and BC = $3a$, joined at B. The axes of the tubes are coincident and the end A is fixed. For the tube AB, the second moment of area of a normal section about a diameter is $2I$; the corresponding quantity for the tube BC is I. Young's modulus for the material of the tubes is E. The cantilever carries three concentrated bodies of masses $2M$, M and M at distances from A of a, $3a$ and $5a$ respectively.

By using any approximate method which is deemed justifiable, obtain an estimate for the fundamental frequency of transverse vibration of the system in the form

$$n = \frac{k}{2\pi}\sqrt{\frac{EI}{Ma^3}}$$

and determine the value of the constant k. (*U. Lond.*) (*Ans.:* 0·579)

24. A uniform beam of length l and total mass M is simply supported at its ends and carries a concentrated mass $2M$ at mid-span. The flexural rigidity of the beam is EI.

Assuming that during free transverse vibration in the first mode the dynamic deflection curve of the beam is half a sine wave, derive an expression for the corresponding frequency of vibration in the form

$$\text{frequency} = \frac{\lambda}{2\pi}\sqrt{\frac{EI}{Ml^3}}$$

and determine the value of λ. (*U. Lond.*) (*Ans.:* 2·156)

25. A uniform beam ABCD is freely supported at A and C and overhangs to D. It carries loads of 2 t at B and 1 t at D. AB = 2·4 m, BC = 2·4 m, CD = 0·9 m.

Find the frequency of transverse vibration allowing approximately for the beam, which has a mass of 320 kg. $I = 4·78 \times 10^{-6}$ m⁴. (*U. Glas.*)

(*Ans.:* 2·065 Hz [y_1 = 63·4 mm; y_2 = −43·85 mm])

26. A shaft ABCDE is supported in swivel bearings at A and D and over-hangs to E. Masses of 400 kg are at B and C and a mass of 100 kg at E. $I = 5.8 \times 10^{-6}$ m^4, AB $= 0.500$ m, BC $= 0.875$ m, CD $= 0.5$ m, DE $= 0.25$ m.

Find the frequency of transverse vibration. The frequency of the shaft may be taken as 56 Hz. (*U. Glas.*)

(*Ans.:* 20.45 Hz [$y_1 = 0.5425$ mm; $y_2 = 0.554$ mm; $y_3 = -0.337$ mm])

27. A beam ABCD is supported at A and D and carries two concentrated loads, each of mass 3 t, at B and C. Using the notation $\alpha_{XY} \equiv$ the deflection at station X produced by a unit load placed at station Y, the elastic properties of the beam are such that

$$\alpha_{BB} = 76.5 \ \mu\text{m per N}$$
$$\alpha_{BC} = \alpha_{CB} = 94.3 \ \mu\text{m per N}$$
$$\alpha_{CC} = 140.4 \ \mu\text{m per N}$$

Neglecting the mass of the beam itself, calculate the two frequencies of free oscillation of the system. (*U. Lond.*)   (*Ans.:* 6.35 and 9.83 Hz)

28. A steel shaft ABC, 50 mm diameter and 1.5 m long, is simply supported at bearings A and B, 1.2 m apart, and overhangs 0.3 m to C. It carries concen-trated masses of 18 kg at the end C and at the centre of span AB. Estimate the frequency of transverse vibration of the shaft, including an approximate allowance for the mass of the shaft, which is 16 kg/m. Use the energy method based on the static deflection curve. $E = 212$ GN/m^2. (*U. Glas.*)

(*Ans.:* 32.9 Hz [$y_1 = 0.225$ mm; $y_2 = -0.238$ mm])

29. A steel shaft overhangs its bearing by 1.2 m, the overhung portion being 50 mm diameter for 0.6 m length next to the bearing and 75 mm diameter for the remaining 0.6 m. Determine the frequency of transverse vibration of the shaft, treating it as a cantilever and using a parabolic arc as an approximation to the static deflection curve. Density of steel $= 7.8$ Mg/m^3. (*U. Glas.*)

(*Ans.:* 18.1 Hz [$\delta_{\max} = 1.21$ mm])

30. A cantilever beam is 3 m long, has a mass of 40 kg/m length, and an EI value of 2.78×10^6 N m^2. It carries a platform on the outer 1 m which, over this portion, increases the load to 136 kg/m and the EI value to 7.42×10^6 N m^2. Using the energy method and assuming the beam to vibrate in a parabolic form, estimate the natural frequency of transverse vibration. (*U. Glas.*)

(*Ans.:* 31 Hz [$\delta_{\max} = 0.376$ mm])

NOTE: In Examples 28 and 29, the mass of the beam has been allowed for by using the formulae of Art. 14.2.

CHAPTER 15

WHIRLING OF SHAFTS

15.1 Whirling of shafts. Let the eccentricity of the centre of gravity G of a disc of mass m attached to a shaft be e, measured from the disc centre O, Fig. 15.1, and let y be the deflection of the shaft axis at the disc, measured from the static deflection position, when rotating at a speed ω rad/s.

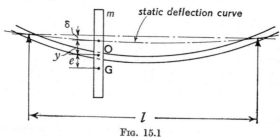

static deflection curve

FIG. 15.1

Then centrifugal force on the shaft $= m\omega^2(y + e)$ and the inward pull exerted by the shaft $= \dfrac{kEI}{l^3}.y$, where k depends upon the position of the load and the end fixing conditions.

Equating these forces, $m\omega^2(y + e) = \dfrac{kEI}{l^3}y$

$$\therefore y = \frac{e}{\dfrac{kEI}{m\omega^2 l^3} - 1} \qquad . \qquad . \qquad . \qquad (15.1)$$

When $\dfrac{kEI}{m\omega^2 l^3} = 1$, the deflection y becomes infinite and whirling takes place. The *whirling* or *critical* speed is therefore

$$\omega_c = \sqrt{\frac{kEI}{ml^3}} = \sqrt{\frac{g}{\delta}} \text{ rad/s}$$

where δ denotes the static deflection of the shaft at the disc,

or $\qquad n_c = \dfrac{1}{2\pi}\sqrt{\dfrac{g}{\delta}} \simeq \dfrac{1}{2\sqrt{\delta}} \text{ rev/s, where } \delta \text{ is in metres} \qquad . \quad (15.2)$

This is the same as the frequency of transverse vibration of the same shaft, and similarly, the whirling speed of the shaft for any other system

354

of loading is identical with the frequency of transverse vibration of the same shaft and loading.

If $\omega_c{}^2$ be substituted for $\dfrac{kEI}{ml^3}$ in equation (15.1), this becomes

$$y = \frac{\omega^2 e}{\omega_c{}^2 - \omega^2} \qquad . \qquad . \qquad . \qquad . \qquad (15.3)$$

This gives the deflection of the shaft from the static deflection position at any speed ω in terms of the critical speed.

When $\omega < \omega_c$, y and e have the same sign, i.e. G lies to the outside of O. When $\omega > \omega_c$, y and e are of opposite sign, i.e. G lies between the centre of the rotating shaft and the static deflection curve. As the speed increases beyond ω_c, $y \to -e$, which means that G finally coincides with the static deflection curve.* Fig. 15.2 shows the positions of O and G relative to the

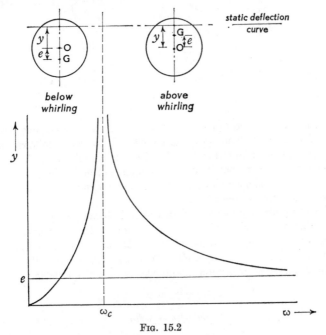

FIG. 15.2

static deflection curve and also the variation in the *numerical* value of y as ω varies from zero to above the whirling speed.

* A full treatment of the problem shows that at the whirling speed the radii to the centre of gravity of the disc and the centre of the rotating shaft are perpendicular and, as the speed increases beyond ω_c, the centre of gravity of the disc moves through a further 90° until it is again the plane of bending. See *Theory of Vibrations*, E. B. Cole, p. 316, or *Applied Mechanics for Engineers*, Prof. Sir Charles Inglis, p. 325.

15.2 Effect of the slope of the disc.

Fig. 15.3 shows a disc attached to a shaft AB, the slope of the disc due to the deflection of the shaft being θ.

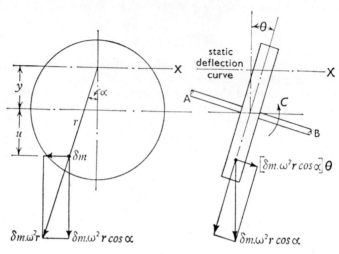

FIG. 15.3

When whirling takes place, the disc rotates about the static deflection curve XX and not about its own axis AB. It therefore follows that the centrifugal force on an element of the disc is perpendicular to XX, so that it has a component perpendicular to the plane of the disc tending to reduce the value of θ. The centrifugal force on such an element, of mass δm and radius r from XX, $= \delta m\, \omega^2 r$.

The horizontal component of this force is balanced by the horizontal force on a corresponding element to the right of the vertical axis.

The vertical component, $\delta m\, \omega^2 r \cos \alpha$, has a component perpendicular to the plane of the disc equal to $(\delta m\, \omega^2 r \cos \alpha) \times \theta$, assuming θ to be small.

The moment of this force about the horizonal diameter

$$= \delta m\, \omega^2 r \cos \alpha \times \theta \times u$$

$$= \delta m\, \omega^2 (y + u)\theta u$$

$\therefore$ total moment about the diameter, $C = \omega^2 \theta \{ y \sum u\, \delta m + \sum u^2\, \delta m \}$

but $\sum u\, \delta m = 0$ and $\sum u^2\, \delta m =$ moment of inertia of the disc about the diameter, I_d

$$\therefore C = I_d \omega^2 \theta* \qquad . \qquad . \qquad . \qquad . \qquad (15.4)$$

---

* This may also be obtained by considering the gyroscopic couple acting on the disc; see Example 7, p. 239.

The effect of C is to reduce the static deflection under the disc, δ, and hence to increase the critical speed. The whirling speed must therefore be calculated first without taking C into account to obtain an approximate value for ω and then a more correct solution obtained by including the effect of this approximate value for C. A second approximation for C could then be obtained, if required.

15.3 Effect of an end-thrust. Assume that the static deflection curve before the application of an end-thrust is a cosine curve. Then if the ordinate at any point on this curve is δ, Fig. 15.4,

$$\delta = \delta_{max} \cos \frac{\pi}{l} x$$

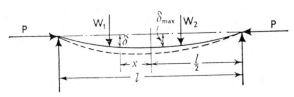

FIG. 15.4

When an end-thrust P is applied, the problem becomes equivalent to the bending of a long column with initial curvature. If the new deflection is Δ, and P_e is the Euler crippling load

$$\left(= \frac{\pi^2 EI}{l^2} \text{ for a simply supported shaft} \right),$$

then

$$\Delta = \frac{P_e}{P_e - P} \delta_{max} \cos \frac{\pi}{l} x \;^*$$

$$= \frac{P_e}{P_e - P} \delta$$

i.e.

$$\frac{\Delta}{\delta} = \frac{P_e}{P_e - P} = \frac{1}{1 - P/P_e} \qquad . \qquad . \qquad . \qquad (15.5)$$

Therefore the effect of P is to increase the static deflection at all points in the ratio :

$$\frac{1}{1 - P/P_e} : 1$$

and hence the whirling speed is reduced in the ratio $\sqrt{1 - P/P_e} : 1$.

* *Strength of Materials*, R. C. Stephens, Art. 7.6, Equation 7.11 (Arnold).

1. *Deduce an equation for the whirling speed of a light shaft which carries a single heavy disc at a given point along its length.*

A shaft 12 mm diameter is supported on spherically seated bearings 300 mm apart and carries a wheel of mass 9 kg at a point 125 mm from one bearing. Find the whirling speed.

If the wheel is mounted with its centre of mass initially 1·25 mm out of alignment with the shaft axis, find the deflection of the shaft when the speed of rotation is 85 rev/s. E = 200 GN/m². (I. Mech. E.)

For a single concentrated load which divides a simply supported beam of length l into two parts of length a and b, $\delta = \dfrac{Wa^2b^2}{3EIl}$

$$\therefore \delta = \frac{9 \times 9{\cdot}81 \times 0{\cdot}125^2 \times 0{\cdot}175^2}{3 \times 200 \times 10^9 \times \dfrac{\pi}{64} \times 0{\cdot}012^4 \times 0{\cdot}3} = 0{\cdot}000231 \text{ m}$$

$$\therefore n_c = \frac{1}{2\sqrt{\delta}} = \frac{1}{2\sqrt{0{\cdot}000\,231}} = \underline{33 \text{ rev/s}}$$

$$y = \frac{\omega^2 e}{\omega_c{}^2 - \omega^2} \qquad . \qquad . \qquad \text{from equation (15.3)}$$

$$= \frac{85^2 \times 1{\cdot}25}{33^2 - 85^2} = \underline{-1{\cdot}47 \text{ mm}}$$

The negative sign signifies that the centre of gravity of the disc lies between the axes of the rotating shaft and the axis through the bearings.

2. *Determine, with the aid of Dunkerley's equation, (a) the first and (b) the second whirling speed of a steel shaft, 25 mm diameter and 0·75 m long, freely supported at both ends, carrying a concentrated mass of 20 kg at the shaft centre.*

The density of the shaft material is 7·8 Mg/m³ and the value of E is 200 GN/m². (U. Lond.)

$$(a)\ n_c = \frac{1}{2\sqrt{\left(\delta + \dfrac{\Delta}{1{\cdot}27}\right)}} \text{ rev/s} \qquad . \qquad . \qquad \text{from equation (14.4)}$$

but $\quad \delta = \dfrac{Wl^3}{48EI} = \dfrac{20 \times 9{\cdot}81 \times 0{\cdot}75^3}{48 \times 200 \times 10^9 \times \dfrac{\pi}{64} \times 0{\cdot}025^4} = 0{\cdot}45 \times 10^{-3} \text{ m}$

and $\quad \Delta = \dfrac{5}{384}\dfrac{\omega l^4}{EI} = \dfrac{5 \times \left(\dfrac{\pi}{4} \times 0{\cdot}025^2 \times 7{\cdot}8 \times 10^3 \times 9{\cdot}81\right) \times 0{\cdot}75^4}{384 \times 200 \times 10^9 \times \dfrac{\pi}{64} \times 0{\cdot}025^4}$

$$= 0{\cdot}0404 \times 10^{-3} \text{ m}$$

$$\therefore n_c = \frac{1}{2\sqrt{\left(0.45 \times 10^{-3} + \dfrac{0.0404 \times 10^{-3}}{1.27}\right)}} = \underline{22.8 \text{ rev/s}}$$

(b) For the second whirling speed the shaft will take up the shape shown in Fig. 15.5. The central load, being located at the node, has no

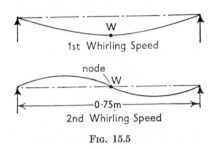

W

1st Whirling Speed

node

W

0.75m

2nd Whirling Speed

FIG. 15.5

effect on the frequency so that the mass of the shaft only has to be considered. The second whirling speed is therefore four times the fundamental frequency of the shaft when only the mass of the shaft is acting (this is the same as the fundamental frequency of a shaft of half the length carrying the same intensity of loading).

i.e.
$$n_c = \frac{1}{2\sqrt{\dfrac{\Delta}{1.27}}} \times 4 = \frac{1}{2\sqrt{\dfrac{0.0404 \times 10^{-3}}{1.27}}} \times 4$$

$$= \underline{355 \text{ rev/s}}$$

3. *A shaft, 12 mm diameter, rotates in spherical bearings with a span of 0·9 m, and carries a disc of mass 12 kg midway between the two bearings.*

Neglecting the mass of the shaft, determine the deflection of the shaft in terms of its speed of rotation in radians per second, if the mass-centre of the disc is 0·25 mm out of centre. $E = 200 \text{ GN/m}^2$.

If the stress in the shaft is not to exceed 100 MN/m², determine the range of speed within which it is unsafe to run the shaft. (U. Lond.)

From equation (15.1), $y = \dfrac{0.000\,25}{\dfrac{48 \times 200 \times 10^9 \times \pi \times 0.012^4}{12 \times \omega^2 \times 0.9^3 \times 64} - 1}$

$$= \frac{0.000\,25}{\dfrac{1117}{\omega^2} - 1} \text{ m} \qquad . \qquad . \qquad . \qquad . \qquad (1)$$

The maximum deflection of the shaft when carrying a central load P is given by

$$y = \frac{Pl^3}{48EI} = \frac{4Pl^3}{3E\pi d^4}$$

and the maximum stress is given by

$$\sigma = \frac{M}{Z} = \frac{Pl/4}{\pi d^3/32} = \frac{8Pl}{\pi d^3}$$

Hence, substituting for P in terms of y,

$$\sigma = \frac{6Ed}{l^2} \times y$$

Therefore when $\sigma = 100$ MN/m²,

$$y = \frac{100 \times 10^6 \times 0 \cdot 9^2}{6 \times 200 \times 10^9 \times 0 \cdot 012}$$

$$= 0 \cdot 005\ 625\ \text{m}$$

Since y and e may be of the same or opposite sign, from equation (1),

$$0 \cdot 005\ 625 = \pm \frac{0 \cdot 000\ 25}{\dfrac{1117}{\omega^2} - 1}$$

from which $\omega = 32 \cdot 7$ or $34 \cdot 2$ rad/s

i.e. $n = 5 \cdot 20$ or $5 \cdot 44$ rev/s

It is therefore unsafe to run the shaft between these speeds.

4. *A 75-mm diameter shaft overhangs the bearing by 500 mm when it enters the boss of a wheel 125 mm wide. Assuming the bearing gives directional restraint and that the outer 125 mm is inflexible, estimate the critical speed of rotation. The distributed mass from A to B is 22·5 kg and from B to C is 135 kg. E = 212 GN/m².*

                                             (U. Glas.)

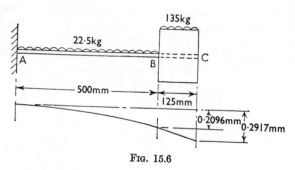

Fig. 15.6

It will be found that
the static deflection at B = 0·2096 mm, Fig. 15.6,
  and that the slope at B = 0·000 656 6 rad

$\therefore$ deflection at $C = 0\cdot2096 + 125 \times 0\cdot000\ 656\ 6 = 0\cdot2917$ mm

(a) Assuming that the vibrating form between A and B is a cosine curve,

then
$$y = 0\cdot2096 \times 10^{-3}\left(1 - \cos\frac{\pi}{1\cdot0}x\right) \text{ m}$$

From B to C the shaft slopes uniformly from $0\cdot2096$ mm to $0\cdot2917$ mm.

$$\int_0^l my\ dx = \int_0^{0\cdot5} \frac{22\cdot5}{0\cdot5} \times 0\cdot2096 \times 10^{-3}\left(1 - \cos\frac{\pi}{1\cdot0}x\right) dx$$
$$+ 135\left(\frac{0\cdot2096 + 0\cdot2917}{2}\right) \times 10^{-3}$$
$$= \left\{9\cdot45\left[x - \frac{1}{\pi}\sin\pi x\right]_0^{0\cdot5} + 33\cdot8\right\} \times 10^{-3}$$
$$= 35\cdot52 \times 10^{-3} \text{ kg m}$$

$$\int_0^l my^2\ dx = \int_0^{0\cdot5} \frac{22\cdot5}{0\cdot5} \times 0\cdot2096^2 \times 10^{-6}\left(1 - \cos\frac{\pi}{1\cdot0}x\right)^2 dx$$
$$+ 135\left(\frac{0\cdot2096 + 0\cdot2917}{2}\right)^2 \times 10^{-6}$$
$$= \left\{1\cdot98\left[\frac{3}{2}x - \frac{2}{\pi}\sin\pi x + \frac{0\cdot25}{\pi}\sin2\pi x\right]_0^{0\cdot5} + 8\cdot47\right\} \times 10^{-6}$$
$$= 8\cdot696 \times 10^{-6} \text{ kg m}^2$$

$$\therefore n_c = \frac{1}{2}\sqrt{\left(\int_0^l my\ dx \Big/ \int_0^l my^2\ dx\right)} \qquad \text{from equation (14.2)}$$
$$= \frac{1}{2}\sqrt{\frac{35\cdot52 \times 10^{-3}}{8\cdot696 \times 10^{-6}}} = \underline{31\cdot9 \text{ rev/s}}$$

(b) Assuming that the vibrating form between A and B is a parabola,

then
$$y = 0\cdot2096 \times 10^{-3}\left\{\frac{x}{0\cdot5}\right\}^2$$

$$\int_0^l my\ dx = \int_0^{0\cdot5} \frac{22\cdot5}{0\cdot5} \times \frac{0\cdot2096 \times 10^{-3}}{0\cdot25}x^2\ dx + 33\cdot8 \times 10^{-3}$$
$$= 35\cdot38 \times 10^{-3} \text{ kg m}$$

$$\int_0^l my^2\ dx = \int_0^{0\cdot5} \frac{22\cdot5}{0\cdot5} \times \frac{0\cdot2096^2 \times 10^{-6}}{0\cdot25^2}x^4\ dx + 8\cdot47 \times 10^{-6}$$
$$= 8\cdot67 \times 10^{-6} \text{ kg m}^2$$

$$\therefore n_c = \frac{1}{2}\sqrt{\frac{35\cdot38 \times 10^{-3}}{8\cdot67 \times 10^{-6}}} = \underline{31\cdot9 \text{ rev/s}}$$

5. *A light-alloy shaft is supported at each end in self-aligning bearings whose centres are 0·5 m apart. The diameter of the shaft at each end is 25 mm and at the mid-span it is 50 mm. The diameter varies symmetrically such that at distance x ($0 \leqslant x \leqslant 0.25$ m) from a bearing centre-line, the second moment of area of the shaft section about a diameter is given by $I_x = I_0/(1 - kx)$ where I_0 is the corresponding second moment of area at the origin and k is a constant. The shaft carries a body of mass 50 kg at mid-span, and the light-alloy has a Young's Modulus $E = 70$ GN/m².*

Neglecting the mass of the shaft, estimate the whirling speed.

(U. Lond.)

The arrangement of the shaft is shown in Fig. 15.7.

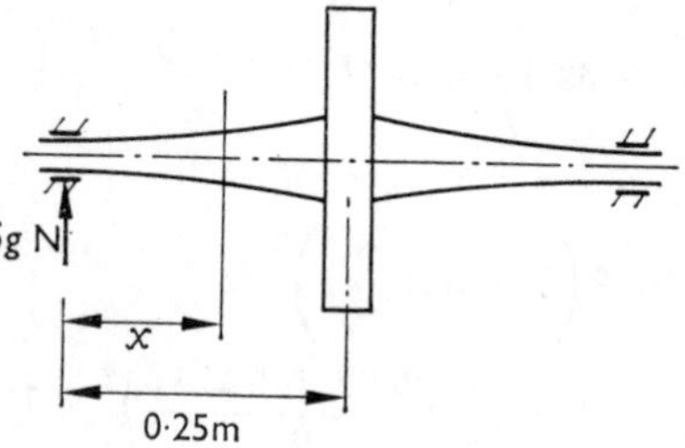

$$I_x = I_0/(1 - kx)$$

$$I_0 = \frac{\pi}{64} \times 0.025^4$$

25g N

x

0·25 m

Fig. 15.7

and

$$I_{0.25} = \frac{\pi}{64} \times 0.05^4$$

$$\therefore \quad 0.05^4 = 0.025^4/(1 - k \times 0.25)$$

$$\therefore \quad k = \frac{15}{4}$$

$$E \frac{d^2y}{dx^2} = \frac{M}{I} = -\frac{25gx}{I_0}(1 - kx) = -\frac{25g}{I_0}\left(x - \frac{15}{4}x^2\right)$$

$$\therefore \quad E \frac{dy}{dx} = -\frac{25g}{I_0}\left(\frac{x^2}{2} - \frac{5}{4}x^3 + A\right)$$

When $x = 0.25$, $\dfrac{dy}{dx} = 0$ $\therefore A = -0.011\,71$

$$\therefore \quad Ey = -\frac{25g}{I_0}\left(\frac{x^3}{6} - \frac{5}{16}x^4 - 0.011\,71x + B\right)$$

When $x = 0$, $y = 0$ $\therefore B = 0$

At $x = 0.25$ m, $y = \dfrac{-25 \times 9.81}{70 \times 10^9 \times \dfrac{\pi}{64} \times 0.025^4}\left(\dfrac{1}{384} - \dfrac{5}{16^3} - \dfrac{0.011\,71}{4}\right)$

$$= 0.000\,283 \text{ m}$$

$$\therefore \quad n = \frac{1}{2\sqrt{0.000\,283}} = \underline{29.7 \text{ rev/s}}$$

6. *A shaft ABC, 150 mm diameter, is carried in three spherical bearings at A, B and C, on a level. AB = 1·8 m and BC = 1·2 m. Wheels of mass 5 t and 3 t are carried in the lengths AB and BC, and may be supposed concentrated at the centres, the masses including an allowance for the shafts.*

Indicate the form of the shaft for the lowest critical speed, and explain how the loads must be set to give the static approximation to this. If in this equivalent static form the reaction at B is 20 kN, find the critical speed. E = 200 GN/m². (U. Glas.)

The form of the shaft at the lowest critical speed will be as shown in Fig. 15.8, and in this form the 3-t load must be set to act upwards. It will

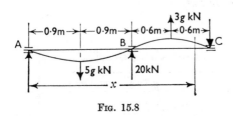

FIG. 15.8

be found that for this setting, the reaction at B is upward and, by taking moments,

$R_A = 20\cdot45$ kN, upward and $R_C = 20\cdot83$ kN, downward

Using Macaulay's Method,

$$EI\frac{d^2y}{dx^2} = -20\cdot45x + 49\cdot1[x - 0\cdot9] - 20[x - 1\cdot8] - 29\cdot43[x - 2\cdot4] \text{ kN m}$$

Integrating twice and inserting the conditions that $y = 0$ when $x = 0$ and $y = 0$ when $x = 1\cdot8$ gives

$$y = \frac{10^3}{200 \times 10^9 \times \frac{\pi}{64} \times 0\cdot15^4}\{-3\cdot41x^3 + 8\cdot18[x - 0\cdot9]^3 - 3\cdot33[x - 1\cdot8]^3 - 4\cdot9[x - 2\cdot4]^3 + 7\cdot73x\} \text{ m}$$

When $x = 0\cdot9$ m $y = 0\cdot9 \times 10^{-3}$ m

and when $x = 2\cdot4$ m, $y = -0\cdot34 \times 10^{-3}$ m

$$\therefore n_c = \frac{1}{2}\sqrt{\frac{\sum my}{\sum my^2}} \quad . \quad . \quad . \quad \text{from equation (14.3)}$$

$$= \frac{1}{2}\sqrt{\left[\frac{(5 \times 0\cdot9 + 3 \times 0\cdot34) \times 10^{-3}}{(5 \times 0\cdot9^2 + 3 \times 0\cdot34^2) \times 10^{-6}}\right]} = \underline{17\cdot7 \text{ Hz}}$$

As in **Example 3**, p. 343, no account is taken of the negative deflection under the 3-t load. In the numerator, both the load and the deflection are negative, giving a positive product, and in the denominator, the negative signs have no significance since the term my^2 represents the kinetic energy of the mass as it passes through its mid-position.

7. *A steel shaft 40 mm diameter is supported in short bearings distant 1·2 m apart. It carries two discs each of mass 20 kg, 0·4 m from each end. Find the whirling speed, neglecting the mass of the shaft.*

If the discs each have a moment of inertia about a diameter of 0·68 kg m², find the whirling speed when allowance is made for the effect of the slope of the discs. E = 200 GN/m².

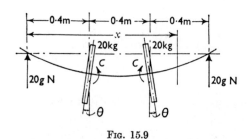

FIG. 15.9

$$EI = 200 \times 10^9 \times \frac{\pi}{64} \times 0{\cdot}04^4 = 8000\pi \text{ N m}^2$$

Using Macaulay's method and ignoring the effect of the couple C (at present unknown),

$$EI \frac{d^2y}{dx^2} = \{-20x + 20[x - 0{\cdot}4] + 20[x - 0{\cdot}8]\}g$$

$$\therefore EI \frac{dy}{dx} = \frac{20g}{2}\{-x^2 + [x - 0{\cdot}4]^2 + [x - 0{\cdot}8]^2 + A\}$$

$$EIy = \frac{20g}{6}\{-x^3 + [x - 0{\cdot}4]^3 + [x - 0{\cdot}8]^3 + 3Ax + B\}$$

When $x = 0$, $y = 0$ $\therefore B = 0$

When $x = 1{\cdot}2$ m, $y = 0$ $\therefore A = 0{\cdot}32$

$$\therefore \frac{dy}{dx} = \frac{20g}{2 \times 8000\pi}\{-x^2 + [x - 0 \cdot 4]^2 + [x - 0 \cdot 8]^2 + 0 \cdot 32\}$$

and $$y = \frac{20g}{6 \times 8000\pi}\{-x^3 + [x - 0 \cdot 4]^3 + [x - 0 \cdot 8]^3 + 0 \cdot 96x\}$$

$\therefore$ when $x = 0 \cdot 4 \; m,$ $\dfrac{dy}{dx} = 0 \cdot 000 \; 624$ rad

and $$y = 0 \cdot 000 \; 416 \; 5 \; m$$

so that $$n_c = \frac{1}{2}\sqrt{\frac{2 \times 20 \times 0 \cdot 000 \; 416 \; 5}{2 \times 20 \times 0 \cdot 000 \; 416 \; 5^2}} = \underline{24 \cdot 8 \text{ rev/s}}$$

Using this approximate whirling speed,

$$C = I_d \omega^2 \theta \qquad . \qquad . \qquad . \qquad . \qquad . \qquad \text{from equation (14.4)}$$

$$= 0 \cdot 68 \times (2\pi \times 24 \cdot 8)^2 \times 0 \cdot 000 \; 624$$

$$= 10 \cdot 34 \text{ N m}$$

Therefore the bending moment equation becomes

$$EI \frac{d^2y}{dx^2} = -20g + 20g[x - 0 \cdot 4] + [10 \cdot 34]^* + 20g[x - 0 \cdot 8] - [10 \cdot 34]\dagger$$

$$\therefore EI \frac{dy}{dx} = -10gx^2 + 10g[x - 0 \cdot 4]^2 + 10 \cdot 34[x - 0 \cdot 4]$$
$$+ 10g[x - 0 \cdot 8]^2 - 10 \cdot 34[x - 0 \cdot 8] + A$$

$$\therefore EIy = -\frac{10gx^3}{3} + \frac{10g}{3}[x - 0 \cdot 4]^3 + 5 \cdot 17[x - 0 \cdot 4]^2$$
$$+ \frac{10g}{3}[x - 0 \cdot 8]^3 - 5 \cdot 17[x - 0 \cdot 8]^2 + Ax + B$$

When $x = 0,$ $y = 0$ $\therefore B = 0$

When $x = 1 \cdot 2 \; m,$ $y = 0$ $\therefore A = 29 \cdot 3$

$$\therefore y = \frac{1}{8000\pi}\left\{-\frac{10}{3}gx^3 + \frac{10}{3}g[x - 0 \cdot 4]^3 + 5 \cdot 17[x - 0 \cdot 4]^2\right.$$
$$\left. + \frac{10}{3}g[x - 0 \cdot 8]^3 - 5 \cdot 17[x - 0 \cdot 8]^2 + 29 \cdot 3x\right\}$$

$\therefore$ when $x = 0 \cdot 4 \; m,$ $y = 0 \cdot 000 \; 383 \; m$

so that $$n_e = \frac{1}{2}\sqrt{\frac{2 \times 20 \times 0 \cdot 000 \; 383}{2 \times 20 \times 0 \cdot 000 \; 383^2}} = \underline{25 \cdot 6 \text{ rev/s}}$$

---

* To be integrated with respect to $[x - 0 \cdot 4]$.
† To be integrated with respect to $[x - 0 \cdot 8]$.

8. *A 125-mm diameter shaft is supported in two bearings with a span of 4·8 m. It carries a load of 500 kg concentrated at mid-span together with 250 kg at each quarter span point, and there is an axial thrust of 35 kN. Neglecting the effect of shaft mass, calculate the critical speed.* $E = 200$ GN/m^2.   (U. Glas.)

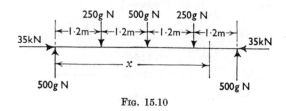

Fig. 15.10

Using Macaulay's method and ignoring the end-thrust,

$$EI \frac{d^2y}{dx^2} = \{-500x + 250[x - 1·2] + 500[x - 2·4] + 250[x - 3·6]\}g$$

Integrating twice and inserting the conditions that $y = 0$ when $x = 0$ and $y = 0$ when $x = 4·8$ m,

$$y = \frac{200 \times 9·81}{6 \times 200 \times 10^9 \times \dfrac{\pi}{64} \times 0·125^4} \{-2x^3 + [x - 1·2]^3 + 2[x - 2·4]^3$$
$$+ [x - 3·6]^3 + 30·25x\} \text{ m}$$

When $x = 1·2$ m, $y = 0·005\ 59$ m and when $x = 2·4$ m, $y = 0·007\ 96$ m

The Euler critical load $= \dfrac{\pi^2 EI}{l^2} = \dfrac{\pi^2 \times 200 \times 10^9 \times \dfrac{\pi}{64} \times 0·125^4}{4·8^2}$

$$= 1·027 \times 10^6 \text{ N}$$

$$\therefore \frac{1}{1 - \dfrac{P}{P_e}} = \frac{1}{1 - \dfrac{35 \times 10^3}{1·027 \times 10^6}} = 1·035 \text{ from equation (15.5)}$$

Hence, when the end thrust is applied, the deflections under the loads become increased in the ratio 1·035 : 1,

i.e. new deflection under 250-kg load $= 1·035 \times 0·005\ 59 = 0·005\ 79$ m

and new deflection under 500-kg load $= 1·035 \times 0·007\ 96 = 0·008\ 24$ m

so that $n_c = \dfrac{1}{2}\sqrt{\left[\dfrac{2 \times 250 \times 0·005\ 79 + 500 \times 0·008\ 24}{2 \times 250 \times 0·005\ 79^2 + 500 \times 0·008\ 24^2}\right]}$

$$= \underline{5·88 \text{ rev/s}}$$

(Unless otherwise stated, take E = 200 GN/m²)

9. A propeller shaft of a motor-driven vehicle consists of a steel tube with universal joints at the ends. Establish formulae for the first and second whirling speeds in terms of the length and the external and internal diameters of the shaft.

Evaluate the first and second whirling speeds in the case of a tube 1·5 m long whose external and internal diameters are 35 and 25 mm respectively. Take density of steel = 7·8 Mg/m³. *(U. Lond.)*

$$(Ans.: \quad \frac{1·77 \times 10^3}{l^2} \sqrt{[D^2 + d^2]} \text{ rev/s}\,; \quad \frac{7·08 \times 10^3}{l^2} \sqrt{[D^2 + d^2]} \text{ rev/s}\,;$$
$$33·8 \text{ rev/s}\,;\ 135·2 \text{ rev/s})$$

10. A light shaft supported in bearings at its ends carries a rotor of mass m at the centre of its length. When displaced laterally the elastic restoring force due to flexure of the shaft is S per unit deflection. If, when at rest, the centre of gravity of the rotor is at distance e from the axis of rotation, show that at a speed ω rad/s the elastic deflection of the shaft at the centre is

$$r = \omega^2 e \Big/ \left(\frac{S}{m} - \omega^2\right)$$

and describe the change in r as the speed is increased from zero to a large value.

If $e = 0·5$ mm and the whirling speed is observed to be 750 rev/min, find the speed range over which the magnitude of r will exceed 1·25 mm. *(I. Mech. E.)*

(Ans.: 633·8 to 968·2 rev/min)

11. A shaft 1·5 m long, supported in flexible bearings at the ends, carries two wheels each of mass 50 kg. One wheel is situated at the centre of the shaft and the other at a distance of 0·375 m from the centre. The shaft is hollow, of external diameter 75 mm, internal diameter 40 mm. The density of the shaft metal is 7·8 Mg/m³.

Find the lowest whirling speed of the shaft, taking into account the mass of the shaft itself. An approximate method may be used. *(U. Lond.)*

(Ans.: 32·8 rev/s [$\delta_1 = 0·121$ mm; $\delta_2 = 0·677$ mm; $\Delta = 0·0558$ mm])

12. A horizontal steel beam supported at the ends carries a load at some point between the supports. Show that, if the mass of the beam is neglected, the period of free vibration of the load is the same as that of a simple pendulum whose length is equal to the static deflection of the beam under the load.

A steel shaft, 60 mm diameter, with its ends freely supported in bearings 0·9 m apart, carries two wheels, each of mass 50 kg and each placed 0·3 m from one of the bearings. Calculate the lowest whirling speed of the shaft, assuming the shaft effect to be equivalent to 17/35ths of its mass concentrated at mid-span. Take $\rho = 7·8$ Mg/m³ for steel. *(U. Lond.)*

(Ans.: 50 rev/s [$y_{50} = 0·0985$ mm; $y_{9·6} = 0·1132$ mm])

13. The rotor of a three-stage centrifugal pump is carried on spherical-seated bearings at 1 m centres. The three impellers of mass 320 kg each are at distances of 0·3, 0·45, 0·6 m from one bearing centre line. The critical speed must be kept above 2500 rev/min. Using the energy method, estimate the minimum permissible shaft diameter, neglecting the mass of the shaft itself. *(U. Glas.)*

$$\left(Ans.: \ 103·5 \text{ mm} \left[y_1 = \frac{14·67 \times 10^{-9}}{d^4}\,; \right. \right.$$
$$\left. \left. y_2 = \frac{17·73 \times 10^{-9}}{d^4}\,;\ y_3 = \frac{16·74 \times 10^{-9}}{d^4} \right] \right)$$

14. A steel shaft of 80 mm diameter is supported in flexible bearings 1·6 m apart. The shaft carries three wheels, each of mass 100 kg, placed symmetrically between the bearings and 0·4 m apart.

(*a*) Neglecting gyroscopic action and the mass of the shaft, calculate approximately the lowest whirling speed of the shaft.

(*b*) Calculate the whirling speed by a different approximate method, using the static deflections under gravity as given below :

Deflection at central wheel 0·510 mm

Deflection at other wheels 0·360 mm

Which of the two methods do you consider the more reliable ? (*U. Lond.*)

(*Ans.:* (*a*) 23·8 rev/s (Dunkerley's method); (*b*) 24·3 rev/s (Energy method). The second answer is more reliable)

15. An initially straight shaft carries a load so heavy that the mass of the shaft is negligible. Show that the critical speed of rotation of the shaft is the same as the frequency of its natural transverse vibration. If such a shaft is to be run at $1\frac{1}{2}$ times its critical speed, and if the centrifugal deflection is not to exceed 0·25 mm, find the greatest permissible displacement of the centre of gravity from the axis of rotation. (*U. Lond.*)   (*Ans.:* 0·139 mm)

16. Write a brief essay on the whirling of rotating shafts, describing this phenomenon and indicating its importance in the design of machinery.

A light shaft carries a single disc of mass 10 kg. The centre of gravity of this disc is at a certain distance e_0 from the axis of the shaft. The whirling speed is 3600 rev/min. At a speed of 3240 rev/min the centre of gravity revolves in a circle of radius 3·8 mm. Calculate the distance e_0.

What is the strain energy of the shaft : (*a*) at 3240 rev/min, (*b*) at 3960 rev/min ? (*U. Lond.*)   (*Ans.:* 0·722 mm; 6·69 J; 12·27 J)

17. A light shaft of 12 mm diameter freely supported in bearings 1·2 m apart, carries a single central load of 0·5 kg. This single load has to be replaced by two loads of m kg each, symmetrically set 0·3 m from the respective bearings and between them.

Determine the magnitude of these two loads for the first whirling speed to be unaltered ; and estimate (*a*) the first whirling speed; (*b*) the second whirling speed for the two load system. (*U. Lond.*)

(*Ans.:* 0·5 kg by energy method ; 0·444 kg by Dunkerley's method ; (*a*) 17 rev/s ; (*b*) 48 rev/s if $m = 0·5$ kg, 50·9 rev/s if $m = 0·444$ kg)

18. Deduce an expression for the whirling speed of a light shaft which carries a single disc at a given point along its length.

The shaft of a small turbine with a single disc is found to have a static deflection of 0·355 mm. Calculate the whirling speed and find what percentage change in the diameter of the shaft will be required in order to raise the whirling speed to 2100 rev/min.

If the initial displacement of the centre of mass of the disc from the axis of the shaft is 0·5 mm compare the deflections of the shaft at 1500 rev/min in the two cases. (*I. Mech. E.*)

(*Ans.:* 1588 rev/min ; 15·07% ; 4·13 mm ; 0·52 mm)

19. In order to calculate the whirling speed of a turbine rotor, the gravitational deflection curve of the rotor was drawn. The length of the rotor was divided into six sections and the mass of each section together with the deflection at the centre of the section were recorded as below :

Section	.	.	1	2	3	4	5	6
Mass (kg)	.	.	16	73	115·5	120	81	20·5
Deflection (mm)	.		0·0325	0·0525	0·0875	0·0875	0·0600	0·0400

Working from first principles, calculate the whirling speed of the rotor and say within what range of speed you would consider it unsafe. (*U. Glas.*)

(*Ans.*: 57·3 rev/s. Unsafe within 30% of whirling speed)

20. Explain fully the strain energy method of finding the whirling speed of a shaft carrying various masses.

Find the whirling speed of a 50-mm diameter steel shaft simply supported at the ends in bearings 1·6 m apart, carrying masses of 75 kg at 0·4 m from one end, 100 kg at the centre and 125 kg at 0·4 m from the other end. Ignore the mass of the shaft. (*U. Lond.*)

(*Ans.*: 9·62 rev/s [$y_1 = 2·268$ mm; $y_2 = 3·25$ mm; $y_3 = 2·36$ mm)]

21. A heavy rotor has a mass of 1200 kg and is fixed at mid-span to a shaft 100 mm diameter and 1·75 m long. The axial length of the rotor is 0·9 m and it may be assumed that over this length the shaft is prevented from bending.

If the ends of the shaft are simply supported, find the whirling speed of the shaft. (*I. Mech. E.*)   (*Ans.*: 40·4 rev/s)

22. A shaft is supported in bearings which do not impose any flexural constraint. The bearings are 1·6 m apart and the shaft carries two equal pulleys between the bearings, each 0·4 m from the nearest bearing. When the shaft is at rest the central deflection is observed to be 0·8 mm. Determine the lowest whirling speed of the shaft. (*U. Lond.*)

(*Ans.*: 20·75 rev/s; 20·4 rev/s assuming parabolic curve)

23. A horizontal shaft, length l and diameter d, is simply supported at the bearings and carries a wheel of mass m at distance a from one bearing. The centre of the wheel is eccentric a distance e to the centre-line of the shaft. Derive the expression for the whirling speed in terms of the above symbols.

When the speed is 10% above the whirling speed, find an expression for the maximum deflection of the shaft at the wheel from its unloaded position.

(*U. Glas.*) $\left(Ans.:\ n = 0·0345 \sqrt{\dfrac{E\pi d^4 l}{ma^2(l-a)^2}}\ \text{rev/s};\ 5·76e\right)$

24. A shaft 150 mm diameter is carried in two spherical bearings at A and B, 1·8 m apart, and overhangs 0·3 m to C, where a pulley of mass 1000 kg is fixed. Find approximately the whirling speed.

For the shaft alone the approximate value of the whirling speed in rev/s $= \dfrac{1}{2\pi} \sqrt{\left(\dfrac{EI}{ml^4} A^4\right)}$ where $A = \pi\left(1 - \dfrac{1}{6}\dfrac{\pi^2 a^3}{l^3}\right)$, l is the length between bearings and a is the overhang. $\rho = 7·8$ Mg/m³. (*U. Glas.*)

(*Ans.*: 40·15 rev/s [$n_1 = 44·8$ rev/s; $n_b = 90·4$ rev/s])

25. A uniform shaft 1·75 m long, supported symmetrically on two short bearings 1·25 m apart, is symmetrically loaded with a central wheel of mass 250 kg, and two equal overhung pulleys at the ends, each of mass 50 kg. Calculate the critical speed, making an *approximate* allowance for the effect of the shaft mass.

Mass of shaft = 100 kg, I of shaft = $5·2 \times 10^{-6}$ m⁴. (*U. Glas.*)

(*Ans.*: 44·85 rev/s [$y_1 = 0·134$ mm; $y_2 = 0·087$ mm])

NOTE: The mass of shaft has been allowed for by using the formulae of Art. 14.2.

26. In a motor-driven centrifuge, a 40-mm diameter shaft has an effective length of 1·2 m from the bearing to the drum, which is carried at the free end of the shaft. The drum has a mass of 70 kg when loaded and has a radius of gyration of 0·3 m. Determine, from first principles, the speed at which whirling will occur, allowing for the rotational inertia of the drum but neglecting the mass of the shaft. (*U. Lond.*)   (*Ans.:* 3·73 rev/s [*see Example* 6, p. 347)]

27. A shaft AB, 150 mm diameter and 3 m long, runs in spherical bearings at A and B, and carries two loads, each 2 t, symmetrically placed at 0·9 m from the ends. Find the whirling speed.

Find also the whirling speed if an end thrust of 1/10 of the Euler critical load is impressed on the shaft. (*U. Glas.*)

$$(\textit{Ans.:} \quad 9\cdot3 \text{ rev/s}; \quad 8\cdot84 \text{ rev/s } [y = 2\cdot88 \text{ mm}])$$

28. Assuming a sine form, obtain an expression for the whirling speed of a uniform shaft supported at its ends in spherical bearings.

Find the whirling speed of a shaft 75 mm diameter and 3 m long. $\rho = 7\cdot8$ Mg/m^3.

If the shaft runs at 70% of this calculated whirling speed but is subject to an occasional end load, find the permissible maximum value of the end load if the running speed is not to exceed 90% of the reduced whirling speed. (*U. Glas.*)

$$\left(\textit{Ans.:} \quad \frac{1}{\sqrt{(\pi y_{max})}}; \quad 16\cdot65 \text{ rev/s}; \quad 134\cdot6 \text{ kN}\right)$$

29. Show that, in a loaded elastic system undergoing free undamped simple harmonic vibrations, the sum of the kinetic and strain energies is constant. Hence establish the rule for the frequency of free transverse vibrations of a loaded beam, neglecting the effect of the beam mass.

A shaft 100 mm diameter, simply supported in end bearings 1·6 m apart, carries equal discs, each of mass 500 kg, symmetrically placed 0·4 m on each side of the centre of the span. The radius of gyration of a disc about a diametral axis is 200 mm. Calculate the critical speed considering disc masses only and then estimate, approximately, the corrections to be made to this for (*a*) the effect of the mass of the shaft, (*b*) the effect of the slope at the discs, and (*c*) the effect of an axial thrust of 50 kN on the shaft. The density of the shaft material is 7·8 Mg/m^3. (*U. Lond.*)

$$(\textit{Ans.:} \quad 24\cdot25 \text{ rev/s}; \quad 23\cdot2 \text{ rev/s}; \quad 28\cdot2 \text{ rev/s}; \quad 24\cdot1 \text{ rev/s}.$$

[For disc masses only, y at each load $= 0\cdot426$ mm])

CHAPTER 16

TORSIONAL VIBRATIONS

16.1 Single-rotor system. If the disc of moment of inertia I attached to the shaft in Fig. 16.1 is given an angular displacement θ and released, the restoring torque, $T = \dfrac{GJ\theta}{l}$.

The angular acceleration of the disc is then given by

$$\frac{GJ\theta}{l} = I\alpha$$

from which

$$\frac{\alpha}{\theta} = \frac{GJ}{Il}$$

$$\therefore n = \frac{1}{2\pi}\sqrt{\frac{GJ}{Il}} \text{ Hz} \quad . \quad . \quad . \quad . \quad (16.1)$$

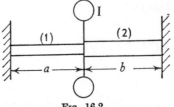

FIG. 16.1   FIG. 16.2

If the rotor is attached to the shaft between two fixed ends, Fig. 16.2, the restoring torque T can be divided into two parts, T_1 and T_2, where T_1 is due to the twisting of length a and T_2 is due to the twisting of length b.

Then

$$T = T_1 + T_2 = G\theta\left(\frac{J_1}{a} + \frac{J_2}{b}\right)$$

so that

$$G\theta\left(\frac{J_1}{a} + \frac{J_2}{b}\right) = I\alpha$$

i.e.

$$\frac{\alpha}{\theta} = \frac{G}{I}\left(\frac{J_1}{a} + \frac{J_2}{b}\right)$$

$$\therefore n = \frac{1}{2\pi}\sqrt{\left[\frac{G}{I}\left(\frac{J_1}{a} + \frac{J_2}{b}\right)\right]} \text{ Hz} \quad . \quad . \quad . \quad (16.2)$$

16.2 Effect of inertia of shaft.

Let the instantaneous angular velocity of the rotor be Ω, Fig. 16.3. Then angular velocity of element at distance x from 0

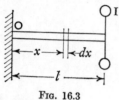

FIG. 16.3

$$= \frac{x}{l}\Omega.$$

If I_s is the polar moment of inertia of the shaft, then K.E. of element of length dx

$$= \tfrac{1}{2}I_s\frac{dx}{l} \times \left(\frac{x}{l}\Omega\right)^2$$

$\therefore$ total K.E. of shaft $= \dfrac{I_s}{2l^3}\Omega^2\displaystyle\int_0^l x^2\,dx$

$$= \frac{I_s}{2}\frac{\Omega^2}{3}$$

$\therefore$ total K.E. of system $= \tfrac{1}{2}\left\{I + \dfrac{I_s}{3}\right\}\Omega^2$

The inertia of the shaft may therefore be allowed for by adding one-third of its inertia to that of the rotor.

16.3 Two-rotor system.

Let K be the node for the vibrations, Fig. 16.4.

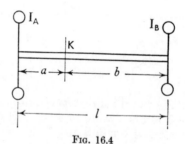

FIG. 16.4

Then

$$n_1 = \frac{1}{2\pi}\sqrt{\frac{GJ}{I_A a}}$$

and

$$n_2 = \frac{1}{2\pi}\sqrt{\frac{GJ}{I_B b}}$$

but

$$n_1 = n_2$$

$$\therefore\ I_A a = I_B b \quad\text{and also}\quad a + b = l$$

Hence a and b can be found and thus the frequency.

16.4 Three-rotor system. There are two possible modes of vibration, having two nodes as shown in Fig. 16.5(a) or only one node as shown in Fig. 16.5(b).

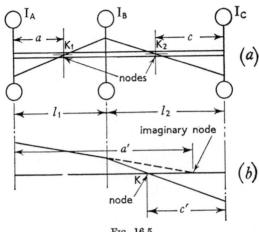

FIG. 16.5

Assuming two nodes, then for rotor A,

$$n_A = \frac{1}{2\pi}\sqrt{\frac{GJ}{I_A a}} \qquad . \qquad . \qquad . \qquad (16.3)$$

and for rotor C,

$$n_C = \frac{1}{2\pi}\sqrt{\frac{GJ}{I_C c}} . \qquad . \qquad . \qquad (16.4)$$

For rotor B,

$$n_B = \frac{1}{2\pi}\sqrt{\left[\frac{GJ}{I_B}\left(\frac{1}{l_1 - a} + \frac{1}{l_2 - c}\right)\right]}, \quad \text{from equation (16.2)} \qquad . \qquad (16.5)$$

The frequency of vibration of A, B and C must be the same so that a and c may be determined by equating n_A, n_B and n_C.

The resulting quadratic will give two values for both a and c; one pair will give the positions of the nodes for the two-node vibration, but of the other pair only one value will give a node lying within the limits of the corresponding part of the shaft.   This will be the node for the single-node vibration.   The other value will give the position where the elastic line* produced intersects the shaft axis.

Substituting these values of a or c in equations (16.3) or (16.4) will then give the two frequencies of vibration.

* The elastic line is a diagram which shows the amplitude of the vibration at all points along the shaft.

16.5   Multi-rotor system.   The analysis of the vibration of a shaft carrying more than three rotors becomes tedious and so a 'trial-and-error' method is adopted.

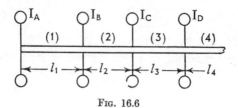

FIG. 16.6

In Fig. 16.6, let the amplitudes of vibration of the rotors be θ_A, θ_B, etc. Then the maximum inertia torques on the rotors are $I_A\omega^2\theta_A$, $I_B\omega^2\theta_B$, etc., where $\omega = 2\pi n$. Since the resultant torque on the shaft must be zero, then $\sum I\omega^2\theta = 0$, i.e. $\sum I\theta = 0$.

Also, the maximum inertia torque transmitted along any part of the shaft is related to the angle of twist of that part of the shaft.

Thus
$$T_A = \frac{GJ_1}{l_1}(\theta_A - \theta_B) = I_A\omega^2\theta_A$$

i.e.
$$\theta_B = \theta_A - \frac{\omega^2 l_1}{GJ_1}I_A\theta_A \quad . \qquad . \qquad . \qquad . \qquad . \quad (16.6)$$

$$T_A + T_B = \frac{GJ_2}{l_2}(\theta_B - \theta_C) = I_A\omega^2\theta_A + I_B\omega^2\theta_B$$

i.e.
$$\theta_C = \theta_B - \frac{\omega^2 l_2}{GJ_2}(I_A\theta_A + I_B\theta_B) \quad . \qquad . \qquad . \quad (16.7)$$

and so on for each rotor.

Take $\theta_A = 1$ radian and assume a suitable value for ω. Then obtain θ_B, θ_C, etc., from equations (16.6) and (16.7) and hence the value of $I_A\theta_A$, $I_B\theta_B$, etc. If the chosen value of ω is correct, then $\sum I\theta$ will be zero ; if not, further values must be tried until this equation is satisfied.

If the value of $\sum I\theta$ is positive, the assumed value of ω is too low. Conversely, if $\sum I\theta$ is negative, the value of ω is too high.

This method, which is known as the Holzer method, is not limited to the determination of the fundamental frequency ; frequencies corresponding to other modes of vibration are obtained in exactly the same way.

Typical examination problems are set out in tabular form in **Examples 6 and 7**.

16.6 Torsionally equivalent shaft. If, in any torsional vibration problem, the shaft is not uniform in section, it may be replaced by an equivalent shaft of uniform section, i.e. by a shaft having the same torsional stiffness. Thus the shaft (2) in Fig. 16.7 may be replaced by a shaft having the same section as shaft (1) if its length l_2' is given by

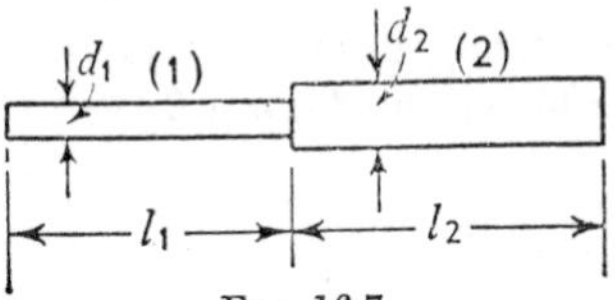

$$\frac{Tl_2}{GJ_2} = \frac{Tl_2'}{GJ_1}$$

i.e. $\quad l_2' = l_2 \times \dfrac{J_1}{J_2}$ $\qquad$. $\quad$. $\quad$ (16.8)

FIG. 16.7

Similarly a coupling or crankshaft of torsional stiffness q may be replaced by a shaft of polar second moment of area J and length l if

$$\frac{GJ}{l} = q \quad . \qquad . \qquad . \qquad . \qquad . \qquad (16.9)$$

16.7 Geared systems. The geared system of Fig. 16.8(a) must first be replaced by an equivalent two-rotor system connected by a single

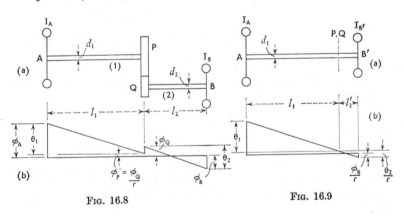

FIG. 16.8 $\qquad$ FIG. 16.9

shaft of uniform diameter, as shown in Fig. 16.9(a).* Rotor B, of moment of inertia I_B, is replaced by an equivalent rotor, of moment of inertia I_B', having the same K.E. of vibration at any instant as rotor B, and shaft (2), with gears P and Q, is replaced by a torsionally equivalent simple shaft, of length l_2' and the same cross-section as shaft (1).

---

* Two systems are equivalent, for vibration purposes, if the maximum K.E. is the same for each and the maximum strain energy is the same for each. For the latter condition to be satisfied, the two systems must have the same torsional stiffness.

Let the gear ratio $\dfrac{n_B}{n_A} = r$. Then since the speed of the equivalent rotor is only $1/r$ of the speed of the original rotor at any instant, its inertia must be increased in the ratio $r^2 : 1$ to give the same K.E.,

i.e.
$$I_B' = r^2 . I_B \qquad . \qquad . \qquad . \qquad . \quad (16.10)$$

If a torque T is applied at A, a corresponding torque T/r must be applied at B. If the angles of twist of shafts (1) and (2) are θ_1 and θ_2 respectively, then

angle of twist of Q relative to B $= \theta_2$

$\therefore$ angle of twist of P relative to B $= \dfrac{\theta_2}{r}$

$\therefore$ angle of twist of A relative to B $= \theta_1 + \dfrac{\theta_2}{r}$

$$= \frac{T l_1}{G J_1} + \frac{\left(\dfrac{T}{r}\right) l_2}{r G J_2}$$

$$= \frac{T}{G J_1}\left\{ l_1 + \frac{l_2 J_1}{r^2 J_2} \right\}$$

The angle of twist of the equivalent two-rotor system under the same torque T applied at A $= \dfrac{T}{G J_1}\{ l_1 + l_2' \}$.

Therefore, for the same twist in the two cases (i.e. for the same torsional stiffness),

$$l_2' = \frac{l_2}{r^2} . \frac{J_1}{J_2} \qquad . \qquad . \qquad . \qquad . \quad (16.11)$$

The elastic lines for the original and equivalent systems are shown in Figs. 16.8(b) and 16.9(b) respectively, no account being taken of the directions of the motions of the rotors. Since $I_B' = r^2 I_B$, the amplitude of B' is $1/r$ of that of B so that the maximum K.E. is the same in the two cases,

i.e.
$$\tfrac{1}{2} I_B (\omega \phi_B)^2 = \tfrac{1}{2} I_B' (\omega \phi_B')^2$$

where ω is the angular, or circular, frequency of the vibration and ϕ is the amplitude.

If the node lies between A and P in the equivalent shaft, its position is the same in the original shaft. If the node lies between Q and B', the corresponding position in the geared system is determined by proportion since the elastic lines for the equivalent and actual shafts are geometrically similar.

When the node is required to coincide with the gears, the gear ratio is irrelevant, except for the calculation of the amplitudes of oscillation.

If the inertia of the gears is to be taken into account, an additional rotor, of moment of inertia $I_P + r^2 I_Q$, must be added at P. The resulting three-rotor system is then analysed as in Art. 16.4.

1. *A steel disc 300 mm diameter, of mass 29 kg, is suspended from the end of a wire 2·5 mm diameter and 1·5 m long which is clamped into a central hole in the disc, the upper end of the wire being rigidly supported. When the disc is set in torsional vibration it is found to make 10 complete oscillations in 78·2 s.*

Find the modulus of rigidity of the wire, and calculate the amplitude of the oscillation which may be allowed if the maximum permissible intensity of shearing stress in the wire is 140 MN/m².

For the wire, $\qquad J = \dfrac{\pi d^4}{32} = \dfrac{\pi \times 2\cdot5^4}{32 \times 10^{12}} = 3\cdot84 \times 10^{-12} \text{ m}^4$

For the disc, $\qquad I = \dfrac{mR^2}{2} = \dfrac{29 \times 0\cdot15^2}{2} = 0\cdot326 \text{ kg m}^2$

From equation (16.1)

$$n = \frac{1}{2\pi}\sqrt{\frac{GJ}{Il}}$$

i.e. $\qquad \dfrac{1 \times 10}{78\cdot2} = \dfrac{1}{2\pi}\sqrt{\left[\dfrac{G \times 3\cdot84 \times 10^{-12}}{0\cdot326 \times 1\cdot5}\right]}$

hence $\qquad G = \underline{82\cdot4 \text{ GN/m}^2}$

If τ is the shear stress in the wire at a radius r, then $\dfrac{\tau}{r} = \dfrac{G\theta}{l}$

i.e. $\qquad \dfrac{140 \times 10^6}{0\cdot001\,25} = \dfrac{82\cdot4 \times 10^9 \times \theta}{1\cdot5}$

$$\therefore \; \theta = \underline{2\cdot04 \text{ rad or } 117°.}$$

2. *A solid shaft AB, 50 mm diameter, is fitted at the end A with a flywheel C, as shown in Fig. 16.10(a). The flywheel C has a mass of 220 kg and a radius of gyration of 450 mm. A concentric hollow shaft DE, 75 mm outside diameter, 55 mm inside diameter and 3 m long, is fitted at D with a flywheel F, of mass 270 kg and radius of gyration 0·6 m. The shaft DE is rigidly fixed to the shaft AB at the point E. AE = 4 m. Except for this fixing at E, the two shafts are free to twist relatively to each other. The modulus of rigidity of the material of the shafts is 80 GN/m².*

(a) If the flywheel C is rotated relatively to the flywheel F and released, find the frequency of the resulting vibrations and the position of the node.

(b) If the maximum angle of rotation of F relative to C is 1·2°, find the corresponding angular movement of F relative to the coincident point G on the shaft AB. (U. Lond.)

N

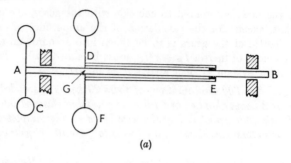

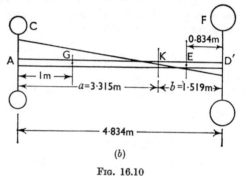

$$(b)$$

Fig. 16.10

The hollow shaft DE is equivalent to a solid shaft of diameter 50 mm and length $= l_{DE} \times \dfrac{J_{AE}}{J_{DE}} = 3 \times \dfrac{50^4}{75^4 - 55^4} = 0.834$ m.

The given system may therefore be replaced by the two-rotor system shown in Fig. 16.10(b), the shaft being 50 mm diameter throughout and 4·834 m long.

If K is the node,

then $\qquad\qquad I_C a = I_F b$

i.e. $\qquad 220 \times 0.45^2 \times a = 270 \times 0.6^2 \times b$

also $\qquad\qquad a + b = 4.834$ m

$\qquad\qquad \therefore\ a = 3.315$ m $\quad$ and $\quad b = 1.519$ m

$$\therefore\ n = \frac{1}{2\pi}\sqrt{\frac{GJ}{I_C a}}$$

$$= \frac{1}{2\pi}\sqrt{\left[\frac{80 \times 10^9 \times \pi \times 0.05^4}{220 \times 0.45^2 \times 32 \times 3.315}\right]}$$

$$= \underline{2.9\ \text{Hz}}$$

The elastic line is shown in Fig. 16.10(b). From similar triangles,

$$\frac{\theta_C}{\theta_F} = \frac{AK}{KD'} = \frac{3\cdot315}{1\cdot519} = 2\cdot18$$

also
$$\theta_C + \theta_F = 1\cdot2°$$

$$\therefore\ \theta_F = 0\cdot377°$$

$$\theta_G = \frac{KG}{KD'} \times \theta_F$$

$$= \frac{2\cdot315}{1\cdot519} \times 0\cdot377 = 0\cdot576°$$

$\therefore$ amplitude of F relative to G $= \theta_F + \theta_G = \underline{0\cdot953°}$

3. *In a two-mass torsional system, two wheels are mounted 1 m apart on a shaft 40 mm diameter. If the moments of inertia are $I_1 = 1\cdot2$ kg m^2; $I_2 = 2$ kg m^2, find the position of the node and the frequency of the free torsional oscillations.*

If the wheel I_2 is connected to the shaft by means of a spring coupling (without change in moment of inertia), as shown in Fig. 16.11, determine the length of the equivalent straight shaft, 40 mm diameter, and find the new position of the node and the new frequency. Each of the four springs has a stiffness of 850 kN/m and they act at a radius of 150 mm. G = 80 GN/m^2.

(U. Glas.)

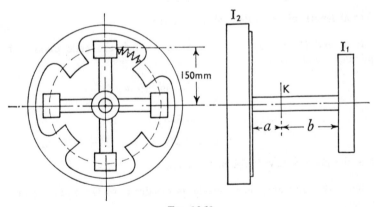

Fig. 16.11

If the node K divides the shaft into two parts of lengths a and b,

then
$$\frac{a}{b} = \frac{I_1}{I_2} = \frac{1\cdot2}{2} \quad \text{and} \quad a + b = 1 \text{ m}$$

$$\therefore\ a = 0{\cdot}375 \text{ m} \quad \text{and} \quad b = 0{\cdot}625 \text{ m}$$

$$\therefore\ n = \frac{1}{2\pi}\sqrt{\frac{GJ}{I_1 b}}$$

$$= \frac{1}{2\pi}\sqrt{\left[\frac{80 \times 10^9 \times \dfrac{\pi}{32} \times 0{\cdot}04^4}{1{\cdot}2 \times 0{\cdot}625}\right]} = \underline{26{\cdot}1 \text{ Hz}}$$

A compression of 1 mm of the springs corresponds to an angular movement of $\frac{1}{150}$ rad,

$$\therefore\ \text{torsional stiffness of spring coupling} = 4 \times \frac{850 \times 0{\cdot}15}{\frac{1}{150}}$$

$$= 76\,500 \text{ N m/rad}$$

If l is the length of the equivalent shaft 40 mm diameter, then

$$\text{torsional stiffness,}\ \frac{T}{\theta} = \frac{GJ}{l}$$

i.e.
$$76\,500 = \frac{80 \times 10^9 \times \dfrac{\pi}{32} \times 0{\cdot}04^4}{l}$$

from which
$$l = 0{\cdot}263 \text{ m}$$

$\therefore$ total length of equivalent shaft $= 1{\cdot}263$ m.

If a' and b' are the new lengths into which the node divides the equivalent shaft, then

$$\frac{a'}{b'} = \frac{1{\cdot}2}{2} \text{ as before}$$

and
$$a' + b' = 1{\cdot}263$$

$$\therefore\ a' = 0{\cdot}474 \text{ m} \quad \text{and} \quad b' = 0{\cdot}789 \text{ m}$$

Thus the node is now $0{\cdot}474$ m from I_1.

Since the frequency is inversely proportional to the square root of the length,

$$\frac{n'}{n} = \sqrt{\frac{a}{a'}}$$

i.e.
$$n' = 26{\cdot}1 \times \sqrt{\frac{0{\cdot}375}{0{\cdot}474}} = \underline{23{\cdot}22 \text{ Hz}}$$

4. *A uniform shaft 85 mm diameter carries three rotors A, B and C having moments of inertia of 17, 40 and 24 kg m² respectively. The distance between A and B is 0·75 m and between B and C 1·35 m. Find the frequencies of the free torsional vibration. If the rotor A has an amplitude of 1° in each case, find the amplitudes of B and C. The modulus of rigidity of the shaft is 80 GN/m².* (U. Lond.)

Referring to Fig. 16.5, $I_A = 17$ kg m², $I_B = 40$ kg m² and $I_C = 24$ kg m² and, by equating (16.3), (16.4) and (16.5),

$$I_A a = I_C c = \frac{I_B}{\dfrac{1}{l_1 - a} + \dfrac{1}{l_2 - c}}$$

$$\therefore \ 17a = 24c \quad \text{from which} \quad a = 1\cdot413c$$

and

$$24c = \frac{40}{\dfrac{1}{0\cdot75 - 1\cdot413c} + \dfrac{1}{1\cdot35 - c}}$$

$$\therefore \ c = 0\cdot342 \text{ m} \quad \text{or} \quad 1\cdot028 \text{ m}$$

and

$$a = 0\cdot483 \text{ m} \quad \text{or} \quad 1\cdot454 \text{ m}$$

The first pair of values locate the nodes for the two-node vibration and, substituting in equation (16.3),

$$n_A = \frac{1}{2\pi}\sqrt{\left[\frac{80 \times 10^9 \times \dfrac{\pi}{32} \times 0\cdot085^4}{17 \times 0\cdot483}\right]} = \underline{35\cdot6 \text{ Hz}}$$

The value $c = 1\cdot028$ m gives the position of the node for the single-node vibration, the value for a being greater than 0·75 m.

Substituting in equation (16.4),

$$n_C = \frac{1}{2\pi}\sqrt{\left[\frac{80 \times 10^9 \times \dfrac{\pi}{32} \times 0\cdot085^4}{24 \times 1\cdot028}\right]} = \underline{20\cdot5 \text{ Hz}}$$

If the amplitude of A is 1°, the other amplitudes may be found from the elastic lines shown in Fig. 16.5.

For the two-node vibration,

$$\theta_B = \frac{l_1 - a}{a} \times 1° = \frac{0\cdot267}{0\cdot483} \times 1° = 0\cdot553°$$

and

$$\theta_C = \frac{c}{l_2 - c} \times 0\cdot553° = \frac{0\cdot342}{1\cdot008} \times 0\cdot553° = 0\cdot188°$$

For the single-node vibration,

$$\theta_B = \frac{a' - l_1}{a'} \times 1° = \frac{0·704}{1·454} \times 1° = 0·484°$$

and $$\theta_C = \frac{c'}{l_2 - c'} \times 0·484° = \frac{1·028}{0·322} \times 0·484° = 1·545°$$

5. *A three-mass torsional system is shown in Fig. 16.12. Determine the torsional stiffness of the shaft BC to make the first natural frequency 6 Hz. Then calculate the corresponding second natural frequency. G = 80 GN/m².*

(U. Glas.)

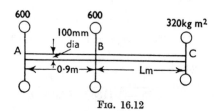

Fig. 16.12

From equation (16.3), $6 = \dfrac{1}{2\pi} \sqrt{\left[\dfrac{80 \times 10^9 \times \dfrac{\pi}{32} \times 0·10^4}{600 \times a}\right]}$

from which $a = 0·921$ m.

From equations (16.3) and (16.4),

$$I_A a = I_C c$$
$$\therefore c = \frac{600}{320} \times 0·921 = 1·727 \text{ m}$$

From equations (16.3) and (16.5),

$$I_A a = \frac{I_B}{\dfrac{1}{l_1 - a} + \dfrac{1}{l_2 - c}} \quad . \quad . \quad . \quad . \quad . \quad . \quad (1)$$

i.e. $$0·921 = \frac{1}{\dfrac{1}{0·9 - 0·921} + \dfrac{1}{L - 1·727}} \quad \text{since } I_A = I_B$$

from which $L = 1·748$ m

$$\therefore \text{torsional rigidity} = \frac{GJ}{L} = \frac{80 \times 10^9 \times \dfrac{\pi}{32} \times 0·10^4}{1·748}$$
$$= \underline{449\ 000 \text{ N m/rad}}$$

For the second frequency,

$$c = \frac{600}{320}a = 1\cdot875a, \quad \text{as before.}$$

$\therefore$ from equation (1),

$$a = \cfrac{1}{\cfrac{1}{0\cdot9 - a} + \cfrac{1}{1\cdot727 - 1\cdot875a}}$$

from which $a = 0\cdot921$ or $0\cdot361$ m

The first value corresponds to the first frequency and the second value corresponds to the second frequency.

The second frequency is given by

$$n = \frac{1}{2\pi}\sqrt{\left[\frac{80 \times 10^9 \times \dfrac{\pi}{32} \times 0\cdot10^4}{600 \times 0\cdot361}\right]} = \underline{9\cdot58 \text{ Hz}}$$

6. *A 150-mm diameter shaft carries four rotors as shown in Fig. 16.13. The natural frequency of torsional oscillations in the first mode is to be adjusted to 20 Hz by adding a flexible coupling and flywheel at end B as indicated. Determine the torsional rigidity of the coupling and the moment of inertia of the flywheel required. The flywheel displacement is to be limited to three times that of end A of the shaft. G = 80 GN/m².*

(U. Glas.)

$$\omega = 2\pi \times 20$$
$$= 40\pi \text{ rad/s}$$

$$\frac{\omega^2}{GJ} = \frac{(40\pi)^2}{80 \times 10^9 \times \dfrac{\pi}{32} \times 0\cdot15^4}$$

$$= 0\cdot003\,97$$

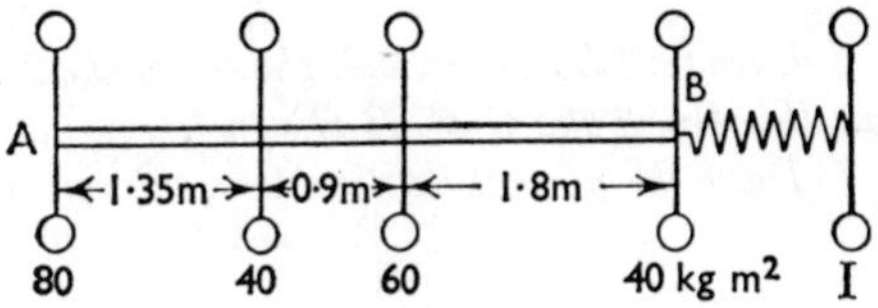

A ← 1·35m → ← 0·9m → ← 1·8m → B

80   40   60   40 kg m²   I

Fig. 16.13

I kg m²	θ rad	$I\theta$	$\Sigma I\theta$	l m	$\dfrac{\omega^2 l}{GJ}$	$\dfrac{\omega^2 l}{GJ}\Sigma I\theta$
80	1·0000	80	80			
				1·35	0·005 36	0·4285
40	0·5715	22·9	102·9			
				0·9	0·003 57	0·3675
60	0·2040	12·25	115·15			
				1·8	0·007 14	0·8220
40	−0·6180	−24·75	90·4			
				L	0·003 97L	0·359 L
I	−0·6180 −0·3590L	$-I\binom{0\cdot6180}{+0\cdot359L}$	$90\cdot4 - I\binom{0\cdot6180}{+0\cdot359L}$			

$$\Sigma \, I\theta = 0,$$

i.e. $$90 \cdot 4 - I\{0 \cdot 618 + 0 \cdot 359L\} = 0 \qquad . \qquad . \qquad . \qquad (1)$$

Also $$\theta_1 = 3 \times \theta_A,$$

i.e. $$0 \cdot 618 + 0 \cdot 359L = 3 \times 1 \cdot 000 \qquad . \qquad . \qquad (2)$$

Hence $$I = \underline{30 \cdot 13 \text{ kg m}^2}$$

From equation (2),

$$L = 6 \cdot 63 \text{ m} = \text{length of equivalent shaft 150 mm diameter}$$

$$\therefore \text{ torsional rigidity of coupling} = \frac{GJ}{L} = \frac{80 \times 10^9 \times \dfrac{\pi}{32} \times 0 \cdot 15^4}{6 \cdot 63}$$

$$= \underline{595\,000 \text{ N m/rad}}$$

7. *A 6-cylinder engine is directly coupled to a centrifugal pump. The equivalent dynamical system consists of six rotors for the six cylinders, each of moment of inertia 12·8 kg m², separated by lengths of 25-mm diameter shafting, each 80 μm long, and a rotor of moment of inertia 355 kg m² for the pump separated from the last cylinder by a 94·8 μm length of 25-mm diameter shaft.*

Using a tabular method, show that the 2-node mode of torsional oscillation is 195 Hz. Take G = 82 GN/m².

Draw the normal elastic curve, taking unit amplitude at the first cylinder.

(U. Lond.)

$$\omega = 2\pi \times 195 \text{ rad/s}$$

$$\therefore \frac{\omega^2}{GJ} = \frac{(2\pi \times 195)^2}{82 \times 10^9 \times \dfrac{\pi}{32} \times 0 \cdot 025^4}$$

$$= 477 \cdot 3$$

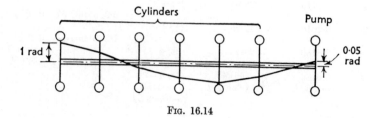

Fig. 16.14

I kg m²	θ rad	$I\theta$	$\Sigma\, I\theta$	$l \times 10^6$ m	$\dfrac{\omega^2 l}{GJ}$	$\dfrac{\omega^2 l}{GJ}\,\Sigma\, I\theta$
12·8	1·000	12·8	12·8			
				80	0·0382	0·489
12·8	0·511	6·54	19·34			
				80	0·0382	0·738
12·8	−0·227	−2·91	16·43			
				80	0·0382	0·627
12·8	−0·854	−10·93	5·50			
				80	0·0382	0·210
12·8	−1·064	−13·63	−8·13			
				80	0·0382	−0·311
12·8	−0·753	−9·63	−17·76			
				94·8	0·0453	−0·804
355	0·051	17·76	0			

Since $\Sigma\, I\theta = 0$, the given frequency of 195 Hz is verified.

The elastic line is shown in Fig. 16.14.

8. *An electric motor running at 2250 rev/min drives a centrifugal pump running at 650 rev/min through a single-stage gear reduction. The motor armature has a moment of inertia of 32 kg m² and the pump impeller one of 84 kg m². The shaft from the pump to the gears is 90 mm diameter and 3·6 m long, and that from the motor to the gears is 0·6 m long.*

What should be the diameter of the shaft from the motor to the gears to ensure that the node for natural torsional vibrations is at the gears? Determine the frequency of these vibrations and the amplitude of the impeller vibrations for an amplitude of one degree at the motor.

The inertia of the shafts and the gears may be neglected. The modulus of rigidity for the steel shafts is 80 GN/m². (U. Lond.)

Referring to Fig. 16·8, let A correspond to the pump and B to the motor. Then $I_A = 84$ kg m², $I_B = 32$ kg m², $d_1 = 90$ mm, $l_1 = 3·6$ m and $l_2 = 0·6$ m.

$$r = \frac{2250}{650} = 3·46$$

Since the node is to be at the gears,

$$n = \frac{1}{2\pi}\sqrt{\frac{GJ_1}{I_A l_1}} = \frac{1}{2\pi}\sqrt{\frac{GJ_2}{I_B l_2}}$$

$$\therefore \frac{J_1}{I_A l_1} = \frac{J_2}{I_B l_2}$$

i.e.
$$\frac{90^4}{84 \times 3·6} = \frac{d_2{}^4}{32 \times 0·6} \quad \therefore\ \underline{d = 45·2 \text{ mm}}$$

$$n = \frac{1}{2\pi}\sqrt{\frac{GJ_1}{I_A l_1}} = \frac{1}{2\pi}\sqrt{\left[\frac{80 \times 10^9 \times \frac{\pi}{32} \times 0.09^4}{84 \times 3.6}\right]}$$

$$= \underline{6.57 \text{ Hz}}$$

Since P does not vibrate, the resultant torque on it must be zero at all times. If θ_A and θ_B are the amplitudes of vibration of A and B respectively,

then   maximum inertia torque on P due to $A = I_A \omega^2 \theta_A$

and   maximum inertia torque on P due to $B = r \times I_B \omega^2 \theta_B$

$$\therefore \ I_A \theta_A = r \times I_B \theta_B$$

since the frequency of vibration is the same for each.

Therefore, since $\theta_B = 1°$, $\theta_A = 3.46 \times \frac{32}{84} = \underline{1.32°}$

9. *Two parallel shafts A and B, of diameters 50 and 75 mm respectively, are connected by a pair of gear wheels, the speed of A being five times that of B. A flywheel of mass 55 kg and radius of gyration 240 mm is mounted on shaft A at a distance of 0.9 m from the gears. Shaft B also carries a flywheel, of mass 90 kg and radius of gyration 430 mm, at a distance of 0.6 m from the gears.*

Neglecting the effect of the shaft masses, find the fundamental frequency of free torsional oscillations, and the position of the node.

Modulus of rigidity = 80 GN/m².   (U. Lond.)

Referring to Fig. 16.8,

$$I_A = 55 \times 0.24^2 = 3.17 \text{ kg m}^2$$

$$I_B = 90 \times 0.43^2 = 16.65 \text{ kg m}^2$$

$$r = \frac{1}{5}$$

$$I_B' = \frac{1}{5^2} \times 16.65 = 0.666 \text{ kg m}^2 \quad \text{from equation (16.10)}$$

$$l_2' = 5^2 \times 0.6 \times \left(\frac{50}{75}\right)^4 = 2.96 \text{ m} \quad \text{from equation (16.11)}$$

The given system therefore reduces to a simple two-rotor system, the shaft connecting I_A and I_B' being 50 mm diameter and 3.86 m long.

If the node divides the shaft into two parts of lengths a and b, then

$$I_A a = I_B' b$$

i.e. $$b = \frac{3.17}{0.666}a = 4.76a$$

also $a + b = 3.86$ m

$$\therefore \ a = 0.67 \text{ m}$$

$$\therefore \ n = \frac{1}{2\pi}\sqrt{\frac{GJ}{I_A a}}$$

$$= \frac{1}{2\pi}\sqrt{\left[\frac{80 \times 10^9 \times \dfrac{\pi}{32} \times 0.05^4}{3.17 \times 0.67}\right]} = \underline{24.2 \text{ Hz}}$$

The node occurs at a distance of 0·67 m from rotor A.

(Unless otherwise stated, take $G = 80$ GN/m²)

10. The upper end of a vertical steel wire 2 mm diameter and 2 m long is held securely. The other end is fixed central to a steel cylinder, 75 mm diameter and of density 7·8 Mg/m³, arranged with its axis horizontal. Find the length of the cylinder to give 0·6 torsional vibration per second.

Calculate the amplitude of the vibrations when the maximum shearing stress is 120 MN/m². (*U. Lond.*)   (*Ans.* : 103·5 mm ; 172°)

11. One end of a shaft of circular cross-section is fixed and a body attached to the other oscillates freely about the axis. Assuming that the angular displacement of each element of the shaft is proportional to its distance from the fixed end, show that the kinetic energy of the shaft at any instant is equivalent to that of a mass with one-third of its moment of inertia moving with the body at the free end.

Establish the expression for the frequency of torsional oscillation of this system. Evaluate this for a shaft 0·9 m long and 25 mm diameter carrying a body 150 mm long and 100 mm diameter, both being of steel for which the density is 7·8 Mg/m³. $G = 85$ GN/m². (*I. Mech. E.*)   (*Ans.* : 84·6 Hz)

12. Two rotors with moment of inertia I_1 and I_2 respectively are fixed at the ends of a shaft with free length L. Show that when they are in free torsional oscillation the stationary section or node in the shaft is at a distance $I_2 L/(I_1 + I_2)$ from the I_1 rotor and determine the expression for the periodic time of the oscillation. Calculate this value if for the first rotor $m = 34$ kg, $k = 115$ mm and for the second $m = 16$ kg, $k = 140$ mm. The shaft is 0·45 m long and 45 mm diameter. (*I. Mech. E.*)   (*Ans.* : 0·01 s)

13. A solid shaft AB, of 100 mm diameter, is rigidly connected at its ends to two hollow shafts AC and BD (both of external diameter 138 mm and internal diameter 106 mm), as shown in Fig. 16.15. At C and D two masses of 130 kg and

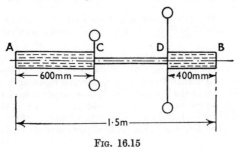

FIG. 16.15

250 kg, and of radius of gyration 175 mm and 375 mm respectively, are attached to the ends of the hollow shafts.

Determine from first principles the frequency of free torsional vibrations, and the position of the node.   (*U. Lond.*)   (*Ans. :* 53·8 Hz; 26·4 mm from B)

14. The flywheel of an engine driving a dynamo has a mass of 135 kg and a radius of gyration of 250 mm ; the armature has a mass of 100 kg and a radius of gyration of 200 mm.   The driving shaft has an effective length of 450 mm and is 50 mm diameter, and a spring coupling is incorporated at one end, having a stiffness of 28 kN m/rad.   Neglecting the inertia of the coupling and shaft, calculate the natural frequency of torsional vibration of the system.

What would be the natural frequency if the spring coupling were omitted ? (*U. Lond.*)   (*Ans. :* 14·43 Hz ; 31·9 Hz)

15. Obtain an expression for the natural frequency of torsional vibration of an elastic shaft carrying two wheels.   For the system shown in Fig. 16.16, calculate the natural frequency and the position of the node.   Specify the dynamic shear stress in the various portions of the shaft corresponding to an amplitude of vibration of 1° at wheel A.   (*U. Glas.*)

(*Ans. :* 7·05 Hz ; 48·9 MN/m² ; 14·7 MN/m²)

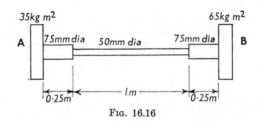

Fɪɢ. 16.16

16. A rotor has a mass of 225 kg and has a radius of gyration of 400 mm.   It is bolted between the ends of two shafts one of which is 75 mm diameter, 0·9 m long and the other is 65 mm diameter, 0·45 m long.   The other ends of the shafts are rigidly fixed in position.   Find the frequency of the natural torsional vibrations of the system.   (*I. Mech. E.*)   (*Ans. :* 20·35 Hz)

17. In a radial engine, the moving parts have a total moment of inertia of 0·8 kg m² and are concentrated in the plane of the single crank pin.   The engine is directly connected to an airscrew of moment of inertia 15 kg m² by a hollow shaft having outer and inner diameters of 75 and 32 mm respectively and an effective length of 250 mm.   The estimated stiffness of the crank throw alone is 2·5 MN m/rad.

Estimate the natural frequency of torsional vibration of the system.   What percentage error is involved if the airscrew mass be assumed infinite ?   (*U. Lond.*)

(*Ans. :* 152 Hz ; 2·56%)

18. Two rotors I_1 and I_2 are mounted on a shaft, 45 mm diameter, with a free length of 530 mm between them.   At a distance of 330 mm from I_1 it is desired to take off a drive which shall be free from oscillating movement and to effect this the remaining 200 mm of shaft adjacent to I_2 is reduced in diameter.   For I_1, the mass is 40 kg and the radius of gyration is 140 mm, and for I_2 the mass is 18 kg and the radius of gyration is 160 mm.   Find the frequency of torsional oscillation for the system and the diameter of the reduced portion of the shaft.   (*I. Mech. E.*)

(*Ans. :* 56·1 Hz ; 34·7 mm)

19. A shaft ABCD, 1·8 m long, has flywheels at its ends A and D. Each wheel has a mass of 55 kg. The radii of gyration are, for the wheel at A, 0·6 m, and for the wheel at D, 0·4 m. The connecting shaft has diameter 55 mm for the portion AB, which is 0·75 m long, and diameter 62 mm for the portion BC, 0·3 m long. Choose the diameter of the remaining portion CD, so that the node of the torsional vibration of the system will be at the centre of the length BC.

Find also the frequency of the torsional vibration. (*U. Lond.*)

(*Ans. :* 44·18 mm ; 10·43 Hz)

20. Derive an expression for the frequency of the torsional vibrations of a shaft fixed at one end and carrying a heavy mass at the free end.

A solid shaft 75 mm in diameter and 1·2 m long is coupled in series with a hollow shaft 80 mm external diameter and 50 mm internal diameter. A wheel of mass 270 kg and radius of gyration 0·6 m is keyed to the free end of the solid shaft and another wheel of mass 360 kg, radius of gyration 0·7 m, is keyed to the free end of the hollow shaft. Neglecting the inertia effect of the shafts and coupling, find the length of the hollow shaft when the node of the torsional vibrations is at the coupling and determine the frequency. (*U. Lond.*)

(*Ans. :* 0·725 m ; 7·35 Hz)

21. The armature of an electric motor, $I = 40$ kg m², is connected to a flywheel, $I = 80$ kg m², by a shaft 1·0 m long and 80 mm diameter. A shaft, 40 mm diameter and 2·0 m long, connects the flywheel to a pulley, $I = 20$ kg m². Find the frequencies of the two fundamental modes of torsional vibration, and the ratio of the angular displacements in the lower mode. (*U. Glas.*)

(*Ans. :* 3·85 Hz ; 17·54 Hz ; 0·1756)

22. In a three-mass case of torsional oscillation, $I_A = I_B = 20$ kg m² and I_C is unknown. AB = BC = 3 m and the shaft is of uniform diameter 80 mm. If it is required to have the node for the lower mode of vibration between B and C and 0·6 m from B, find the value of I_C and the frequency of the lower mode. (*U. Glas.*)

(*Ans. :* 31 kg m², 10·48 Hz)

23. A shaft 50 mm diameter and 1·2 m long carries two flywheels at the ends and is driven by a gear wheel fixed to the shaft between the flywheels.

The left-hand flywheel has a mass of 350 kg with a radius of gyration of 0·6 m, whilst the right-hand flywheel has a mass of 550 kg with a radius of gyration of 0·75 m. The gear wheel, which has a mass of 225 kg with a radius of gyration of 0·4 m, is 0·45 m from the left-hand flywheel.

Determine the two frequencies of torsional vibrations which the system may have. Neglect the moment of inertia of the shaft. (*U. Lond.*)

(*Ans. :* 3·33 Hz ; 11·74 Hz)

24. A light elastic shaft AB of uniform diameter, supported freely in bearings, carries a wheel at each end, and it is found that the natural frequency of torsional vibration is 40 Hz. A third wheel is mounted on the shaft at a point C such that AC = ¾AB. If all the wheels have the same moment of inertia, determine the natural frequencies of torsional vibration. (*U. Lond.*)

(*Ans. :* 38 Hz ; 84·17 Hz)

25. As a first approximation in the determination of the natural frequencies of torsional oscillation, a four cylinder engine, flywheel and a rigidly connected generator are equivalent to three rotors mounted on a shaft of diameter 150 mm. The engine is equivalent to a rotor of moment of inertia 35 kg m² and the generator to one of moment of inertia 40 kg m² at a distance of 2 m. The flywheel has a moment of inertia 300 kg m², and is between the engine and the generator at a distance of 0·9 m from the engine rotor. Find the two frequencies of natural torsional oscillation. (*U. Lond.*)

(*Ans. :* 50 Hz ; 60·6Hz)

26. A torsional vibration system consists of four masses, A, B, C and D, carried on a steel shaft of 50 mm diameter. The moments of inertia are respectively 0·6, 0·2, 0·1 and 0·25 kg m² and the lengths of shafts are AB 0·5 m, BC 0·25 m and CD 1 m. An estimate of the first natural frequency gives 62·5 Hz. Check this estimate and determine whether it is high or low. Sketch the mode of vibration at this frequency and find the position of the node.   (*U. Glas.*)

                                       (*Ans.:* Low; 0·5287 m from A)

27. Details of a marine engine installation are shown in Fig. 16.17. Torsiograph measurements taken on the shaft give an amplitude of 1·8° at position A

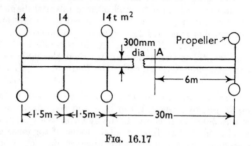

FIG. 16.17

when the system vibrates at its natural frequency of 2·5 Hz. The moment of inertia of the propeller itself is 10 t m²; estimate the inertia of the water entrained in the propeller and the maximum vibratory torque transmitted by the shaft under these conditions.   (*U. Glas.*)   (*Ans.:* 0·41 t m² ; 106·6 kN m)

28. Use the tabular method to find the length of shaft (L), between the rotors C and D of the system shown in Fig. 16.18, so that the natural frequency of torsional oscillations is 11·25 Hz.

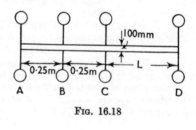

FIG. 16.18

$$I_a = 80 \text{ kg m}^2$$
$$I_b = 80 \text{ kg m}^2$$
$$I_c = 1350 \text{ kg m}^2$$
$$I_d = 550 \text{ kg m}^2$$

                                      U. Lond.)
                                  (*Ans.:* 0·285 m)

29. A four-cylinder engine drives a generator and for the purpose of investigating torsional oscillations the assembly may be regarded as equivalent to the six-rotor system as shown in Fig. 16.19. The relevant data are as follows:

Rotor	Description	Moment of Inertia (kg m²)	Torsional stiffness of connecting shaft (N m/rad)
1	Engine crank	4·72	
			800 000
2	,, ,,	4·72	
			800 000
3	,, ,,	4·72	
			800 000
4	,, ,,	4·72	
			575 000
5	Engine flywheel	I	
			480 000
6	Generator	63	

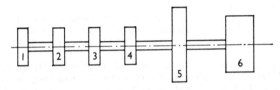

FIG. 16.19

(a) Determine the value of I, the moment of inertia of the engine flywheel, if the second mode of torsional oscillation is to have a frequency of 22·5 Hz.

(b) With this flywheel in position and with the system oscillating with the second mode frequency, the amplitude of the motion of No. 1 crank being 0·1 degree, determine the maximum torque induced in the shaft and the section in which it occurs. (*U. Lond.*)

(*Ans. :* 157 kg m² ; 478 N m between rotors 4 and 5)

30. The propeller of a motor launch is driven by a three-cylinder engine. For the purpose of investigating torsional oscillations the complete arrangement can be reduced to the following equivalent five-rotor system as shown in Fig. 16.20.

Item	Description	Moment of inertia (kg m²)
A, B, C	Equivalent rotors for the three cylinders	7 (each)
D	Engine flywheel	48
E	Propeller	17

The shaft can be regarded as being uniform with a diameter of 125 mm.

Determine the length L of the tail-shaft between the flywheel and the propeller, if the fundamental frequency of free torsional oscillation of the system is to be 40 Hz. (*U. Lond.*)

(*Ans. :* 2·1 m)

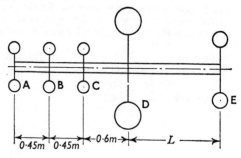

Fig. 16.20

31. An electric motor is to drive a centrifuge, running at four times the motor speed, through a spur gear and pinion. The steel shaft from motor to gear wheel is 54 mm diameter, and L long; the shaft from pinion to centrifuge is 40 mm diameter and 0·4 m long. The masses and radii of gyration of motor and centrifuge are respectively 35 kg, 100 mm, and 27 kg, 140 mm; the inertia effect of the gears may be neglected.

Find the value of L if the gears are to be at the node for torsional vibration of the system, and determine the frequency of torsional vibration. Establish any formulae used. (*U. Lond.*) (*Ans. :* 2·01 m; 49 Hz)

32. A torsional system consists of a flywheel of mass 180 kg and radius of gyration 250 mm connected to a shaft of diameter 50 mm and length 350 mm which drives a second shaft through a step-up gear having a velocity ratio of 1 : 4. The second shaft has a diameter of 25 mm for a length of 150 mm and a diameter 40 mm for a further length of 640 mm.

A second flywheel of 70 kg mass and a radius of gyration of 125 mm is connected to the end of the second shaft. The inertia of the gear wheels can be neglected.

Derive the equivalent dynamic system consisting of two flywheels coupled by a uniform shaft of diameter 40 mm. (*U. Glas.*)

(*Ans. :* 0·703 kg m²; 1·094 kg m²; 3·917 m *or* 11·25 kg m²; 17·5 kg m²; 0·2448 m)

33. A motor drives a centrifugal pump through gearing, the pump speed being one-third that of the motor. The shaft from the motor to the pinion is 55 mm diameter and 250 mm long, the motor having a moment of inertia of 40 kg m². The impeller shaft is 100 mm diameter and 0·6 m long, the moment of inertia of the impeller being 140 kg m².

Working from first principles, determine the frequency of torsional oscillations of the system. Neglect the inertia and flexibility of the gear and the inertia of the shaft. (*U. Lond.*) (*Ans. :* 14·8 Hz)

34. A turbine rotor is connected to a pinion B by a shaft AB of uniform diameter and the pinion meshes with a gear wheel C. The moments of inertia of A, B and C are in the ratio of 15 : 1 : 36 and the gear ratio is 5 : 1. If the system is in torsional vibration find the position of the node along the shaft AB.

The moment of inertia of A is 140 kg m² and the torsional stiffness of the shaft AB is 5·5 kN m per degree of twist. Find the frequency of the torsional vibrations of the system. (*I. Mech. E.*) (*Ans. :* 20·2 Hz)

35. A rotor A is mounted on one end of a shaft and a pinion B on the other end, and the pinion is in mesh with a gear wheel C, giving a speed reduction ratio of 5. The moments of inertia of A, B and C are 120, 20 and 200 kg m² respectively, and the torsional stiffness of the shaft is 2·7 kN m per degree of twist.

Derive an expression for the natural frequency of torsional vibration of the equivalent two-mass system, and find its value. (*U. Lond.*)

(*Ans. :* 13·13 Hz)

36. Fig. 16.21 shows a turbine, *T*, driving a propeller, *P*, through a double reduction gear. Determine the natural frequencies of torsional oscillation of the system given the following data :

Moment of inertia of turbine *T*	800 kg m²
Moment of inertia of propeller *P*	200 kg m²
Moment of inertia of gear 1	100 kg m²
Moment of inertia of gear 2	400 kg m²
Moment of inertia of gear 3	160 kg m²
Moment of inertia of gear 4	600 kg m²
Number of teeth in gear 1	45
Number of teeth in gear 2	145
Number of teeth in gear 3	52
Number of teeth in gear 4	220

The shaft connecting gears 2 and 3 is rigid.

Stiffness of shaft *A*	22 kN m/rad
Stiffness of shaft *B*	75 kN m/rad

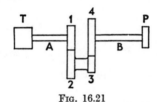

Fig. 16.21

(*U. Lond.*)
(*Ans. :* 2·055 Hz and 3·08 Hz)

NOTE. The equivalent three-rotor system for the above system is as shown in Fig. 16.22.

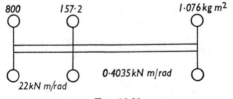

Fig. 16.22

CHAPTER 17

DAMPED VIBRATIONS

17.1 Viscous and Coulomb damping. When a vibrating body is subjected to a damping force, this force may be due to either (*a*) viscous friction (as obtained with a dash-pot), in which the resistance is assumed to be proportional to the velocity, or (*b*) Coulomb friction (as obtained between two dry surfaces), in which the resistance is assumed to be independent of the velocity.

17.2 Linear vibrations with viscous damping. If a body of mass m, Fig. 17.1, is acted upon by a restoring force S per unit displacement from the equilibrium position and also by a damping force c per unit velocity, the equation of motion is

$$m\frac{d^2x}{dt^2} = -Sx - c\frac{dx}{dt}$$

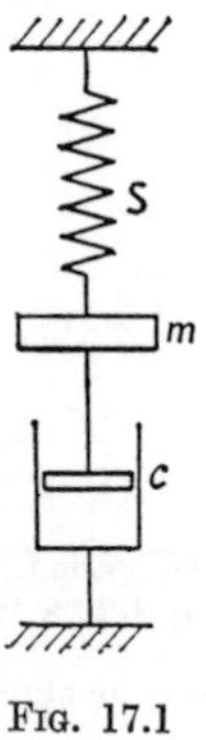

The quantity c is called the *damping coefficient* and the force $c\dfrac{dx}{dt}$ is negative since its direction is opposite to that of the velocity, $\dfrac{dx}{dt}$.

This equation may be written

$$\frac{d^2x}{dt^2} + 2\mu\frac{dx}{dt} + \omega^2 x = 0 \qquad . \qquad . \quad (17.1)$$

Fig. 17.1

where
$$2\mu = \frac{c}{m} \quad \text{and} \quad \omega^2 = \frac{S}{m}$$

or
$$(D^2 + 2\mu D + \omega^2)x = 0 \quad \text{where } D \text{ represents } \frac{d}{dt}$$

so that
$$D = -\mu \pm \sqrt{(\mu^2 - \omega^2)}$$

(i) $\underline{\mu > \omega}$ $\quad x = C_1 e^{[-\mu + \sqrt{(\mu^2 - \omega^2)}]t} + C_2 e^{[-\mu - \sqrt{(\mu^2 - \omega^2)}]t}$

The coefficient of t is negative in each term, so that x is the sum of two vanishing exponential terms. The motion is non-oscillatory and the mass, when disturbed from rest, will slowly return to its equilibrium position. The damping is described as *heavy* and the motion is termed *aperiodic*.

(ii) $\underline{\mu = \omega}$ $x = (C_1 + C_2 t)e^{-\mu t}$ *

The motion is again aperiodic but the damping coefficient now has the least value which will produce such motion. The damping in this case is said to be *critical*.

The damping coefficient which makes the damping critical is denoted by c_0 and the ratio c/c_0 is called the *damping ratio*.

(iii) $\underline{\mu < \omega}$.

$$x = C_1 e^{[-\mu + \sqrt{(\mu^2 - \omega^2)}]t} + C_2 e^{[-\mu - \sqrt{(\mu^2 - \omega^2)}]t}$$
$$= e^{-\mu t}\{C_1 e^{i\sqrt{(\omega^2 - \mu^2)}t} + C_2 e^{-i\sqrt{(\omega^2 - \mu^2)}t}\}$$
$$= e^{-\mu t}\{C_1[\cos\sqrt{(\omega^2 - \mu^2)}t + i\sin\sqrt{(\omega^2 - \mu^2)}t]$$
$$\qquad\qquad + C_2[\cos\sqrt{(\omega^2 - \mu^2)}t - i\sin\sqrt{(\omega^2 - \mu^2)}t]\}$$
$$= e^{-\mu t}\{A\cos\sqrt{(\omega^2 - \mu^2)}t + B\sin\sqrt{(\omega^2 - \mu^2)}t\} \qquad . \qquad . \quad (17.2)$$

where $A = (C_1 + C_2)$ and $B = i(C_1 - C_2)$.

This equation represents an oscillatory motion in which the amplitude decreases exponentially, as shown in Fig. 17.2.

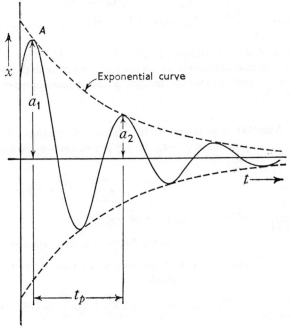

Fig. 17.2

* *Differential Equations*, H. T. H. Piaggio, Art. 27.

The periodic time, $t_p = \dfrac{2\pi}{\sqrt{(\omega^2 - \mu^2)}}$ s $\qquad . \qquad . \qquad .$ (17.3)

or frequency, $\qquad n = \dfrac{1}{2\pi} \sqrt{(\omega^2 - \mu^2)}$ Hz $\qquad . \qquad .$ (17.4

If a_1 and a_r are the first and rth amplitudes on the same side of the equilibrium position, then, measuring the time from the point A so that $t = 0$ when $x = a_1$, and $t = (r - 1)t_p$ when $x = a_r$,

$$\frac{a_1}{a_r} = \frac{e^0\{A \cos 0 + B \sin 0\}}{e^{-\mu(r-1)t_0}\{A \cos \sqrt{\omega^2 - \mu^2}(r-1)t_p + B \sin \sqrt{\omega^2 - \mu^2}(r-1)t_p\}}$$

$$= e^{\mu(r-1)t_p}\frac{A}{\{A \cos (r-1)2\pi + B \sin (r-1)2\pi\}}$$

$$= e^{\mu(r-1)t_p} \qquad . \qquad . \qquad . \qquad . \qquad . \qquad . \qquad . \qquad .$$ (17.5)

Thus the ratio of successive amplitudes is $e^{\mu t_p}$

or $\qquad \log_e \dfrac{a_1}{a_2} = \mu t_p = \dfrac{2\pi\mu}{\sqrt{(\omega^2 - \mu^2)}}$ $\qquad . \qquad . \qquad .$ (17.6)

The term $\dfrac{2\pi\mu}{\sqrt{(\omega^2 - \mu^2)}}$ is referred to as the *logarithmic decrement*.

When $\mu < \omega$, the damping is described as *light*. This is the case which most commonly arises in engineering problems and usually μ is so small in comparison with ω that the periodic time approximates very closely to $2\pi/\omega$.

17.3 Angular vibrations with viscous damping.

The corresponding equation of motion of damped angular vibrations is

$$I\frac{d^2\theta}{dt^2} + c\frac{d\theta}{dt} + q\theta = 0$$

where I = moment of inertia

c = damping torque per unit angular velocity

and $\quad q$ = restoring torque per unit angular displacement

This can be written $\dfrac{d^2\theta}{dt^2} + 2\mu\dfrac{d\theta}{dt} + \omega^2\theta = 0$

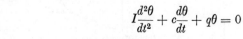

FIG. 17.3

where $\qquad\qquad 2\mu = \dfrac{c}{I}$ and $\omega^2 = \dfrac{q}{I}$ $\qquad . \qquad . \qquad .$ (17.7)

The solution of this equation is exactly as for linear vibrations.

When $\mu < \omega$, $$n = \frac{1}{2\pi} \sqrt{(\omega^2 - \mu^2)} \text{ Hz} \qquad . \qquad . \qquad . \quad (17.8)$$

and $\dfrac{\phi_1}{\phi_r} = e^{\mu(r-1)t_p}$ where ϕ_1 and ϕ_r are the first and rth amplitudes   (17.9)

17.4 Linear vibrations with Coulomb damping. Let a body of mass m rest on a horizontal plane, Fig. 17.4, which exerts a constant friction

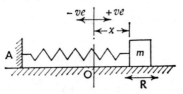

Fig. 17.4

force R upon it and be connected by a spring of stiffness S to a fixed point A. Then, if O is the point at which there is no force in the spring, the restoring force Sx is opposite in direction to the displacement x and the friction force R is opposite in direction to the velocity $\dfrac{dx}{dt}$.

When x is $+$ve and $\dfrac{dx}{dt}$ is $+$ve, $m\dfrac{d^2x}{dt^2} = -S(+x) - R$ . . (17.10)

When x is $+$ve and $\dfrac{dx}{dt}$ is $-$ve, $m\dfrac{d^2x}{dt^2} = -S(+x) + R$ . . (17.11)

When x is $-$ve and $\dfrac{dx}{dt}$ is $-$ve, $m\dfrac{d^2x}{dt^2} = +S(-x) + R$ . . (17.12)

When x is $-$ve and $\dfrac{dx}{dt}$ is $+$ve, $m\dfrac{d^2x}{dt^2} = +S(-x) - R$ . . (17.13)

It will be observed that equations (17.10) and (17.13) are identical and that equations (17.11) and (17.12) are identical.

Equations (17.10) and (17.13) may be written

$$\frac{d^2x}{dt^2} + \omega^2 x = -\frac{R}{m} \qquad . \qquad . \qquad . \quad (17.14)$$

and equations (17.11) and (17.12) may be written

$$\frac{d^2x}{dt^2} + \omega^2 x = +\frac{R}{m} \qquad . \qquad . \qquad . \quad (17.15)$$

where $\omega^2 = \dfrac{S}{m}$.

The solutions are: $\qquad x = A \cos \omega t + B \sin \omega t - \dfrac{R}{S}$ . . (17.16)

and $\qquad\qquad\qquad x = A \cos \omega t + B \sin \omega t + \dfrac{R}{S}$ . . (17.17)

respectively, the constants A and B being different for the two cases. Thus there are different equations of motion for each half-period, one for motion to the right and another for motion to the left.

The periodic time, $t_p = \dfrac{2\pi}{\omega}$, as in undamped vibration.

Let successive amplitudes of the motion at each half-period be a_0, a_1, a_2, . . . a_r.

Then, for $0 \leqslant t \leqslant \dfrac{\pi}{\omega}$, $\qquad x = A \cos \omega t + B \sin \omega t + \dfrac{R}{S}$

When $t = 0$, $\qquad\qquad\qquad\qquad x = a_0 \quad \text{and} \quad \dfrac{dx}{dt} = 0$

$$\therefore \ B = 0 \quad \text{and} \quad a_0 = A + \frac{R}{S}$$

so that $A = a_0 - \dfrac{R}{S}$, $\qquad \therefore \ x = \left(a_0 - \dfrac{R}{S}\right) \cos \omega t + \dfrac{R}{S}$

When $t = \dfrac{\pi}{\omega}$, $\qquad\qquad\qquad\qquad x = -a_1$

$$\therefore \ -a_1 = \left(a_0 - \frac{R}{S}\right) \cos \pi + \frac{R}{S}$$

$$\therefore \ a_1 = a_0 - \frac{2R}{S}$$

For $\dfrac{\pi}{\omega} \leqslant t \leqslant \dfrac{2\pi}{\omega}$, $\qquad\qquad x = A \cos \omega t + B \sin \omega t - \dfrac{R}{S}$

When $t = \dfrac{\pi}{\omega}$, $\qquad\qquad\qquad x = -a_1 \quad \text{and} \quad \dfrac{dx}{dt} = 0$

$$\therefore \ B = 0 \quad \text{and} \quad -a_1 = -A - \frac{R}{S}$$

so that $\qquad\qquad\qquad\qquad A = a_1 - \dfrac{R}{S} = a_0 - \dfrac{3R}{S}$

$$\therefore \ x = \left(a_0 - \frac{3R}{S}\right) \cos \omega t - \frac{R}{S}$$

When $t = \dfrac{2\pi}{\omega}$, $\quad x = a_2 \quad \therefore \ a_2 = \left(a_0 - \dfrac{3R}{S}\right) \cos 2\pi - \dfrac{R}{S}$

$$= a_0 - \frac{4R}{S}$$

Thus the amplitudes of successive half-periods are decreasing at the uniform rate of $2\dfrac{R}{S}$ so that $a_r = a_0 - r \times 2\dfrac{R}{S}$ . . . (17.18)

The variation of displacement with time is shown in Fig. 17.5.

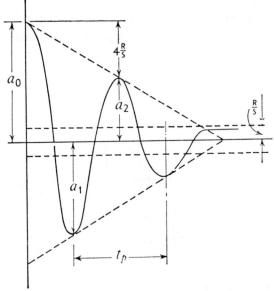

Fig. 17.5

Motion ceases when $a_r \leqslant \dfrac{R}{S}$ since the restoring force exerted by the spring is then equal to or less than R.

The number of half-periods which elapse before motion ceases is therefore given by $a_0 - r \times 2\dfrac{R}{S} \leqslant \dfrac{R}{S}$,

i.e. $$r \geqslant \frac{a_0 - \dfrac{R}{S}}{2\dfrac{R}{S}}$$ . . . (17.19)

Motion ceases at the first extreme position when the amplitude of the body becomes less than $\dfrac{R}{S}$.

17.5　Angular vibrations with Coulomb damping. The equations of motion for angular vibrations corresponding to equations (17.10) and (17.13) and equations (17.11) and (17.12) are

$$I\frac{d^2\theta}{dt^2} + q\theta = -T \quad \text{and} \quad I\frac{d^2\theta}{dt^2} + q\theta = +T$$

where　I = moment of inertia

　　　　T = constant damping torque

and　　q = restoring torque per unit angular displacement

These equations may be written

$$\frac{d^2\theta}{dt^2} + \omega^2\theta = -\frac{T}{I} \quad . \qquad . \qquad . \qquad . \quad (17.20)$$

and

$$\frac{d^2\theta}{dt^2} + \omega^2\theta = +\frac{T}{I} \quad . \qquad . \qquad . \qquad . \quad (17.21)$$

where

$$\omega^2 = \frac{q}{I}$$

The solutions are similar to those for linear vibrations.

Thus

$$t_p = \frac{2\pi}{\omega}$$

$$\phi_r = \phi_0 - r \times 2\frac{T}{q} \quad . \qquad . \qquad . \quad (17.22)$$

and motion ceases when $r \geqslant \dfrac{\phi_0 - \dfrac{T}{q}}{2\dfrac{T}{q}}$. $\quad . \qquad . \qquad . \quad (17.23)$

1. *A mass of 5 kg hangs from a spring and makes damped oscillations. The time of 50 complete oscillations is found to be 20 s, and the ratio of the first downward displacement to the sixth is found to be 2·25.*

Find the stiffness of the spring and the damping force.　　(U. Lond.)

$$t_p = \frac{20}{50} = 0.4 \text{ s}$$

$$\frac{a_1}{a_6} = e^{5\mu t_p} \quad . \qquad . \qquad . \qquad . \qquad \text{from equation (17.5)}$$

i.e.　　　　$2 \cdot 25 = e^{5\mu \times 0 \cdot 4}$

$$\therefore \mu = 0 \cdot 4055 \text{ rad/s}$$

$$t_p = \frac{2\pi}{\sqrt{(\omega^2 - \mu^2)}} \qquad . \qquad . \qquad \text{from equation (17.3)}$$

i.e.
$$0 \cdot 4 = \frac{2\pi}{\sqrt{(\omega^2 - 0 \cdot 4055^2)}}$$

$$\therefore \omega^2 = 246 \cdot 96 = \frac{S}{m} \qquad . \qquad . \qquad \text{from equation (17.1)}$$

$$\therefore S = 246 \cdot 96 \times 5 = \underline{1235 \text{ N/m}}$$

$$\mu = \frac{c}{2m} \qquad . \qquad . \qquad . \qquad . \qquad \text{from equation (17.1)}$$

$$\therefore c = 2 \times 5 \times 0 \cdot 4055 = \underline{4 \cdot 055 \text{ N s/m}}$$

2. *A body of mass 6 kg is hung on a spring of stiffness 1 kN/m. It is pulled down 50 mm below its static equilibrium position and released. There is a frictional resistance which is proportional to the velocity, and which is 36 N when the velocity is 1 m/s.*

Write down the differential equation of the motion, and its solution, evaluating the constants.

Calculate the time which elapses, and the distance which the body moves, from the instant of release until it is again at rest at the highest part of its travel. (U. Lond.)

The equation of motion is

$$6\frac{d^2x}{dt^2} + 36\frac{dx}{dt} + 1000x = 0$$

or
$$\frac{d^2x}{dt^2} + 6\frac{dx}{dt} + 166 \cdot 7x = 0$$

The solution is $x = e^{-3t}\{A \cos 12 \cdot 56t + B \sin 12 \cdot 56t\}$. . . (1)

$$\therefore \frac{dx}{dt} = e^{-3t}\{12 \cdot 56[-A \sin 12 \cdot 56t + B \cos 12 \cdot 56t]$$
$$- 3[A \cos 12 \cdot 56t + B \sin 12 \cdot 56t]\} \qquad (2)$$

when $t = 0$, $x = 0 \cdot 05$ $\therefore A = 0 \cdot 05$ . . . from equation (1)

when $t = 0$, $\frac{dx}{dt} = 0$, $\therefore 0 = 12 \cdot 56B - 3A$ . . from equation (2)

$$\therefore B = 0 \cdot 011 \text{ 95}$$

$$\therefore x = \underline{e^{-3t}\{0 \cdot 05 \cos 12 \cdot 56t + 0 \cdot 011 \text{ 95} \sin 12 \cdot 56t\}}$$

$$t_p = \frac{2\pi}{12 \cdot 56} = 0 \cdot 5 \text{ s}$$

$\therefore$ time to reach highest point = $\underline{0 \cdot 25 \text{ s}}$

$$\therefore x = e^{-3 \times 0 \cdot 25}\{0 \cdot 05 \cos (12 \cdot 56 \times 0 \cdot 25) + 0 \cdot 011 \text{ 95} \sin (12 \cdot 56 \times 0 \cdot 25)\}$$
$$= -0 \cdot 05e^{-0 \cdot 75} = -0 \cdot 0236 \text{ m}$$

Downward displacements have been considered positive and hence the negative sign indicates that the displacement is now measured upwards from the mean position.

$$\therefore \text{ total distance travelled} = 0{\cdot}05 + 0{\cdot}0236 = \underline{0{\cdot}0736 \text{ m}}$$

3. *A rod is hinged at one end and supported by a spring of stiffness S at the other end. A mass m is attached at $\frac{1}{3}$ of the length from the hinge and a dash-pot having a damping coefficient c is attached at $\frac{2}{3}$ of the length from the hinge. Find the equivalent mass and damping coefficient at the spring and derive an expression for the frequency of the damped free vibrations of the system.*   (U. Lond.)

Neglecting the mass of the rod, dash-pot plunger and spring, $I_A = ma^2$, Fig. 17.6.

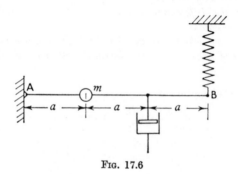

Fig. 17.6

Let x be the instantaneous displacement of the end B.

Then   restoring force exerted by spring $= Sx$

$$\therefore \text{ restoring moment about A} = Sx \times 3a$$

$$\text{Velocity of dash-pot plunger} = \frac{2}{3}\frac{dx}{dt}$$

$$\therefore \text{ damping force} = c \times \frac{2}{3}\frac{dx}{dt}$$

$$\therefore \text{ damping moment about A} = c \times \frac{2}{3}\frac{dx}{dt} \times 2a$$

$$= \frac{4ac}{3}\frac{dx}{dt}$$

$$\text{Angular acceleration of rod about A} = \left(\frac{d^2x}{dt^2}\right)\Big/ 3a$$

Therefore the equation motion of the rod is

$$\frac{ma^2}{3a}\left(\frac{d^2x}{dt^2}\right) + \frac{4ac}{3}\frac{dx}{dt} + Sx \times 3a = 0$$

i.e.

$$\frac{m}{9}\frac{d^2x}{dt^2} + \frac{4c}{9}\frac{dx}{dt} + Sx = 0$$

$$\therefore \text{ equivalent mass at spring} = \frac{m}{9}$$

and equivalent damping coefficient at spring $= \dfrac{4c}{9}$

The equation of motion can be written

$$\frac{d^2x}{dt^2} + 2\mu\frac{dx}{dt} + \omega^2 x = 0 \quad \text{where} \quad 2\mu = \frac{4c}{m} \quad \text{and} \quad \omega^2 = \frac{9S}{m}$$

$$\therefore n = \frac{1}{2\pi}\sqrt{(\omega^2 - \mu^2)} = \frac{1}{2\pi}\sqrt{\left(\frac{9S}{m} - \frac{4c^2}{m^2}\right)}$$

$$= \underline{\frac{1}{2\pi m}\sqrt{(9Sm - 4c^2)}\ \text{Hz}}$$

4. *The disc of a torsional pendulum has a moment of inertia of 0·068 kg m² and is immersed in a viscous fluid. The brass shaft (G = 40 GN/m²) attached to it is of 10 mm diameter and 380 mm long. When the pendulum is vibrating the observed amplitudes on the same side of the rest position for successive angles are 5°, 3° and 1·8°.*

Determine (i) the logarithmic decrement, (ii) the damping torque at unit velocity, (iii) the periodic time of the vibration.

What would be the frequency of the vibrations if the disc were removed from the viscous fluid? (U. Lond.)

Ratio of successive amplitudes $= \dfrac{5}{3} = \dfrac{3}{1\cdot8} = e^{\mu t_p}$

$$\therefore \text{ logarithmic decrement} = \mu t_p = \log_e 1\tfrac{2}{3} = 0\cdot511$$

$$\omega^2 = \frac{q}{I} = \frac{GJ}{Il} = \frac{40 \times 10^9 \times \dfrac{\pi}{32} \times (0\cdot01)^4}{0\cdot068 \times 0\cdot38} \quad \text{from equation (17.7)}$$

$$= 1520\ (\text{rad/s})^2$$

$$t_p = \frac{2\pi}{\sqrt{(\omega^2 - \mu^2)}} \quad . \quad . \quad . \quad . \quad \text{from equation (17.3)}$$

i.e.

$$\frac{0\cdot511}{\mu} = \frac{2\pi}{\sqrt{(1520 - \mu^2)}}$$

from which $\mu = 3{\cdot}16 = \dfrac{c}{2I}$ . . . . from equation (17.7)

$$\therefore c = 3{\cdot}16 \times 2 \times 0{\cdot}068 = \underline{0{\cdot}43 \text{ N m s/rad}}$$

$$t_p = \frac{2\pi}{\sqrt{(1520 - 3{\cdot}16^2)}} = \underline{0{\cdot}1615 \text{ s}}$$

If the disc is removed from the fluid,

$$t_p = \frac{2\pi}{\omega}$$

i.e. $\qquad n = \dfrac{\omega}{2\pi} = \dfrac{1}{2\pi} \times \sqrt{1520} = \underline{6{\cdot}2 \text{ H z}}$

5. *The rotor of a galvanometer is controlled by a spiral hairspring and a viscous torsional damper. When the rotor is in its normal position of equilibrium, the pointer indicates zero on the scale of the instrument. The scale is graduated in equal divisions. The pointer reads 100 divisions on the scale when the rotor has turned 40°, and the torque required to maintain the rotor in this position is 1·25 μN m. When the rotor is held in this position and then released, the pointer swings to a reading of −10 divisions and then to +1 division, the time for each swing being 2 s.*
Calculate :

(a) the moment of inertia of the rotor ;

(b) the periodic time of the oscillation, if there were no damping ;

(c) the ratio in which the damping would need to be increased, in order to make the motion aperiodic. (U. Lond.)

$$\frac{\text{1st angular displacement}}{\text{2nd angular displacement}} = \frac{100}{1} = e^{\mu t_p} \quad . \quad \text{from equation (17.9)}$$

but $\qquad t_p = 4 \text{ s} \quad \therefore \mu = 1{\cdot}151 \text{ rad/s}$

also $\qquad\qquad\qquad\qquad t_p = \dfrac{2\pi}{\sqrt{(\omega^2 - \mu^2)}}$

i.e. $\qquad\qquad\qquad\qquad 4 = \dfrac{2\pi}{\sqrt{(\omega^2 - 1{\cdot}151^2)}}$

$$\therefore \omega^2 = 3{\cdot}79 \text{ (rad/s)}^2$$

$$q = \frac{1{\cdot}25 \times 10^{-6}}{40} \text{ N m/degree} = 1{\cdot}79 \times 10^{-6} \text{ N m/rad}$$

$$\therefore 3{\cdot}79 = \frac{1{\cdot}79 \times 10^{-6}}{I} \qquad \text{from equation (17.7)}$$

$$\therefore I = \underline{0{\cdot}472 \times 10^{-6} \text{ kg m}^2}$$

Without damping, $t_p = \dfrac{2\pi}{\omega} = \dfrac{2\pi}{\sqrt{3 \cdot 79}} = \underline{3 \cdot 225 \text{ s}}$

For the motion to be aperiodic, $u \geqslant \omega$

$$\geqslant \sqrt{3 \cdot 79}$$

Therefore, since the damping torque is proportional to μ, it must be increased in the ratio $\dfrac{\sqrt{3 \cdot 79}}{1 \cdot 151} = \underline{1 \cdot 693 : 1.}$

6. *A compound pendulum consists of a concentrated bob of mass 10 kg attached to the lower end of a thin uniform rod 1·25 m long, of mass 7 kg, and suspended from a knife-edge 100 mm below the upper end of the rod. Working from first principles, find its time of free swing.*

The pendulum is now subjected to a damping torque which is proportional to its angular velocity, and it is found that the angle of swing diminishes to one eighth of its initial value after one complete double swing. Calculate the time of the damped swing, and the magnitude of the damping torque when the angular velocity is 1 rad/s. (U. Lond.)

For a *small* angular displacement, θ, Fig. 17.7,

restoring torque $= 10g \times 1 \cdot 15\theta + 7g \times 0 \cdot 525\theta = 15 \cdot 175g\theta$ N m

I of bob about $O = 10 \times 1 \cdot 15^2 = 13 \cdot 23$ kg m^2

I of rod about $O = \dfrac{7 \times 1 \cdot 25^2}{12} + 7 \times 0 \cdot 525^2 = 2 \cdot 84$ kg m^2

$\therefore$ total I about $O = 16 \cdot 07$ kg m^2

$\therefore 15 \cdot 175 \times 9 \cdot 81\theta = 16 \cdot 07\alpha$

$\therefore \dfrac{\theta}{\alpha} = 0 \cdot 108$

$\therefore t_p = 2\pi\sqrt{0 \cdot 108} = \underline{2 \cdot 065 \text{ s}}$

When damped,

$$\frac{\text{1st angular displacement}}{\text{2nd angular displacement}} = 8 = e^{\mu t_p}$$

$$\therefore 2 \cdot 079 = \mu t_p = \mu \times \frac{2\pi}{\sqrt{(\omega^2 - \mu^2)}}$$

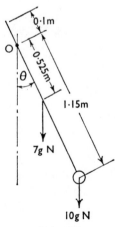

0·1m

O

0·525m

θ

1·15m

7g N

10g N

Fig. 17.7

But from equation (5.7),

$$\omega^2 = \frac{\text{restoring torque/rad displacement}}{I_0}$$

$$= \frac{15 \cdot 175 \times 9 \cdot 81}{16 \cdot 07} = 9 \cdot 26 \ (\text{rad/s})^2$$

$$\therefore \ 2 \cdot 079 = \frac{2\pi\mu}{\sqrt{(9 \cdot 26 - \mu^2)}}$$

from which $\quad \mu = 0 \cdot 955 \ \text{rad/s}$

$$\therefore \ t_p = \frac{2\pi}{\sqrt{(9 \cdot 26 - 0 \cdot 955^2)}} = \underline{2 \cdot 175 \ \text{s}}$$

Damping torque, $c = 2\mu \times I = 2 \times 0 \cdot 955 \times 16 \cdot 07$

$$= \underline{30 \cdot 7 \ \text{N m s/rad}}$$

7. *A vertical spring of stiffness 8·5 kN/m supports a mass of 32 kg. There is a friction force of 45 N which always resists the vertical motion whether upwards or downwards. The mass is released from a position in which the total extension of the spring is 125 mm.*

Determine :

(a) the time which elapses before the mass comes finally to rest,

(b) the final extension of the spring. (U. Lond.)

Static deflection of spring due to mass

$$= \frac{32 \times 9 \cdot 81}{8 \cdot 5 \times 10^3} = 0 \cdot 0369 \ \text{m}$$

$$\therefore \ a_0 = 0 \cdot 125 - 0 \cdot 0369 = 0 \cdot 0881 \ \text{m}$$

$$r \geqslant \frac{a_0 - \dfrac{R}{S}}{2\dfrac{R}{S}} \quad . \quad . \quad . \quad \text{from equation (17.19)}$$

$$\geqslant \frac{0 \cdot 0881 - \dfrac{45}{8 \cdot 5 \times 10^3}}{2 \times \dfrac{45}{8 \cdot 5 \times 10^3}}$$

$$\geqslant 7 \cdot 81$$

Hence motion ceases after 8 half-cycles.

$$t_p = 2\pi\sqrt{\frac{m}{S}} = 2\pi\sqrt{\frac{32}{8 \cdot 5 \times 10^3}} = 0 \cdot 385 \ \text{s}$$

$$\therefore \ \text{time taken} = 4 \times 0 \cdot 385 = \underline{1 \cdot 54 \ \text{s}}$$

After 8 half-cycles,

$$a = 0.0881 - 8 \times \frac{2 \times 45}{8.5 \times 10^3} = 0.0034 \text{ m} \qquad \text{from equation (17.18)}$$

This is of the same sign as the initial displacement and so the mass settles 0.0034 m below the frictionless equilibrium position.

Hence, final extension of spring = $0.0369 + 0.0034 = \underline{0.0403 \text{ m}}$

8. *A flywheel of moment of inertia 1·6 kg m² is attached to the lower end of a vertical shaft 1·5 m long and 25 mm diameter, the upper end of which is rigidly fixed. A band brake is fitted to the flywheel and exerts upon it a constant frictional torque of 25 N m. If the flywheel is displaced through an angle of 10° and released, find:*

(a) *the frequency,*

(b) *the number of oscillations performed before coming to rest, and*

(c) *the final settling position.*

$$G = 80 \text{ GN/m}^2$$

(a)
$$q = \frac{GJ}{l} = \frac{80 \times 10^9 \times \dfrac{\pi}{32} \times 0.025^4}{1.5}$$

$$= 2045 \text{ N m/rad}$$

$$n = \frac{1}{2\pi}\sqrt{\frac{q}{I}} \qquad . \qquad . \qquad . \qquad \text{from equation (13.16)}$$

$$= \frac{1}{2\pi}\sqrt{\frac{2045}{1.6}} = \underline{5.69 \text{ Hz}}$$

(b) Motion ceases when

$$r \geqslant \frac{\phi_0 - \dfrac{T}{q}}{2\dfrac{T}{q}} \qquad . \qquad . \qquad . \qquad \text{from equation (17.23)}$$

$$\frac{T}{q} = \frac{25}{2045} = 0.012\ 23 \text{ rad}$$

and
$$\phi_0 = 10 \times \frac{\pi}{180} = 0.1745 \text{ rad}$$

$$\therefore r \geqslant \frac{0.1745 - 0.012\ 23}{2 \times 0.012\ 23}$$

$$\geqslant 6.633$$

Hence motion ceases after $\underline{7}$ half-cycles.

(c) After 7 half-cycles,

$$\phi = 0.1745 - 7 \times 2 \times 0.012\ 23 \qquad . \qquad \text{from equation (17.22)}$$
$$= 0.003\ 28\ \text{rad} = 0.188°$$

Hence the flywheel settles at $0.188°$ on the same side of the equilibrium position as the initial displacement.

9. A mass suspended from a helical spring vibrates in a viscous medium whose resistance varies directly with the speed. It is observed that the frequency of the damped vibrations is 1.5 Hz and that the amplitude decreases to 20% of its initial value in one complete vibration. Find the frequency of the free undamped vibrations of the system. (*I. Mech. E.*)   (*Ans.* : 15.487 Hz)

10. A machine of mass 70 kg is mounted on springs and is fitted with a dash-pot to damp out vibrations. There are three springs each of stiffness 9 kN/m and it is found that the amplitude of the vibrations diminishes from 36 mm to 6 mm in two complete vibrations. Assuming that the damping force varies directly as the velocity, determine the resistance of the dash-pot at unit velocity and compare the frequency of the damped vibrations with the frequency when the dash-pot is not in operation. (*U. Lond.*)

(*Ans.* : 388 N s/m ; 3.095 Hz ; 3.125 Hz)

11. A machine of mass 100 kg is supported on springs which deflect 20 mm under the load. The vibration of the machine on the springs is constrained to be linear and vertical, and a dash-pot is fitted in order to reduce the amplitude to one-quarter of its initial value in two complete oscillations.

Find the magnitude of the damping force required at unit speed and compare the frequency of the damped and the free vibrations of the system. (*U. Lond.*)

(*Ans.* : 485 N s/m ; 3.505 Hz ; 3.525 Hz)

12. Obtain the equation of motion, and its solution, for the vibration of a body of mass m when acted on by a restoring force of S per unit displacement from the position of static equilibrium, and subjected to a damping force of c per unit velocity.

Given that $m = 14$ kg and $S = 8.4$ kN/m, and that the amplitude of the vibration diminishes to one-tenth of its original value in 2 complete vibrations, find the frequency of vibration and the value of the damping force c. (*U. Lond.*)

(*Ans.* : 3.83 Hz ; 123.6 N s/m)

13. (a) A mass of 100 kg is suspended from a vertical coil spring of stiffness 20 kN/m. Find the frequency of free vibration.

(b) The system is now damped. The mass is pulled downwards a definite distance, and when released, the vibration amplitude is reduced to 0.10 of the original in four complete oscillations.

Determine the frequency of the damped oscillation and the value of the damping force. (*U. Lond.*)   (*Ans.* : 2.25 Hz ; 2.24 Hz ; 258 N s/m)

14. A mass $m = 50$ kg is suspended from a spring whose stiffness $S = 16$ kN/m. When vibrating freely the amplitude falls to half its initial value after 10 complete vibrations. The motion may be assumed to be given by $x = Ce^{-\frac{1}{2}\mu t}\cos(kt - \alpha)$, where $k = \sqrt{(p^2 - \frac{1}{4}\mu^2)}$, $p^2 = S/m$ and $\mu = c/m$, c being the viscous damping force and S the stiffness of the spring.

Find (a) the value of c and (b) the ratio between this value of c and the minimum value of c required to prevent free vibration. (*I. Mech. E.*)

(*Ans.* : 19.74 N s/m ; 1/91)

15. Write down the differential equation and obtain its solution for the free damped vibration of a body of mass m, if the restoring force is S per unit displacement from the neutral position and the damping force is c per unit velocity.

In such a system a mass of 25 kg is suspended from a helical spring of stiffness 15 kN/m, the motion being controlled by a dash-pot such that the amplitude of the vibration decreases to one-fifth of its original value after 2 complete vibrations.

Find (a) the value of the damping force,
(b) the frequency of the vibration. (*U. Lond.*)
(*Ans. :* 155·6 N s/m ; 3·87 Hz)

16. Show from first principles that no free oscillatory motion can be obtained in a simple system consisting of a mass m suspended from an elastic member of negligible mass and of stiffness k and experiencing damping proportional to the velocity of motion, when the damping coefficient is equal to $2\sqrt{km}$. Discuss briefly where such a condition of damping is applicable.

Derive a relationship between the logarithmic decrement and the ratio c/c_0 for such a simple system, where c is the damping coefficient and c_0 the critical damping coefficient. (*U. Lond.*)

$$\left(Ans.:\ 2\pi\frac{c}{c_0}\Big/\sqrt{\left[1-\left(\frac{c}{c_0}\right)^2\right]}\simeq 2\pi\frac{c}{c_0}\ \text{when}\ \frac{c}{c_0}\ \text{is small}\right)$$

17. A mass is suspended from a helical spring and allowed to vibrate under the action of a damper in which the resistance is directly proportional to velocity. It is observed that the time for one complete oscillation is 0·8 s and that the amplitude decreases by one-half in each complete oscillation. Find the error involved in assuming the natural frequency to be the same for both the damped and the undamped vibration of the system. (*U. Lond.*)

(*Ans. :* Damped frequency = 1·25 Hz ; undamped frequency = 1·2573 Hz ; error = 0·58%)

18. A mass of 2·4 kg hanging from the lower end of a vertical spring is pulled downward through a definite distance and then released. If the resulting motion is controlled by viscous damping such that the ratio of the first downward displacement to the third is 4 : 1 and five vibrations are completed in 4 s, find the stiffness of the spring and the damping force. (*U. Lond.*)

(*Ans.:* 150 N/m ; 4·16 N s/m)

19. A transmission dynamometer is fitted with a dash-pot to damp out vibrations. It is observed that, with a certain oil in the dash-pot, the amplitudes of successive swings are 40 mm and 35 mm. It is desired to replace this oil with a new oil such that the amplitudes of successive swings are 40 mm and 25 mm approximately. Determine the ratio of the viscosities of the two oils. (*U. Lond.*)

(*Ans. :* 3·512:1)

20. A light uniform bar of mass m and length l is hinged at one end while the other end is carried by a spring of stiffness S so that in its rest position the bar is horizontal. Half-way along the bar a dash-pot is attached which produces a damping force of c per unit velocity.

Write down the equation of motion obtained by taking moments about the hinge, and give an expression for the time period. (*U. Lond.*)

$$\left(Ans.:\ \frac{16\pi m}{\sqrt{(192Sm-9c^2)}}\right)$$

o

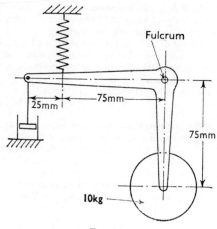

FIG. 17.8

21. The motion of a 10 kg pendulum mass is controlled by a spring and dash-pot, as shown in Fig. 17.8. The spring stiffness is 7 kN/m and the dash-pot piston has a mass of 0·75 kg. The bell-crank lever has a mass of 2·5 kg and a radius of gyration of 50 mm about its centre of gravity which is at the fulcrum. When the mass is displaced from its equilibrium position and released, the amplitude of the resulting vibration is reduced by one-half over three vibrations. Determine the damping coefficient at the dash-pot, and the value required to give critical damping. (*U. Glas.*) (*Ans.:* 13·28 N s/m; 362 N s/m)

22. A disc which has a moment of inertia of 0·6 kg m² is fitted to the end of a shaft with a torsional stiffness of 5 N m/rad, the other end of the shaft being fixed. A damping torque of 0·3 N m s/rad acts on the disc. Find the frequency of the damped vibration of the disc and the ratio of successive amplitudes on the same side of the equilibrium position. (*U. Lond.*) (*Ans.:* 0·4575 Hz; 1·726)

23. A torsional pendulum consists of a wire, 0·75 m long, 6 mm diameter, fixed at its upper end and attached at its lower end to a disc having a moment of inertia of 0·045 kg m². The disc is immersed in a viscous liquid and set in vibration when successive amplitudes on the same side of the equilibrium position are found to be in the ratio of 1·5 : 1. The modulus of rigidity of the wire is 40 GN/m².

Derive an expression for the oscillation of the pendulum and find :

(*a*) the damping coefficient of the system,

(*b*) the frequency of the damped vibration. (*U. Lond.*)

(*Ans.:* 0·0711 N m s/rad; 1·95 Hz)

24. A flywheel of mass 10 kg, suspended in frictionless bearings, makes rotational oscillations under the control of a torsion spring which exerts a torque of 4 N m for each radian of angular displacement of the flywheel. The periodic time of the oscillations is 2·5 s. Calculate the radius of gyration of the flywheel.

When a viscous damper is fitted to the system, the ratio of successive amplitudes of oscillation in the same sense is 0·10. Find the periodic time of the damped oscillation.

If, in this damped system, the flywheel is displaced by one radian from its position of equilibrium and being held at rest is then released, what is the angular velocity of the flywheel at the end of the first quarter period ? (*U. Lond.*)

(*Ans. :* 0·2515 m ; 2·665 s ; −1·56 rad/s)

25. A flywheel is mounted in bearings at which there is viscous damping, with its axis horizontal. A uniform disc 100 mm radius and of mass 1 kg is fastened on to the face of the wheel so that its centre of gravity is 250 mm from the axis of the wheel. The time of a small complete oscillation of the system is found to be 7 s. It is also observed that a point on the circumference of the wheel is displaced 120 mm for the first swing and 75 mm for the following swing, both being measured on the same side of the rest position.

Determine (*i*) the moment of inertia of the flywheel,
(*ii*) the damping torque at unit velocity due to bearings. (*U. Lond.*)

(*Ans. :* 2·96 kg m² ; 0·4065 N m s/rad)

26. A rotor of mass 180 kg with radius of gyration 380 mm is fitted to one end of a solid shaft 50 mm diameter, 3 m long, the other end of which is fixed. A torsional damper is fitted which exerts a resisting torque proportional to the angular velocity and which causes the amplitude to diminish by 30% in each complete vibration. Find from first principles the frequency of the damped vibrations and the damping torque per rad/s. $G = 80$ GN/m². (*U. Lond.*)

(*Ans. :* 3·98 Hz ; 73·9 N m s/rad)

27. Show that the frequency of vibration of a system damped by a constant resisting force is the same as that of the system when undamped. Show that such a vibration as this consists of successive half-cycles of S.H.M.

In a particular instance a mass of 250 kg is suspended from a spring which has a stiffness of 10 kN/m. The damping force is constant at 250 N.

If the mass is pulled downward a distance of 225 mm below its frictionless equilibrium position and then released, determine the final settling position and the number of complete vibrations. Show in a sketch how the amplitude of the vibration dies away. (*U. Lond.*) (*Ans. :* 0·025 m below equilibrium position ; 2)

28. A simple system consisting of a mass of 7 kg controlled by a spring of stiffness 1·2 kN/m and experiencing Coulomb friction of constant magnitude equal to 1 N, is given an initial displacement of 80 mm from the mid-position, i.e. the position for which the spring exerts no force.

Determine from first principles the time which will elapse for motion to cease and the displacement from the mid-position at that time. (*U. Lond.*)

(*Ans. :* 11·52 s ; 0)

29. Obtain an expression for the frequency of natural vibrations of a system subject to a damping force constant in value (unaffected by velocity) and show that this frequency is the same as that for the undamped system.

A mass of 30 kg is suspended from a spring of stiffness 6 kN/m and its motion is resisted by a constant force of 60 N. If the mass is pulled downwards a distance of 75 mm below the frictionless equilibrium position and then released, find the settling position of the mass and the number of complete oscillations performed. (*U. Lond.*) (*Ans. :* 5 mm above equilibrium position ; 2)

CHAPTER 18

FORCED VIBRATIONS

18.1 Forced linear vibrations. If a body of mass m, Fig. 18.1, is acted upon by a restoring force S per unit displacement from the equilibrium position and also by an external harmonic force $P \cos pt$, the equation of motion is

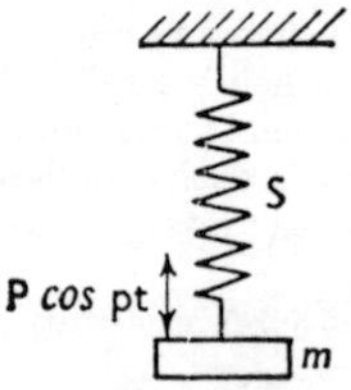

$$m\frac{d^2x}{dt^2} = -Sx + P \cos pt$$

or

$$\frac{d^2x}{dt^2} + \omega^2 x = \frac{P}{m} \cos pt \qquad . \qquad . \quad (18.1)$$

where

$$\omega^2 = \frac{S}{m}$$

Fig. 18.1

The complementary function, i.e. the solution of $\dfrac{d^2x}{dt^2} + \omega^2 x = 0$, is

$$x = A \cos \omega t + B \sin \omega t \qquad \text{as in equation (13.18)}$$

To obtain the particular integral, equation (18.1) may be written

$$(D^2 + \omega^2)x = \frac{P}{m} \cos pt, \quad \text{where } D \text{ represents } \frac{d}{dt}$$

$$\therefore \ x = \frac{1}{(D^2 + \omega^2)} \cdot \frac{P}{m} \cos pt$$

$$= \frac{P \cos pt}{m(\omega^2 - p^2)}$$

since $\qquad f(D^2) \cos pt = f(-p^2) \cos pt$*

The complete solution is therefore

$$x = A \cos \omega t + B \sin \omega t + \frac{P \cos pt}{m(\omega^2 - p^2)}$$

The first two terms represent the free vibration of the body, which dies out due to damping effects, leaving

$$x = \frac{P \cos pt}{m(\omega^2 - p^2)} \qquad . \qquad . \qquad . \qquad . \qquad . \quad (18.2)$$

to represent the steady-state vibration.

* *Differential Equations*, H. T. H. Piaggio, Art. 33.

This is a harmonic motion of frequency $p/2\pi$ Hz and amplitude

$$a = \frac{P}{m(\omega^2 - p^2)} \qquad . \qquad . \qquad . \qquad . \quad (18.3)$$

In cases where $p > \omega$, x becomes negative, showing that the body is 180° out of phase with the disturbing force.

The actual amplitude is, in all cases, the numerical value only of $\dfrac{P}{m(\omega^2 - p^2)}$ regardless of sign, i.e. $\left| \dfrac{P}{m(\omega^2 - p^2)} \right|$.

When $p = \omega$, the amplitude becomes infinite and *resonance* occurs.

The above formulae are applicable whether P is constant or is proportional to p^2, as in the case of the inertia force on a rotating or reciprocating mass. The graphs of a against ω in the two cases are shown in Figs. 19.2 and 19.3, μ being zero for forced vibrations without damping.

18.2 Periodic force transmitted to support. If a body is attached to a spring, the other end of which is attached to a rigid support, the maximum *dynamic* force transmitted to the support

= spring stiffness × amplitude of vibration

Maximum *total* force = weight of body + maximum dynamic force

18.3 Forced angular vibrations. The corresponding equation of motion for forced angular vibrations, in which the body is subjected to an external harmonic torque $T \cos pt$, Fig. 18.2, is

$$I\frac{d^2\theta}{dt^2} + q\theta = T \cos pt$$

where I = moment of inertia

and q = restoring torque per unit angular displacement

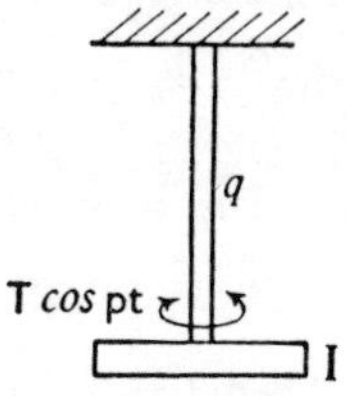

This can be written

FIG. 18.2

$$\frac{d^2\theta}{dt^2} + \omega^2\theta = \frac{T}{I} \cos pt \qquad . \qquad . \qquad . \quad (18.4)$$

where

$$\omega^2 = \frac{q}{I}$$

The solution of this equation is similar to that for linear vibrations.

18.4 Periodic movement of the support. In some cases the body may be subjected to a disturbing force due to the movement of the spring support rather than by a force applied directly to the body.

Suppose that the support for the system shown in Fig. 18.3 is vibrating such that $y = h \cos pt$.

Then   change in spring length = $x - y$

so that   restoring force = $S(x - y)$

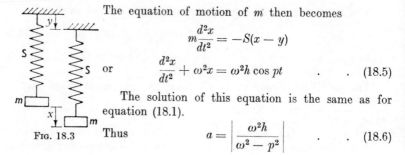

FIG. 18.3

The equation of motion of m then becomes

$$m\frac{d^2x}{dt^2} = -S(x - y)$$

or

$$\frac{d^2x}{dt^2} + \omega^2 x = \omega^2 h \cos pt \qquad . \qquad . \qquad (18.5)$$

The solution of this equation is the same as for equation (18.1).

Thus

$$a = \left| \frac{\omega^2 h}{\omega^2 - p^2} \right| \qquad . \qquad . \qquad (18.6)$$

1. *The time of free vibration of a mass hung from the end of a helical spring is 0·8 s. When the mass is stationary the upper end is made to move upwards with a displacement y mm such that $y = 45 \sin 2\pi t$, where t is the time in seconds measured from the beginning of the motion.*

Neglecting the mass of the spring and any damping effects, determine the vertical distance through which the mass is moved in the first 0·3 s.

(U. Lond.)

$$\omega = \frac{2\pi}{t_p} = \frac{2\pi}{0·8} = 2·5 \text{ rad/s}$$

The equation of motion (from equation 18.5) is

$$\frac{d^2x}{dt^2} + \omega^2 x = \omega^2 h \sin pt \quad \text{where} \quad \omega^2 = \frac{S}{m}$$

The solution is

$$x = A \cos \omega t + B \sin \omega t + \frac{\omega^2 h}{\omega^2 - p^2} \sin pt$$

When $t = 0$, $x = 0$, $\therefore A = 0$

$$\therefore \frac{dx}{dt} = \omega B \cos \omega t + \frac{p\omega^2 h}{\omega^2 - p^2} \cos pt$$

When $t = 0$, $\frac{dx}{dt} = 0$, $\therefore 0 = \omega B + \frac{p\omega^2 h}{\omega^2 - p^2}$

$$\therefore B = - \frac{p\omega h}{\omega^2 - p^2}$$

$$\therefore x = - \frac{p\omega h}{\omega^2 - p^2} \sin \omega t + \frac{\omega^2 h}{\omega^2 - p^2} \sin pt$$

$$= - \frac{2\pi \times 2·5\pi \times 45}{(2·5\pi)^2 - (2\pi)^2} \sin 2·5\pi t + \frac{(2·5\pi)^2 \times 45}{(2·5\pi)^2 - (2\pi)^2} \sin 2\pi t$$

$$= -100 \sin 2·5\pi t + 125 \sin 2\pi t$$

$\therefore$ when $t = 0·3$ s, $x = -100 \sin 0·75\pi + 125 \sin 0·6\pi = \underline{\underline{48·18 \text{ mm}}}$

2. *In Fig. 18.4, a pair of uniform parallel beams AB, 1·2 m long and together of mass 8 kg, are hinged at A and supported at B by a single spring of stiffness 4·5 kN/m. The beams carry a flywheel D of mass 28 kg in bearings 0·9 m from A. When in static equilibrium, AB is horizontal. Find the natural frequency of vibration of the system.*

If the centre of gravity of the flywheel is 3 mm from its axis of rotation, find the total vertical movement of B when the flywheel rotates at 200 rev/min.

(U. Lond.)

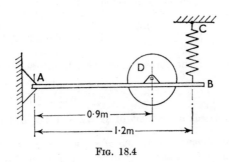

FIG. 18.4

Moment of inertia of system about $A = 8 \times \dfrac{1·2^2}{3} + 28 \times 0·9^2$

$$= 26·52 \text{ kg m}^2$$

Let B be displaced downwards a distance x m and let the corresponding angular displacement of the beams be θ.

Restoring force due to spring $= 4500x$ N

$\therefore$ restoring moment about $A = 4500x \times 1·2$

$$= 5400x \text{ N m}$$

$$= 6480\theta \quad \text{since } x = 1·2 \times \theta$$

The equation of motion is

$$26·52\frac{d^2\theta}{dt^2} + 6480\theta = 0$$

i.e.

$$\frac{d^2\theta}{dt^2} + 244\theta = 0$$

$$\therefore \omega = \sqrt{244} = 15·62 \text{ rad/s}$$

$$\therefore n = \frac{1}{2\pi} \times 15·62 = \underline{2·485 \text{ Hz}}$$

At 200 rev/min, $p = \dfrac{2\pi}{60} \times 200 = 20.94$ rod/s

centrifugal force on shaft bearings $= 28 \times 20.94^2 \times 0.003$
$$= 36.85 \text{ N}$$

The moment of this force about A $= 36.85 \times 0.9 = 33.1$ N m

The equation of motion therefore becomes

$$26.52\dfrac{d^2\theta}{dt^2} + 6480\theta = 33.1 \cos\left(\dfrac{2\pi}{60} \times 200\right)t$$

or
$$\dfrac{d^2\theta}{dt^2} + 244 = 1.248 \cos 20.94t$$

The amplitude of the forced vibrations, from equation (18.3),

$$= \left|\dfrac{1.248}{244 - 20.94^2}\right| = 0.006\,42 \text{ rad}$$

$\therefore$ total vertical movement of B $= 2 \times 1.2 \times 0.00642 = \underline{0.0154 \text{ m}}$

3. *A mass m_1 hangs by a spring of stiffness s_1 from another mass m_2 which in turn hangs by a spring of stiffness s_2 from a fixed point. Write down the equations of motion for the masses and hence find an expression for the periods of free oscillation.*

If the previously fixed point be now compelled to execute vertical oscillations $h \cos pt$, find the amplitudes of the vibrations of the masses. (U. Lond.)

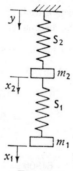

Fig. 18.5

When the support is fixed, the equations of motion are for m_1 :

$$m_1 \dfrac{d^2x_1}{dt^2} = -s_1(x_1 - x_2) \qquad . \qquad . \qquad . \qquad . \qquad (1)$$

and for m_2 : $$m_2 \dfrac{d^2x_2}{dt^2} = -s_2x_2 + s_1(x_1 - x_2) \qquad . \qquad . \qquad . \qquad (2)$$

Assuming that $x_1 = a_1 \cos \omega t$ and $x_2 = a_2 \cos \omega t$, equation (1) becomes

$$-m_1\omega^2 a_1 \cos \omega t = -s_1(a_1 \cos \omega t - a_2 \cos \omega t)$$

i.e.
$$m_1\omega^2 = s_1\left(1 - \frac{a_2}{a_1}\right)$$

i.e.
$$\frac{a_2}{a_1} = 1 - \frac{m_1\omega^2}{s_1} \qquad . \qquad . \qquad . \qquad . \qquad (3)$$

and equation (2) becomes

$$-m_2 a_2 \omega^2 \cos \omega t = -s_2 a_2 \cos \omega t + s_1(a_1 \cos \omega t - a_2 \cos \omega t)$$

i.e.
$$m_2 \frac{a_2}{a_1}\omega^2 = s_2 \frac{a_2}{a_1} - s_1\left(1 - \frac{a_2}{a_1}\right)$$

i.e.
$$\frac{a_2}{a_1} = \frac{s_1}{s_1 + s_2 - m_2\omega^2} \qquad . \qquad . \qquad . \qquad . \qquad (4)$$

Therefore from equations (3) and (4)

$$1 - \frac{m_1\omega^2}{s_1} = \frac{s_1}{s_1 + s_2 - m_2\omega^2}$$

i.e.
$$m_1 m_2 \omega^4 - [m_1(s_1 + s_2) + m_2 s_1]\omega^2 + s_1 s_2 = 0$$

This expression gives the values of ω from which the periodic times can be found.

When the support is vibrating, let the instantaneous displacement of the support be y such that $y = h \cos pt$.

Then equations (1) and (2) become

$$m_1 \frac{d^2x_1}{dt^2} = -s_1(x_1 - x_2) \qquad . \qquad . \qquad . \qquad . \qquad (5)$$

and
$$m_2 \frac{d^2x_2}{dt^2} = -s_2(x_2 - y) + s_1(x_1 - x_2) \qquad . \qquad . \qquad (6)$$

The free vibration of the system dies out, leaving a forced vibration only, having the same frequency as that of the support.

Assuming that $x_1 = a_1 \cos pt$ and $x_2 = a_2 \cos pt$, then, as in equation (3), equation (5) becomes

$$\frac{a_2}{a_1} = 1 - \frac{m_1 p^2}{s_1} \qquad . \qquad . \qquad . \qquad . \qquad (7)$$

and equation (6) becomes

$$-m_2 p^2 a_2 \cos pt = -s_2(a_2 \cos pt - h \cos pt) + s_1(a_1 \cos pt - a_2 \cos pt)$$

i.e.
$$\frac{a_2}{a_1}[-m_2 p^2 + (s_1 + s_2)] - \frac{s_2 h}{a_1} - s_1 = 0$$

$$\therefore \; a_1 = \frac{s_2 h}{\left(1 - \dfrac{m_1 p^2}{s_1}\right)[-m_2 p^2 + (s_1 + s_2)] - s_1}$$

substituting for $\dfrac{a_2}{a_1}$ from equation (7)

$$= \frac{s_1 s_2 h}{m_1 m_2 p^4 - [m_1(s_1 + s_2) + m_2 s_1]p^2 + s_1 s_2}$$

$$a_2 = \left(1 - \frac{m_1 p^2}{s_1}\right) a_1 \qquad . \qquad . \qquad \text{from equation (7)}$$

$$= \frac{s_2 h(s_1 - m_1 p^2)}{m_1 m_2 p^4 - [m_1(s_1 + s_2) + m_2 s_1]p^2 + s_1 s_2}$$

4. *An engine of mass 180 kg is to be supported on four helical springs. When the engine speed is 900 rev/min there is a primary vertical periodic disturbing force of maximum value 300 N due to the unbalanced reciprocating weights. Assuming that the engine vibrates in the vertical direction with no horizontal or angular movement, find the stiffness of each spring to limit the maximum total periodic force on the foundations to 20 N. What will be the amplitude of vibration of the engine when its speed is 600 rev/min?*

(U. Lond.)

From equation (18.3), $a = \left| \dfrac{P}{m(\omega^2 - p^2)} \right|$ where $\omega^2 = \dfrac{S}{m}$

When the additional force on the foundations is 20 N, the additional compression of the springs is $20/S$, S being the combined stiffness of the springs. This is the amplitude of the forced vibration, so that

$$\frac{20}{S} = \left| \frac{300}{180 \left[\dfrac{S}{180} - \left(\dfrac{2\pi}{60} \times 900 \right)^2 \right]} \right|$$

i.e. $\qquad\qquad (1 \pm 15)S = 1{\cdot}6 \times 10^6$

$$\therefore \; S = 10^5 \text{ N/m (taking the } +\text{ve sign)}$$

$$\therefore \text{ stiffness of each spring} = \frac{10^5}{4} = \underline{25 \text{ kN/m}}$$

At 600 rev/min, primary disturbing force $= 300 \times \left(\dfrac{600}{900}\right)^2 = \dfrac{400}{3}$ N

$$\therefore \; a = \left| \frac{400}{3 \times 180 \left[\dfrac{10^5}{180} - \left(\dfrac{2\pi}{60} \times 600 \right)^2 \right]} \right| = 0{\cdot}000218 \text{ m}$$
$$\text{or } \underline{0{\cdot}218 \text{ mm}}$$

5. *A small-scale model of a horizontal single-cylinder reciprocating engine is suspended by vertical flexible cords from an overhead support, so that the model is free to swing as a pendulum. The length of each cord is 750 mm. The mass of the model is 5 kg, the reciprocating mass is 0·8 kg, the crank radius is 20 mm and the length of the connecting rod is 80 mm.*

The crankshaft is rotated at 60 rev/min and the model begins to oscillate under the effect of the inertia of its reciprocating parts. Calculate the amplitudes of the forced oscillations set up (a) by the primary inertia force, and (b) by the secondary force. State also the travel of the model in each swing of the combined oscillation. (U. Lond.)

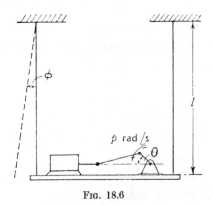

Fig. 18.6

The inertia force on the reciprocating parts when the crank makes an angle θ with the inner dead centre, Fig. 18.6,

$$= mp^2r\left(\cos\theta + \frac{r}{l}\cos 2\theta\right)$$

$$= 0{\cdot}8 \times \left(\frac{2\pi}{60} \times 60\right)^2 \times 0{\cdot}02 \times \left(\cos\theta + \frac{20}{80}\cos 2\theta\right)$$

$$= 0{\cdot}632(\cos\theta + \tfrac{1}{4}\cos 2\theta)$$

$$\therefore \text{ primary inertia force} = 0{\cdot}632\cos\theta \text{ N}$$

and $\qquad\qquad$ secondary inertia force $= 0{\cdot}158\cos 2\theta$ N

The equation of motion of the model is

$$ml^2 \frac{d^2\phi}{dt^2} + mgl\phi = Pl\cos pt, \quad \text{assuming } \phi \text{ to be small}$$

i.e. $\qquad \dfrac{d^2\phi}{dt^2} + \omega^2\phi = \dfrac{P}{ml}\cos pt \quad \text{where} \quad \omega^2 = \dfrac{g}{l}$

$\therefore$ amplitude of $\phi = \left| \dfrac{P}{ml(\omega^2 - p^2)} \right|$

For the primary inertia force,

$$\text{amplitude of } \phi = \left| \frac{0{\cdot}632}{5 \times 0{\cdot}75 \left[\dfrac{9{\cdot}81}{0{\cdot}75} - \left(\dfrac{2\pi}{60} \times 60 \right)^2 \right]} \right|$$

$$= 0{\cdot}006\ 38 \text{ rad}$$

$\therefore$ amplitude of horizontal movement $= 750 \times 0{\cdot}006\ 38 = \underline{4{\cdot}785 \text{ mm}}$

For the secondary inertia force,

$$\text{amplitude of } \phi = \left| \frac{0{\cdot}158}{5 \times 0{\cdot}75 \left[\dfrac{9{\cdot}81}{0{\cdot}75} - \left(\dfrac{2\pi}{60} \times 120 \right)^2 \right]} \right| = 0{\cdot}000\ 290\ 5 \text{ rad}$$

$\therefore$ amplitude of horizontal movement $= 750 \times 0{\cdot}000\ 290\ 5 = \underline{0{\cdot}218 \text{ mm}}$

Horizontal movement of engine

$$= 4{\cdot}785 \cos \theta + 0{\cdot}218 \cos 2\theta \text{ mm}$$

$\therefore$ amplitude of combined motion, Fig. 18.7,

$$= 4{\cdot}785 + 0{\cdot}218 \quad \text{when } \theta = 0,\ 2\pi,\ \text{etc.}$$

$$= \underline{5{\cdot}003 \text{ mm.}}$$

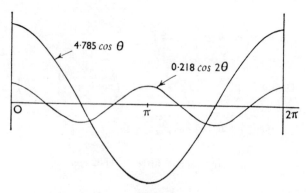

FIG. 18.7

6. *A small vertical two-cylinder in-line engine with cranks at 180°, is bolted to floor beams in a building. The reciprocating parts per line have a mass of 2·4 kg, the crank radius is 75 mm and the connecting rods are 300 mm long. The total mass of the engine and frame is 350 kg. Resonant vibration is produced by the secondary forces due to the reciprocating parts at a speed of 1200 rev/min and the engine speed is limited to 1100 rev/min at which, however, the vibration is still severe.*

It is proposed to reduce the vibration by interposing special elastic packing between the engine and floor beams and by limiting the speed to 1000 rev/min. If the elastic force of the packing is 8·5 MN/m compare, neglecting damping effects, the vibration amplitudes and dynamical reactions for the two cases.

(U. Lond.)

The primary cranks are 180° out of phase and rotate at 1200 rev/min. The secondary cranks are in phase and rotate at 2400 rev/min. The inertia force on the reciprocating parts per cylinder

$$= mp^2r\left\{\cos\theta + \frac{r}{l}\cos 2\theta\right\} \qquad \text{from Art. 7.1.}$$

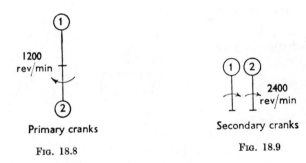

1200 rev/min

Primary cranks

Fig. 18.8

2400 rev/min

Secondary cranks

Fig. 18.9

∴ maximum secondary force

$$= mp^2\frac{r^2}{l} = m(2p)^2\frac{r^2}{4l}$$

∴ for the two cylinders, maximum secondary force

$$= 2 \times 2\cdot4 \times \left(\frac{2\pi}{60} \times 2400\right)^2 \times \frac{0\cdot075^2}{4 \times 0\cdot3}$$

$$= 1420 \text{ N}$$

At resonance, $\omega = p = \dfrac{2\pi}{60} \times 2400 = 251\cdot4$ rad/s

If S_1 = stiffness of the floor beams, then

$$\omega^2 = \frac{S_1}{m} \qquad . \qquad . \qquad . \qquad \text{from equation (18.1)}$$

$$\therefore \; S_1 = 251 \cdot 4^2 \times 350 = 22 \cdot 13 \times 10^6 \text{ N/m}$$

At 1100 rev/min, maximum secondary force

$$= 1420 \times \left(\frac{1100}{1200}\right)^2 = 1194 \text{ N}$$

and

$$p^2 = \left(\frac{2\pi}{60} \times 2200\right)^2 = 53\,080 \text{ (rad/s)}^2$$

$$\therefore \; a = \left| \frac{P}{m(\omega^2 - p^2)} \right| \qquad . \qquad . \qquad \text{from equation (18.3)}$$

$$= \left| \frac{1194}{350(251 \cdot 4^2 - 53\,080)} \right| = \underline{0 \cdot 000\,339 \text{ m}}$$

Dynamic reaction on floor $= 22 \cdot 13 \times 10^6 \times 0 \cdot 000\,339 = 7500 \text{ \underline{N}}$

If S_2 = stiffness of packing,

$$\text{deflection of floor, } x_1 = \frac{P}{S_1}$$

and   deflection of packing, $x_2 = \dfrac{P}{S_2}$

$$\therefore \; x_1 + x_2 = P\left(\frac{1}{S_1} + \frac{1}{S_2}\right) = P\left(\frac{S_1 + S_2}{S_1 S_2}\right)$$

$$\therefore \text{ equivalent stiffness, } \frac{P}{x_1 + x_2} = \frac{S_1 S_2}{S_1 + S_2}$$

$$= \frac{22 \cdot 13 \times 10^6 \times 8 \cdot 5 \times 10^6}{22 \cdot 13 \times 10^6 + 8 \cdot 5 \times 10^6} = 6 \cdot 14 \times 10^6 \text{ N/m}$$

$$\therefore \text{ new } \omega^2 = \frac{6 \cdot 14 \times 10^6}{350} = 17\,540 \text{ (rad/s)}^2$$

At 1000 rev/min,

$$\text{maximum secondary force} = 1420 \times \left(\frac{1000}{1200}\right)^2 = 986 \text{ N}$$

and

$$p^2 = \left(\frac{2\pi}{60} \times 2000\right)^2 = 43\,870 \text{ (rad/s)}^2$$

$$\therefore \; a = \left| \frac{986}{350(17\,540 - 43\,870)} \right| = \underline{0 \cdot 000\,107 \text{ m}}$$

Dynamic reaction on floor $= 6 \cdot 14 \times 10^6 \times 0 \cdot 000\,107 = \underline{657 \text{ N}}$

7. *A rotor A has a gear wheel with 40 teeth attached to it, the total mass being 20 kg and radius of gyration 150 mm. This gear meshes with a pinion on a rotor B with a parallel axis, the rotating parts having a mass of 22·5 kg, with radius of gyration 100 mm. A torque of 10 cos 15t Nm is applied to the rotor A, t being the time in seconds.*

(a) If the pinion on rotor B has 20 teeth, find the amplitude of movement of B.

(b) Find how many teeth the pinion on the rotor B should have, in order that the amplitude of the movement of B may be a maximum, and find this maximum amplitude. It may be assumed that the moment of inertia of the rotor B is not affected by any alteration in the size of the pinion.

(U. Lond.)

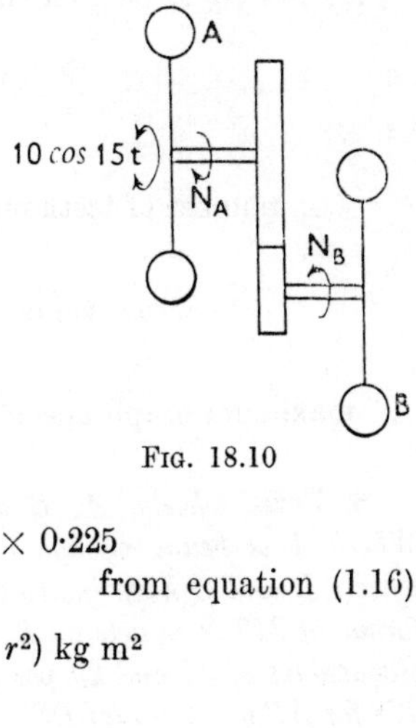

$$I_A = 20 \times 0.15^2 = 0.45 \text{ kg m}^2$$

$$I_B = 22.5 \times 0.1^2 = 0.225 \text{ kg m}^2$$

Let $\quad r = $ gear ratio, $\dfrac{N_B}{N_A}$.

Fig. 18.10

Then   equivalent inertia of A $= 0.45 + r^2 \times 0.225$

from equation (1.16)

$$= 0.225(2 + r^2) \text{ kg m}^2$$

Therefore the equation of motion of A is

$$0.225(2 + r^2)\frac{d^2\theta_A}{dt^2} = 10 \cos 15t$$

$$\therefore \frac{d^2\theta_A}{dt^2} = \frac{44.4}{(2 + r^2)} \cos 15t$$

$$\therefore \frac{d\theta_A}{dt} = \frac{2.96}{(2 + r^2)} \sin 15t + C_1$$

When $t = 0$, $\dfrac{d\theta_A}{dt} = 0$, $\therefore C_1 = 0$

$$\therefore \theta_A = -\frac{0.1975}{(2 + r^2)} \cos 15t + C_2$$

When $t = 0$, $\theta_A = 0$, $\therefore\ C_9 = \dfrac{0 \cdot 1975}{(2 + r^2)}$

$$\therefore\ \theta_A = \frac{0 \cdot 1975}{(2 + r^2)}\, (1 - \cos 15t)$$

$$\therefore\ \text{amplitude of } A,\ \phi_A = \frac{0 \cdot 1975}{(2 + r^2)}\ \text{rad} = \frac{11 \cdot 3\ \text{deg}}{(2 + r^2)}$$

$$\therefore\ \text{amplitude of B},\ \phi_B = \frac{11 \cdot 3r\ \text{deg}}{(2 + r^2)}$$

(a) $r = \dfrac{40}{20} = 2$, $\therefore\ \phi_B = \dfrac{11 \cdot 3 \times 2}{6} = \underline{3 \cdot 77^\circ}$

(b) For ϕ_B to be a maximum, $\dfrac{d\phi_B}{dr} = 0$

i.e. $(2 + r^2) \times 11 \cdot 3 = 11 \cdot 3r \times 2r$

i.e. $r = \sqrt{2}$

$$\therefore\ \text{number of teeth on B} = \frac{40}{\sqrt{2}} = 28\ \text{(to the nearest whole number)}$$

$$\therefore\ \text{actual value of } r = \frac{40}{28} = 1 \cdot 428$$

$$\therefore\ \text{maximum amplitude of B} = \frac{11 \cdot 3 \times 1 \cdot 428}{2 + 1 \cdot 428^2} = \underline{3 \cdot 99^\circ}$$

8. *Three wheels, A, B and C, form a system in torsional vibration. Wheel A is heavy enough to justify assuming no variation in its speed; wheels B and C have inertias of 4 and 0·8 kg m² respectively. A harmonic torque of 240 N m acts on B. Find the amplitudes of B and C at impressed frequencies of 25 and 29 per second. The shaft stiffnesses in kN m/rad are 128 for AB and 25·6 for BC.* (U. Glas.)

The end A of the shaft can be considered as rigidly fixed, Fig. 18.11.

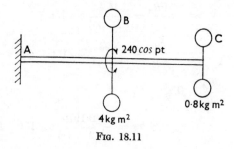

Fig. 18.11

Let θ_b and θ_c be the instantaneous angular displacements of B and C respectively, assumed to be in the same direction.

Then the equation of motion of C is

$$0{\cdot}8\frac{d^2\theta_c}{dt^2} = -25\,600(\theta_c - \theta_b) \qquad . \qquad . \qquad . \qquad . \qquad . \tag{1}$$

and the equation of motion of B is

$$4\frac{d^2\theta_b}{dt^2} = -128\,000\theta_b + 25\,600(\theta_c - \theta_b) + 240\cos pt \tag{2}$$

where $p = 2\pi \times$ impressed frequency.

The free vibration of the system dies out, leaving a forced vibration only, having the same frequency as the impressed torque.

Assuming $\theta_b = \phi_b \cos pt$ and $\theta_c = \phi_c \cos pt$, equation (1) becomes

$$-0{\cdot}8p^2\phi_c \cos pt = -25\,600(\phi_c \cos pt - \phi_b \cos pt)$$

i.e.
$$p^2\phi_c = 32\,000(\phi_c - \phi_b)$$

or
$$\frac{\phi_b}{\phi_c} = 1 - \frac{p^2}{32\,000} \qquad . \qquad . \qquad . \qquad . \tag{3}$$

Equation (2) becomes

$$-4p^2\phi_b \cos pt = -128\,000\phi_b \cos pt$$
$$+ 25\,600(\phi_c \cos pt - \phi_b \cos pt) + 240\cos pt$$

i.e.
$$-p^2\phi_b = -32\,000\phi_b + 6400(\phi_c - \phi_b) + 60$$

or
$$-p^2 = -38\,400 + \frac{6400}{\dfrac{\phi_b}{\phi_c}} + \frac{60}{\phi_b}$$

$$= -38\,400 + \frac{6400}{1 - \dfrac{p^2}{32\,000}} + \frac{60}{\phi_b} \qquad . \qquad . \tag{4}$$

substituting for $\dfrac{\phi_b}{\phi_c}$ from equation (3).

When $p = 2\pi \times 25$, $\phi_b = -0{\cdot}004\,22$ rad $= -0{\cdot}242°$ from equation (4)

and
$$\frac{\phi_b}{\phi_c} = 0{\cdot}229 \qquad . \qquad . \qquad . \qquad \text{from equation (3)}$$

so that $\underline{\phi_c = -1{\cdot}055°}$

When $p = 2\pi \times 29$, $\phi_b = +0{\cdot}000\,341$ rad $= +0{\cdot}0195°$

and
$$\frac{\phi_b}{\phi_c} = -0{\cdot}0375$$

so that $\cdot\,\underline{\phi_c = -0{\cdot}52°}$

The node is at A in the first case and is between B and C in the second case.

P

9. *A balancing machine frame is equivalent to a uniform rigid slender bar AB of length 2a and mass 50 kg, supported by a spring at each end, Fig. 18.12. The stiffness of each spring is 90 kN/m and in the position of static equilibrium the bar is horizontal.*

A vertical harmonic exciting force $P = P_0 \sin pt$ N is applied to the bar at a distance a/10 from the centre of mass of the bar.   Determine the amplitudes of vibration of A and B if the amplitude of the exciting force is 65 N at a frequency of 17·5 Hz.   Neglect damping and consider small amplitudes of oscillation for the steady-state forced vibration only.   (U. Lond.)

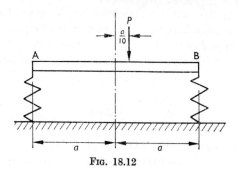

FIG. 18.12

Let the linear displacement of the centre of the beam be x and the angular displacement of the beam be θ, Fig. 18.13.

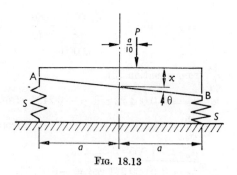

FIG. 18.13

Then the equation of linear motion is

$$m\frac{d^2x}{dt^2} = -S(x - a\theta) - S(x + a\theta) + P_0 \sin pt$$

$$= -2Sx + P_0 \sin pt \qquad . \qquad . \qquad . \qquad . \qquad (1)$$

and the equation of angular motion is

$$m\frac{a^2}{3}\frac{d^2\theta}{dt^2} = S(x - a\theta)a - S(x + a\theta)a + P_0 \sin pt \times \frac{a}{10}$$

$$= -2Sa^2\theta + \frac{a}{10}P_0 \sin pt$$

$$m\frac{d^2\theta}{dt^2} = -6S\theta + \frac{3P_0}{10a} \sin pt \qquad . \qquad . \qquad . \qquad (2)$$

Assuming that $x = k \sin pt$ and $\theta = \phi \sin pt$, equations (1) and (2) become

$$-p^2 mk = -2Sk + P_0$$

and $\qquad -p^2 m\phi = -6S\phi + \dfrac{3P_0}{10a}$

from which $\qquad k = \dfrac{P_0}{2S - p^2 m} = \dfrac{65}{2 \times 90 \times 10^3 - (35\pi)^2 \times 50}$

$$= -0.000\ 154 \text{ m}$$

and $\qquad \phi a = \dfrac{3P_0}{10(6S - p^2 m)} = \dfrac{3 \times 65}{10(6 \times 90 \times 10^3 - (35\pi)^2 \times 50)}$

$$= -0.000\ 302 \text{ m}$$

$\therefore$ amplitude of A $= 0.000\ 302 - 0.000\ 154 = \underline{0.000\ 148 \text{ m}}$

and amplitude of B $= 0.000\ 302 + 0.000\ 154 = \underline{0.000\ 456 \text{ m}}$

10. A body of mass m is suspended from the lower end of a spring of stiffness S and the upper end of the spring is given a periodic displacement $y = y_{\max} \sin pt$. Determine the displacement x of the body from its equilibrium position when the motion has become established. Hence, show that the motion of the body relative to the support has an amplitude $\dfrac{p^2}{\omega^2 - p^2}$ times that of the support, where $\omega^2 = \dfrac{S}{m}$. (*I. Mech. E.*)

11. A mass $m = 35$ kg is supported by a spring with a stiffness $S = 25$ kN/m, and acted on by a disturbing force of amplitude $P = 40$ N and frequency $f = 5$ Hz. Find the amplitude of the forced vibration if there is no damping. Work from first principles and establish the formulae required. (*I. Mech. E.*)
(*Ans.* : 4·185 mm)

12. A light helical spring carries a mass of 5 kg and the upper end of the spring is attached to a pin which moves in a vertical path with S.H.M., with a total stroke of 40 mm. When the frequency of oscillation is 200 cycles/min, the total movement of the mass is observed to be 30 mm. Find (*a*) the strength of the spring, (*b*) the maximum spring force, (*c*) the natural frequency of the mass-spring system. (*U. Lond.*)
(*Ans.* : 940 N/m ; 82 N ; 2·18 Hz)

13. A mass of 100 kg is suspended from a spring of stiffness 4·5 kN/m. The upper end of the spring is given S.H.M. in a vertical direction by means of a crank

6 mm long. Determine the total movement, up and down, of the suspended mass and also its maximum velocity, if the crank is driven at 20 rad/s. (*U. Lond.*)

(*Ans.*: 1·52 mm; 15·2 mm/s)

14. A mass suspended from a spring has a natural frequency n for vertical oscillations. When a vertical force P_0 is applied to the mass it is deflected statically through a distance δ. If the system is set into forced oscillations (undamped) due to the action of a vertical force P on the mass, where $P = P_0 \sin 2\pi ft$, show that the amplitude of these vibrations will be

$$\delta\left\{\frac{1}{1 - \left(\frac{f}{n}\right)^2}\right\}.$$

In a particular instance an undamped vibrating system is set in motion by a sinusoidal force of constant amplitude, but variable frequency. The vibration amplitude is 25 mm at 10 Hz and decreases *continuously* to 2·5 mm at 20 Hz. Determine the critical frequency of this system, and hence the static deflection of the suspension due to the dead weight of the load. (*U. Lond.*)

(*Ans.*: 8·165 Hz; 3·73 mm)

15. An undamped vibrating system excited by a sinusoidal force of constant magnitude but variable frequency has a travel between extreme positions of 100 mm at 200 cycles/min and 50 mm at 600 cycles/min, and it is known that the critical frequency lies somewhere between these two rates. Determine the natural frequency of the system and hence, for a suspension stiffness of 7 kN/m, the mass of the suspended system. (*U. Lond.*)

(*Ans.*: 6·38 Hz; 4·375 kg)

16. A mass of 240 kg, supported by an elastic structure giving a restoring force proportional to displacement and having a stiffness equal to 1·75 MN/m, is acted upon by a simple harmonic disturbing force of maximum value 50 N and frequency 35 Hz. The amplitude of this system is to be reduced to zero by attaching a dynamic absorber to the mass. The absorber is to consist of a mass of 0·2 kg on a spring of stiffness S. The system experiences no damping. Find from first principles the numerical value of the stiffness S, neglecting the masses of the elastic member of the system and of the spring of the absorber. (*U. Lond.*)

(*Ans.*: 9·67 kN/m)

17. An electric motor rests on spring supports which allow vertical motion and the deflection under its own weight is 0·5 mm. When running at 1400 rev/min the observed amplitude of forced vibration is 0·25 mm. Find the amount of out of balance in the machine expressed as kg at 1 mm radius. The mass of the motor is 500 kg.

It may be assumed that the amplitude of the oscillation is $1\left/\left\{1 - \left(\frac{p}{\omega}\right)^2\right\}\right.$ times the displacement which the disturbing force would produce if statically applied, (p/ω) being the ratio of the disturbing to the natural frequency. (*I. Mech. E.*)

(*Ans.*: 10·88 kg)

18. A single-cylinder engine of total mass 180 kg is to be mounted on an elastic support which permits vibratory movement in the vertical direction only. The piston of the engine has a mass of 3 kg and has a vertical reciprocating motion (which may be assumed to be simple harmonic) with stroke 150 mm. It is desired that the maximum vibratory force transmitted through the elastic support to the foundation shall be 500 N when the engine speed is 800 rev/min and less than this at all higher engine speeds.

(a) Find the necessary stiffness of the elastic support and the amplitude of vibration of the engine at 800 rev/min.

(b) If the speed of the engine is reduced below 800 rev/min, at what speed will the transmitted force again become 500 N ?   (*U. Lond.*)

(*Ans.* : 304 kN/m ; 1·645 mm ; 296 rev/min)

19. A sensitive instrument which requires to be insulated from vibrations is to be used in a laboratory where reciprocating machinery is in use. The vibrations of the floor of the laboratory may be assumed to be simple harmonic motion having a frequency in the range 1000 to 3000 cycles /min.   The instrument is to be mounted on a small platform and supported on three springs of equal stiffness and arranged to carry equal loads. If the combined mass of the instrument and supporting table is 5 kg, calculate suitable values for the stiffness of the springs if the amplitude of transmitted vibrations is to be less than 15% of the floor vibrations over the given frequency range. Damping is to be neglected and any formula used should be derived. (*U. Lond.*)   (*Ans.* : 2·387 kN/m)

20. A machine of mass m rests on a foundation of stiffness S. Its moving parts set up a disturbing force in the vertical direction with amplitude P and fixed frequency $p/2\pi$ Hz. Assuming that the amplitude of the forced oscillation is $\dfrac{1}{\omega^2 - p^2}\dfrac{P}{m}$, where $\omega^2 = \dfrac{S}{m}$, obtain an expression for the amplitude of the disturbing force transmitted to the foundation in terms of P and the frequency ratio ω/p. Draw to scale a curve showing the value of this force for ω/p between 0 and 3.

If, in an example, $m = 120$ kg and the disturbing force makes 2400 cycles/min, find S to limit the transmitted force to half its initial value, and the corresponding statical deflection of the machine when placed on the foundation. (*I. Mech. E.*)   (*Ans.* : 2·528 MN/m ; 0·466 mm)

21. A machine, fixed to the floor of a workshop, produces a static deflection of 2 mm immediately under the machine. When the machine is working, there is an unbalanced mass which produces a vertical alternating force whose frequency is equal to the speed of the driving shaft on the machine. When the speed of this shaft is 240 rev/min, the amplitude of the forced vibration of the floor is 1·25 mm. If the floor is assumed to be elastic and damping is neglected, what will be the amplitude of the forced vibration when the speed is 480 rev/min ? At what speed will resonance occur ? (*U. Lond.*)

(*Ans.* : 8·98 mm ; 668·8 rev/min)

22. An engine rests on an elastic foundation which deflects 0·85 mm under the dead load. Find the frequency of free vertical vibration.

If the engine has a mass of 1·25 t and when running at 450 rev/min there is an out of balance force of this frequency and amplitude 2·4 kN, find the amplitude of the forced vibration. Find also the maximum force exerted on the foundation. (*I. Mech. E.*)   (*Ans.* : 17·15 Hz ; 0·206 mm ; 15·22 kN)

23. A load of 500 kg is supported on a beam which deflects 2·5 mm statically at the load. A fluctuating vertical force is applied to the load by a mass of 5 kg rotating at 200 mm radius at a speed of 500 rev/min. Neglecting the mass of the beam, derive an expression for the amplitude of the forced vibration of the load and find its value. To what speed should the rotation of the mass be *increased* in order to reduce the amplitude of vibration to 2·5 mm ? (*U. Lond.*)

(*Ans.* : 4·66 mm ; 1336 rev/min)

24. A single-cylinder vertical engine of total mass 250 kg is mounted on springs having an effective stiffness of 150 kN/m. Revolving masses provide primary balance for 60% of the reciprocating parts which have a total mass of 18 kg. The stroke is 200 mm.

Neglecting the obliquity of the connecting rod, determine the amplitude of the steady-state vertical vibrations when the engine is running at 190 rev/min. At what speed of running would trouble probably be experienced due to resonance? (*U. Lond.*)   (*Ans.* : 5·58 mm ; 234 rev/min)

25. A small vertical engine is supported on a steel joist at mid-span. The mass of the engine including an allowance for the mass of the joist is 100 kg and the static deflection of the joist is 4 mm. The stroke of the engine is 100 mm and the mass of the reciprocating parts is 1 kg. If the reciprocating parts are assumed to have S.H.M., find the amplitude of the forced vertical vibrations when the speed of rotation of the engine crank is 800 rev/min. (*I. Mech. E.*)

(*Ans.* : 0·769 mm)

26. An extension spindle, 32 mm diameter, is rigidly held in a bracket attached to an engine frame and protrudes therefrom by 0·6 m. It carries a wheel of mass 18 kg at the free end. The engine when running at 420 rev/min sets up a transverse vibration at the wheel of twice this frequency and of amplitude 8 mm. Determine the amplitude of vibration of the bracket and estimate the reduction of overhang necessary to reduce the wheel vibration to $\frac{1}{10}$ of its observed value. $E = 200\ \text{GN/m}^2$. (*U. Lond.*)   (*Ans.* : 0·2065 mm ; 0·052 m)

27. A small high-speed steam turbine has a single wheel of mass 6 kg mounted at the mid-point of a 10-mm diameter steel shaft. The bearing span is 450 mm. Owing to slight manufacturing inaccuracies the centre of gravity of the wheel is 0·025 mm from the centre of rotation.

If the turbine rotates at 3000 rev/min, determine the amplitude of the steady state forced vibration, the dynamic load transmitted to the self-aligning bearings and the stress in the shaft due to this load if the mass of the shaft is neglected. $E = 200\ \text{GN/m}^2$. (*U. Lond.*)

(*Ans.* : 0·0274 mm ; 0·709 N (each) ; 1·625 MN/m²)

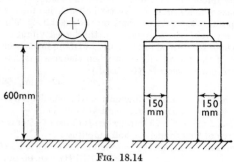

Fig. 18.14

28. A motor with a total mass of 17 kg is carried on four vertical legs as shown in Fig. 18.14. The motor has an unbalanced rotating mass of 1·8 kg at 25 mm radius. Calculate the range of speed over which the stress in the legs due to horizontal vibration of the motor exceeds 150 MN/m². Each leg is a steel flat 150 mm by 12 mm and is 600 mm long. The legs may be treated as cantilevers, fixed in direction at the bottom and hinged at the top. $E = 200\ \text{GN/m}^2$. (*U. Glas.*)   (*Ans.* : 1046 to 1252 rev/min)

29. A load of 700 kg is supported at the middle of a simply-supported beam of 6 m span. The second moment of area of the cross-section about the neutral axis is 88×10^{-6} m⁴. Neglecting the mass of the beam itself, find the natural frequency of vibration of the load.

A fluctuating force is applied to the load by a mass of 5 kg rotating at a radius of 200 mm at a speed of 400 rev/min. Calculate the amplitude of forced vibration of the load.

(For the above conditions the deflection at the load is $Wl^3/48EI$.)
$E = 200$ GN/m². (*U. Lond.*)   (*Ans. :* 11·94 Hz ; 0·647 mm)

30. A motor is mounted on a rigid rectangular plate, of mass 12 kg, which is hinged along one side, A, and is supported on the opposite side, B, by two helical springs, placed one at each corner as shown in Fig. 18.15. The motor has a mass

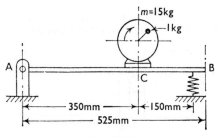

Fig. 18.15

of 15 kg and is fitted with a crank of 50 mm radius, which rotates at 400 rev/min. The mass acting at the crankpin is 1 kg. The stiffness of each spring is 15 kN/m of compression. Assuming that the applied vertical forces act at a point C, 350 mm from the hinge, and that the forces on the springs are equal, calculate (*a*) the natural frequency of vibration of the system, and (*b*) the maximum amplitude of C due to the varying load. Neglect horizontal forces. (*U. Lond.*)

(*Ans. :* 8·03 Hz ; 4·6 mm)

31. Two shafts A and B are geared together with speed ratio $\dfrac{\omega_b}{\omega_a} = r$. The shafts carry rotors with moments of inertia I_a and I_b respectively. If, with the system initially at rest, a torque $T \cos pt$ is applied to shaft A, find the value of r in terms of I_a and I_b for maximum amplitude of oscillation of shaft B. (*U. Lond.*)

(*Ans. :* $\sqrt{(I_a/I_b)}$)

32. An engine, having an effective moment of inertia of 8 kg m², is coupled to a flywheel, of moment of inertia 16 kg m² by means of a solid shaft 2 m long and 80 mm diameter. The engine torque varies by $400 \cos 2\Omega t$ N m, where Ω is the speed of the engine. When this speed is 900 rev/min, calculate (*i*) the amplitude of oscillation of the flywheel, and (*ii*) the maximum vibration stress in the shaft. $G = 80$ GN/m². (*Ans. :* 0·1517° ; 14·96 MN/m²)

33. A flywheel of moment of inertia 12 kg m² is keyed to one end of a shaft 40 mm diameter and 1 m long. The other end of the shaft is rigidly fixed. Find the frequency of the natural torsional vibrations of the system.

If a sinusoidal torque which varies between the limits of ± 65 N m is applied to the flywheel at a frequency of 5 Hz, find the amplitude of the forced vibration of the flywheel and the corresponding maximum vibration stress in the shaft material. $G = 80$ GN/m². *(I. Mech. E.)*

(*Ans.* : 6·52 Hz ; 0·45° ; 12·6 MN/m²)

34. Two flywheels A and B are mounted, one at each end, on a shaft 4 m long, 50 mm diameter, which is supported in bearings of negligible friction. A has a mass of 240 kg with radius of gyration 0·6 m. B has a mass of 360 kg with radius of gyration 0·8 m. The modulus of rigidity of the shaft is 80 GN/m². Find the frequency of free torsional vibrations and the position of the node.

A torque of 700 cos 20*t* N m is now applied to flywheel A. Calculate the amplitude of the forced vibrations of each flywheel and the new position of the node. *(U. Lond.)*

(*Ans.* : 2·225 Hz ; 1·09 m from B ; −1·965° ; +0·302° ; 0·533 m from B)

35. A motor M drives a centrifugal pump impeller C through gearing as shown in Fig. 18.16. The shaft between the motor and gear wheel W is very short and

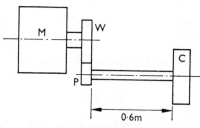

Fig. 18.16

may be regarded as being torsionally rigid. The shaft between the pinion P and the pump impeller is 0·6 m long and is of solid section having a diameter of 32 mm. The modulus of rigidity is 80 GN/m².

The moment of inertia of the motor shaft assembly, including gear wheel W, is 1·4 kg m². The pinion P on the pump shaft has a pitch circle diameter of 120 mm, its moment of inertia is 0·007 kg m² and it runs at twice the speed of wheel W. The moment of inertia of the pump impeller, including an allowance for the entrained water, is 0·175 kg m². Damping is assumed to be negligible.

During running conditions the motor torque contains a harmonic 16 sin 446*t* N m, where *t* is the time in seconds. By working from first principles, determine the amplitude of torsional vibration of the impeller and the maximum tangential force at the pitch circle of the gears due to this harmonic. *(U. Lond.)*

(*Ans.* : 0·00621° ; 704 N)

36. A rotor of mass 100 kg, of radius of gyration 100 mm, is fixed at the end A of a shaft AB, which is 12 mm diameter and 600 mm long. The modulus of rigidity is 80 GN/m².

(*a*) If the end B of the shaft is fixed, find the frequency of free torsional vibrations of the rotor.

(*b*) If the end B is given rotational simple harmonic motion of frequency 100 vib/min and maximum displacement 2° on either side of its original position, find the corresponding maximum displacement of the rotor and the maximum torque in the shaft. *(U. Lond.)*   (*Ans.* : 2·625 Hz ; 3·35° ; 6·4 N m)

37. A trailer consists of a body supported by springs upon an axle having two wheels. The body has a mass of 400 kg and the axle and wheel assembly has a mass of 100 kg. The total stiffness of the suspension springs is 80 kN/m and the total stiffness of the tyres is 200 kN/m.

The trailer is drawn along a level road which has a sinusoidal surface of wavelength 3 m. Determine the critical speeds of the trailer. (*U. Lond.*)

(*Ans.* : 5·65 m/s ; 25·55 m/s)

38. A system of two masses and three springs in series is arranged between fixed supports. The elements are, in order, a spring of stiffness k_1, a mass m_1, a spring of stiffness k_2, a mass m_2, and a spring stiffness k_3. A force $P = P_0 \cos pt$ acts on mass 1. Obtain expressions for the amplitudes of the two masses in sustained vibration. Show that, if k_2 is very much less than k_1 and k_3, the natural frequencies of the system are approximately the same for mass 1 on spring 1 alone and for mass 2 on spring 3 alone. (*U. Lond.*)

$$\left(Ans. : \ a_1 = \frac{(k_2 + k_3 - m_2 p^2)P_0}{m_1 m_2 p^4 - \{(m_1 + m_2)k_2 + m_1 k_3 + m_2 k_1\}p^2 + (k_1 k_2 + k_2 k_3 + k_3 k_1)}\right.$$

$$a_2 = \frac{k_2 P_0}{m_1 m_2 p^4 - \{(m_1 + m_2)k_2 + m_1 k_3 + m_2 k_1\}p^2 + (k_1 k_2 + k_2 k_3 + k_3 k_1)}$$

39. A girder of mass 250 kg, supported by an elastic structure giving a restoring force proportional to displacement and of stiffness 1·4 MN/m, is acted upon by a simple harmonic disturbing force of maximum value 50 N and frequency 1400 cycles/min. The amplitude of the girder movement is to be reduced to zero by attaching to it a dynamic absorber comprising a mass of 0·35 kg on a spring of stiffness S ; there is no damping.

Find, *from first principles*, the value of S, neglecting the masses of the elastic member of the system and absorber spring. (*U. Lond.*)

(*Ans.* : 7·52 kN/m)

40. A horizontal beam of negligible inertia supports at its mid-point a motor of mass 50 kg. The mid-point of the beam deflects vertically 1 mm under a load of 100 N. It is observed that, due to the presence of rotor unbalance, the mid-point of the beam vibrates with simple harmonic motion of amplitude 5 mm when the motor is running at its normal speed of 440 rev/min. It is proposed to eliminate the vibration by the addition of a light spring attached to the beam at its mid-point, the lower end of the spring supporting a body of mass of 2 kg.

Deduce *from first principles* the required spring stiffness and the amplitude of vibration of the added body. Friction may be neglected. (*U. Lond.*)

'(*Ans.* : 4·24 kN/m ; 118 mm)

41. A light inextensible belt passes around two pulleys, P_1 and P_2, which are free to rotate about the parallel axes O_1 and O_2 respectively. Inserted in each of the straight portions of the belt is a spring ; one, S_1, is of stiffness 2 kN/m and the other, S_2, is of stiffness 4 kN/m. The centre distance $O_1 O_2$ is adjusted so that the static force in each spring is 80 N. The effective diameter, mass and radius of gyration of pulley P_1 are 120 mm, 2 kg and 50 mm respectively. The corresponding quantities for pulley P_2 are 180 mm, 4 kg and 80 mm. Pulley P_1 is subjected to a torque which varies sinusoidally, having an amplitude of 1·2 N m and a frequency of 15 Hz. The belt does not slip on the pulleys.

By considering only the steady-state forced vibration of the system, determine the maximum value of the ratio $\dfrac{\text{force in spring } S_2}{\text{force in spring } S_1}$. (*U. Lond.*) (*Ans.* : 1·467)

CHAPTER 19

FORCED-DAMPED VIBRATIONS

19.1 Forced-damped linear vibrations. If a body of mass m, Fig. 19.1, is acted upon by a restoring force S per unit displacement, a damping force c per unit velocity and also by an external harmonic force $P \cos pt$, the equation of motion is

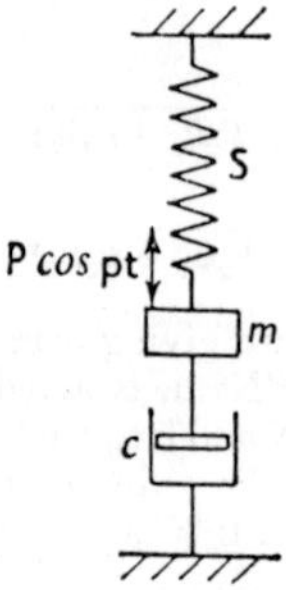

$$m\frac{d^2x}{dt^2} + c\frac{dx}{dt} + Sx = P \cos pt$$

or

$$\frac{d^2x}{dt^2} + 2\mu\frac{dx}{dt} + \omega^2 x = \frac{P}{m}\cos pt \tag{19.1}$$

where

$$2\mu = \frac{c}{m} \quad \text{and} \quad \omega^2 = \frac{S}{m}$$

Fig. 19.1

The complementary function, i.e. the solution of

$$\frac{d^2x}{dt^2} + 2\mu\frac{dx}{dt} + \omega^2 x = 0$$

is, as in Chapter 17,

$$x = C_1 e^{[-\mu + \sqrt{(\mu^2 - \omega^2)}]t} + C_2 e^{[-\mu - \sqrt{(\mu^2 - \omega^2)}]t} \qquad \text{if } \mu > \omega$$

or

$$x = (C_1 + C_2 t)e^{-\mu t} \qquad \text{if } \mu = \omega$$

or

$$x = e^{-\mu t}\{A \cos \sqrt{(\omega^2 - \mu^2)}t + B \sin \sqrt{(\omega^2 - \mu^2)}t\} \qquad \text{if } \mu < \omega$$

Equation (19.1) may be written

$$(D^2 + 2\mu D + \omega^2)x = \frac{P}{m}\cos pt$$

so that the particular integral is given by

$$x = \left(\frac{1}{D^2 + 2\mu D + \omega^2}\right)\frac{P}{m}\cos pt$$

$$= \left(\frac{1}{2\mu D + (\omega^2 - p^2)}\right)\frac{P}{m}\cos pt \quad \text{since } f(D^2)\cos pt = f(-p^2)\cos pt$$

$$= \left(\frac{2\mu D - (\omega^2 - p^2)}{4\mu^2 D^2 - (\omega^2 - p^2)^2}\right)\frac{P}{m}\cos pt$$

$$= \frac{P}{m}\left(\frac{-2\mu p \sin pt - (\omega^2 - p^2)\cos pt}{-4\mu^2 p^2 - (\omega^2 - p^2)^2}\right)$$

434

$$= \frac{P\sqrt{[4\mu^2p^2 + (\omega^2 - p^2)^2]} \cos{(pt - \alpha)}}{m[4\mu^2p^2 + (\omega^2 - p^2)^2]}$$

where $\qquad\qquad \alpha = \tan^{-1}\dfrac{2\mu p}{\omega^2 - p^2}$ . . . . (19.2)

i.e. $\qquad\qquad x = \dfrac{P\cos{(pt - \alpha)}}{m\sqrt{[4\mu^2p^2 + (\omega^2 - p^2)^2]}}$ . . . (19.3)

The complementary function (or *transient solution*) represents the damped free vibration of the body which dies out, leaving

$$x = \frac{P\cos{(pt - \alpha)}}{m\sqrt{[4\mu^2p^2 + (\omega^2 - p^2)^2]}}$$

to represent the steady-state vibration. This is a harmonic motion of frequency $\dfrac{p}{2\pi}$ Hz and amplitude

$$a = \frac{P}{m\sqrt{[4\mu^2p^2 + (\omega^2 - p^2)^2]}} \qquad . \qquad . \qquad . \qquad (19.4)$$

lagging the disturbing force by an angle α.

When $p = \omega$, the amplitude is $\dfrac{P}{2\mu mp}$ instead of infinity, as in the case of undamped motion.

The maximum amplitude occurs when $\dfrac{da}{dp} = 0$.

If P is constant, then $p = \sqrt{(\omega^2 - 2\mu^2)}$ for maximum amplitude. When μ is small, the maximum amplitude occurs when $p \simeq \omega$, but as μ increases, the maximum amplitude occurs at lower values of p, reaching the limiting value of $p = 0$ when $\mu = \dfrac{\omega}{\sqrt{2}}$.

If $P \propto p^2$, then $p = \dfrac{\omega^2}{\sqrt{(\omega^2 - 2\mu^2)}}$ for maximum amplitude. When μ is small, the maximum amplitude occurs when $p \simeq \omega$, but as μ increases, the maximum amplitude occurs at higher values of p, reaching the limiting value of $p = \infty$ when $\mu = \dfrac{\omega}{\sqrt{2}}$.

Figs. 19.2 and 19.3 show the variation in a as p varies for the two cases of P constant and $P \propto p^2$ respectively. In each case, the graphs have been drawn for various values of $\dfrac{\mu}{\omega}\left(= \dfrac{c}{c_0}\right)$, from $\dfrac{\mu}{\omega} = 0$ to $\dfrac{1}{\sqrt{2}}$.

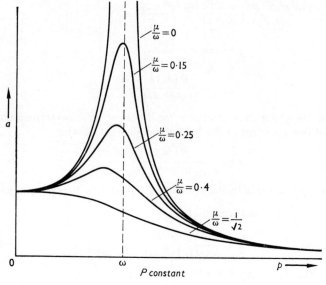

Fig. 19.2

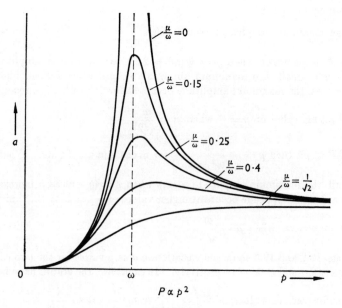

Fig. 19.3

The angle of lag (or *phase angle*), α, is given by equation (19.2) and Fig. 19.4 shows how α varies with p for the same values of $\dfrac{\mu}{\omega}$.

When μ is small, the effect upon the amplitude and phase angle is small, except near the resonant frequency.

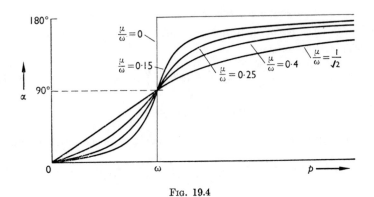

FIG. 19.4

19.2 Dynamic magnifier. The amplitude may be written in the form

$$a = \frac{P}{m\omega^2 \sqrt{\left[\frac{4\mu^2 p^2}{\omega^4} + \left(1 - \left[\frac{p}{\omega}\right]^2\right)^2\right]}}$$

but $\dfrac{P}{m\omega^2} = \dfrac{P}{S} =$ deflection under a static load $P = \Delta$

$$\therefore \ a = K\Delta \quad \text{where} \quad K = \frac{1}{\sqrt{\left[\left(\frac{2\mu p}{\omega^2}\right)^2 + \left(1 - \left[\frac{p}{\omega}\right]^2\right)^2\right]}} \quad . \quad (19.5)$$

Thus K is the ratio of the vibration amplitude due to the force $P \cos pt$ to the deflection under a static load P and is called the *dynamic magnifier.*

For undamped forced vibrations, $\mu = 0$, so that

$$K = 1 \Big/ \left[1 - \left(\frac{p}{\omega}\right)^2\right] \quad . \quad . \quad . \quad (19.6)$$

19.3   Forced-damped angular vibrations.   The corresponding equation of motion for forced-damped angular vibrations, in which a body is subjected to an external harmonic torque $T \cos pt$, Fig. 19.5, is

$$I\frac{d^2\theta}{dt^2} + c\frac{d\theta}{dt} + q\theta = T \cos pt$$

where $I =$ moment of inertia

              $c =$ damping torque per unit angular velocity

and $q =$ restoring torque per unit angular displacement

This can be written

$$\frac{d^2\theta}{dt^2} + 2\mu\frac{d\theta}{dt} + \omega^2\theta = \frac{T}{I}\cos pt \quad . \quad . \quad . \quad (19.7)$$

where $2\mu = \dfrac{c}{I}$ and $\omega^2 = \dfrac{q}{I}$

The solution of this equation is similar to that for linear vibrations.

Thus, the amplitude $\phi = \dfrac{T}{I\sqrt{[4\mu^2 p^2 + (\omega^2 - p^2)^2]}}$.   .   (19.8)

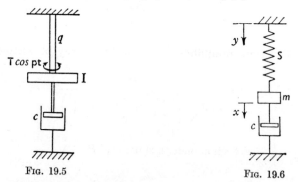

FIG. 19.5   FIG. 19.6

19.4   Periodic movement of support.   If the vibration is produced by a periodic movement of the support, Fig. 19.6, such that the displacement of the support, $y = h \cos pt$, the equation of motion is

$$m\frac{d^2x}{dt^2} + c\frac{dx}{dt} + S(x - y) = 0$$

i.e. $\dfrac{d^2x}{dt^2} + 2\mu\dfrac{dx}{dt} + \omega^2 x = \omega^2 h \cos pt.$.   (19.9)

The solution of this equation is the same as for equation (19.1), so that

$$a = \frac{\omega^2 h}{\sqrt{[4\mu^2 p^2 + (\omega^2 - p^2)^2]}} \quad . \quad . \quad . \quad (19.10)$$

If the 'fixed' member of the dash-pot is attached to the moving support, Fig. 19.7, the damping is proportional to the *relative* velocity between the mass and the support.

The equation of motion then becomes

$$m\frac{d^2x}{dt^2} + c\left(\frac{dx}{dt} - \frac{dy}{dt}\right) + S(x - y) = 0 \qquad . \qquad . \quad (19.11)$$

If the relative displacement between the mass and the support is z, then

$$x - y = z$$

$$\frac{dx}{dt} - \frac{dy}{dt} = \frac{dz}{dt}$$

and

$$\frac{d^2x}{dt^2} - \frac{d^2y}{dt^2} = \frac{d^2z}{dt^2}$$

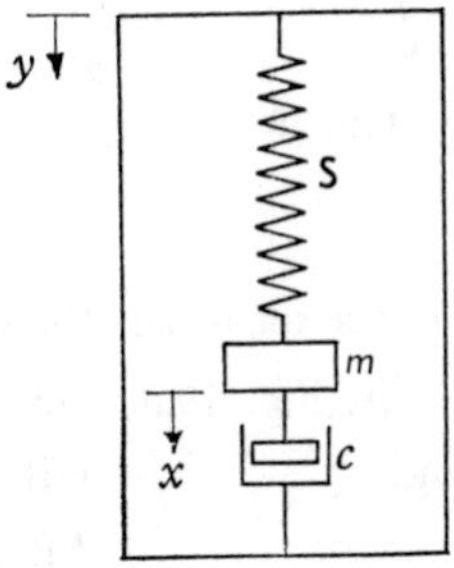

Equation (19.11) can therefore be written

$$m\left(\frac{d^2z}{dt^2} + \frac{d^2y}{dt^2}\right) + c\frac{dz}{dt} + Sz = 0$$

FIG. 19.7

or

$$m\frac{d^2z}{dt^2} + c\frac{dz}{dt} + Sz = -m\frac{d^2y}{dt^2} = mp^2h \cos pt$$

$$\therefore \quad \frac{d^2z}{dt^2} + 2\mu\frac{dz}{dt} + \omega^2 z = p^2h \cos pt \qquad . \qquad . \qquad . \quad (19.12)$$

Equation (19.12) is similar to equation (19.10) but the amplitude obtained from this equation is that of the *relative* motion. The amplitude of the *absolute* motion of m must be obtained from a vector diagram, since the motions of the mass and the support are not in phase.

Similar equations may be deduced for angular motion.

1. *A mass of 18 kg is carried on a spring of stiffness 7500 N/m from a support which has a vertical harmonic motion of amplitude ± 4 mm at a frequency of 4 Hz. The motion of the mass is opposed by a force proportional to its absolute velocity, and of amount 75 N s/m. Calculate the amplitude of the steady motion of the mass and its phase relative to the motion of the support. Find also the maximum instantaneous extension of the spring.* (U. Glas.)

$$a = \frac{\omega^2 h}{\sqrt{[4\mu^2 p^2 + (\omega^2 - p^2)^2]}} \qquad \text{from equation (19.10)}$$

$$\omega^2 = \frac{S}{m} = \frac{7500}{18} = 416 \cdot 7 \text{ (rad/s)}^2$$

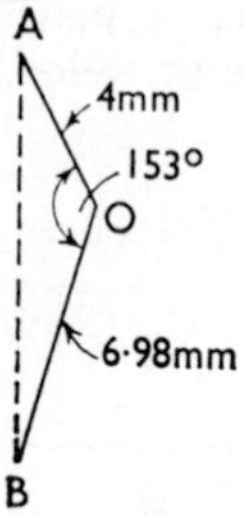

$$2\mu = \frac{c}{m} = \frac{75}{18} = 4\cdot167 \text{ rad/s}$$

$$p^2 = (2\pi \times 4)^2 = 631\cdot7 \text{ (rad/s)}^2$$

$$\therefore a = \frac{416\cdot7 \times 4}{\sqrt{[4\cdot167^2 \times 631\cdot7 + (416\cdot7 - 631\cdot7)^2]}}$$

$$= \underline{6\cdot98 \text{ mm}}$$

Fig. 19.8

$$\alpha = \tan^{-1} \frac{2\mu p}{\omega^2 - p^2} \qquad . \qquad . \qquad \text{from equation (19.2)}$$

$$= \tan^{-1} \frac{4\cdot167 \times 8\pi}{416\cdot7 - 631\cdot7} = 153°$$

i.e. the mass lags the disturbing force by 153°.

In Fig. 19.8, OA represents the amplitude of the disturbing force and OB represents the amplitude of the mass

∴ AB represents the maximum extension of the spring.

Using the cosine rule,

$$AB = \sqrt{(4^2 + 6\cdot98^2 - 2 \times 4 \times 6\cdot98 \times \cos 153°)}$$

$$= \underline{10\cdot7 \text{ mm}}$$

2. *A periodic torque, having a maximum value of 0·5 N m at a frequency corresponding to 4 rad/s, is impressed on a flywheel suspended from a wire. The wheel has a moment of inertia of 0·12 kg m² and the wire has a stiffness of 1 N m/rad. A viscous dash-pot applies a damping couple of 0·4 N m at an angular velocity of 1 rad/s.*

Calculate (a) the maximum angular displacement from the rest position, and (b) the maximum couple applied to the dash-pot.   (U. Lond.)

From equation (19.8), $\phi = \dfrac{T}{I\sqrt{[4\mu^2 p^2 + (\omega^2 - p^2)^2]}}$

$$2\mu = \frac{c}{I} = \frac{0\cdot4}{0\cdot12} = 3\cdot33 \text{ rad/s}$$

$$\omega^2 = \frac{q}{I} = \frac{1}{0\cdot12} = 8\cdot33 \text{ (rad/s)}^2$$

$$p^2 = 4^2 = 16 \text{ (rad/s)}^2$$

$$\therefore \phi = \frac{0\cdot5}{0\cdot12\sqrt{[3\cdot33^2 \times 16 + (8\cdot33 - 16)^2]}} = \underline{0\cdot271 \text{ rad}}$$

Maximum angular velocity of flywheel $= p\phi = 4 \times 0\cdot271$

$$= 1\cdot084 \text{ rad/s}$$

∴ maximum couple on dash-pot $= 1\cdot084 \times 0\cdot4 = \underline{0\cdot4336 \text{ N m}}$

3. *Fig. 19.9 shows an instrument for recording the vertical movements of the body on which it stands. The mass m is suspended by a spring of stiffness S and the motion is damped by a dash-pot having a viscous damping*

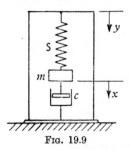

FIG. 19.9

coefficient c. The movement of m relative to the frame of the instrument is recorded on a drum. If the frequency of free undamped oscillations of the mass is $\frac{\omega}{2\pi}$, if the value of c is $1/\sqrt{2}$ of the value for critical damping and if the frame is given a vertical oscillation of $h \cos pt$, show that the amplitude recorded by the instrument is $\dfrac{hp^2}{\sqrt{(\omega^4 + p^4)}}$. (U. Lond.)

Both the mass and the frame are oscillating, so that the damping of the mass is proportional to the relative velocity between them.

Hence, if x and y are the instantaneous displacements of the mass and the frame respectively, the equation of motion of m is

$$m\frac{d^2x}{dt^2} + c\left(\frac{dx}{dt} - \frac{dy}{dt}\right) + S(x - y) = 0 \qquad . \qquad . \qquad (1)$$

If the relative displacement between the mass and machine is z, then

$$\frac{d^2z}{dt^2} + 2\mu\frac{dz}{dt} + \omega^2 z = -\frac{d^2y}{dt^2} = p^2 h \cos pt \qquad . \quad \text{from equation (19.12)}$$

where $2\mu = \dfrac{c}{m}$ and $\omega^2 = \dfrac{S}{m}$.

$\therefore$ amplitude of $z = \dfrac{p^2h}{\sqrt{[4\mu^2p^2 + (\omega^2 - p^2)^2]}}$ from equation (19.10)

but $\qquad\qquad c = \dfrac{1}{\sqrt{2}} \times$ critical value of c

$$\therefore \mu = \frac{1}{\sqrt{2}} \times \text{critical value of } \mu = \frac{\omega}{\sqrt{2}}$$

$\therefore$ amplitude of $z = \dfrac{p^2h}{\sqrt{[2\omega^2p^2 + (\omega^2 - p^2)^2]}} = \dfrac{p^2h}{\sqrt{(\omega^4 + p^4)}}$

4. *A periodic force* $P = 40 \cos 10t$ *N is applied to the system shown in Fig. 19.10. Calculate from first principles the work done in one cycle.*

(U. Lond.)

From equation (19.3)

$$x = \frac{P \cos (pt - \alpha)}{m\sqrt{[4\mu^2 p^2 + (\omega^2 - p^2)^2]}}$$

$$= a \cos (pt - \alpha)$$

where

$$\alpha = \tan^{-1} \frac{2\mu p}{\omega^2 - p^2}$$

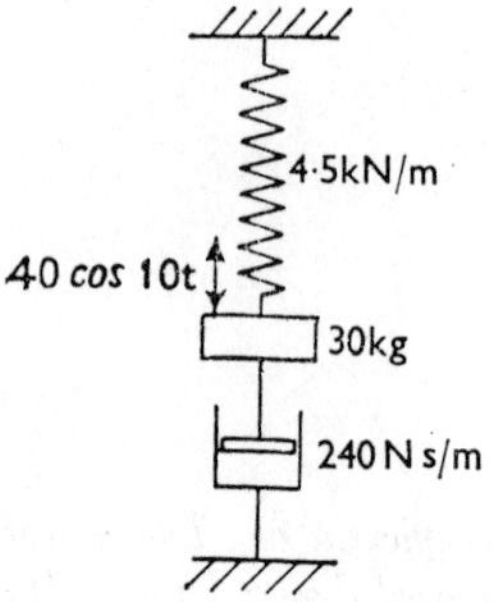

$$\therefore \frac{dx}{dt} = -ap \sin (pt - \alpha)$$

Fig. 19.10

Work done per cycle

$$= \int (P \cos pt) \, dx$$

$$= \int_0^{\frac{2\pi}{p}} (P \cos pt)\{-ap \sin (pt - \alpha)\} \, dt$$

$$= -Pap \int_0^{\frac{2\pi}{p}} \{\cos pt \sin pt \cos \alpha - \cos^2 pt \sin \alpha\} \, dt$$

$$= -Pap\left\{-\frac{\sin \alpha}{2} \times \frac{2\pi}{p}\right\} = \pi Pa \sin \alpha$$

$$2\mu = \frac{c}{m} = \frac{240}{30} = 8 \text{ rad/s}$$

$$\omega^2 = \frac{S}{m} = \frac{4 \cdot 5 \times 10^3}{30} = 150 \text{ (rad/s)}^2$$

$$\therefore a = \frac{40}{30\sqrt{[8^2 \times 100 + (150 - 100)^2]}}$$

$$= 0 \cdot 014 \ 14 \text{ m}$$

$$\alpha = \tan^{-1} \frac{8 \times 10}{150 - 100} = 58°$$

$$\therefore \text{ work done per cycle} = \pi \times 40 \times 0 \cdot 014 \ 14 \times \sin 58°$$

$$= \underline{1 \cdot 501 \text{ J}}$$

5. *A machine is liable to perform vibrations of the type, displacement* $= h \cos pt$. *A mass m is connected to the machine by a spring so that it can move in the direction of the vibrations. The mass is controlled by the spring which is of stiffness S units of force per unit deflection and a damper which produces a resisting force of c units per unit velocity of the mass relative to the machine. Find an expression for the mean power absorbed by the damper, and hence the value of c, in terms of S, p and m, which will make the power absorbed a maximum.* (U. Lond.)

Let the instantaneous displacements of the mass and the machine be x and y respectively and let z be the relative displacement.

Then, as in Example 3, the equation of motion reduces to

$$\frac{d^2z}{dt^2} + 2\mu\frac{dz}{dt} + \omega^2 z = p^2 h \cos pt \quad \text{where} \quad 2\mu = \frac{c}{m} \quad \text{and} \quad \omega^2 = \frac{S}{m}$$

Therefore from equation (19.2),

$$z = \frac{p^2 h \cos(pt - \alpha)}{\sqrt{[4\mu^2 p^2 + (\omega^2 - p^2)^2]}}$$

$$= a \cos(pt - \alpha) \quad \text{where} \quad \alpha = \tan^{-1}\frac{2\mu p}{\omega^2 - p^2}$$

and

$$\frac{dz}{dt} = -ap \sin(pt - \alpha)$$

Damping force $= c\dfrac{dz}{dt} = -cap \sin(pt - \alpha)$

$\therefore$ energy absorbed per cycle

$$= \int \left(c\frac{dz}{dt}\right) dz$$

$$= \int_0^{\frac{2\pi}{p}} \{-cap \sin(pt - \alpha)\}\{-ap \sin(pt - \alpha)\}\, dt$$

$$= ca^2 p^2 \int_0^{\frac{2\pi}{p}} \sin^2(pt - \alpha)\, dt = ca^2 p^2 \times \frac{\pi}{p} = \pi ca^2 p$$

$\therefore$ mean power absorbed $= \dfrac{\pi ca^2 p}{t_p}$

$$= \frac{\pi cp}{\left(\dfrac{2\pi}{p}\right)} \times \frac{h^2 p^4}{4\mu^2 p^2 + (\omega^2 - p^2)^2}$$

$$= \frac{ch^2 p^6}{2\left\{\dfrac{c^2 p^2}{m^2} + \left(\dfrac{S}{m} - p^2\right)^2\right\}} = \frac{ch^2 p^6 m^2}{2\{c^2 p^2 + (S - mp^2)^2\}}$$

Differentiating with respect to c and equating to zero for maximum power absorbed,

$$2\{c^2p^2 + (S - mp^2)^2\} \times h^2p^6m^2 = ch^2p^6m^2 \times 4cp^2$$

i.e.
$$(S - mp^2)^2 = c^2p^2$$

$$\therefore c = \underline{\frac{S - mp^2}{p}}$$

6. *Some machinery is mounted on a bedplate which is supported on four elastic members, having a stiffness each of 3·5 MN/m. The total mass to be supported is 1 t. It is estimated that the total damping force exerted on the system is 20 per cent of the critical, it being assumed that the damping is proportional to the velocity of motion.*

Measurements of the motion of the bedplate show that when the speed of rotation of the machine is 2000 rev/min, the maximum amplitude of vertical motion of the bedplate is 0·06 mm.

Calculate the total maximum force transmitted through each mounting to the ground, assuming the motion to be simple harmonic.   (U. Lond.)

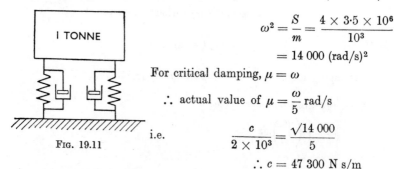

FIG. 19.11

$$\omega^2 = \frac{S}{m} = \frac{4 \times 3 \cdot 5 \times 10^6}{10^3}$$

$$= 14\,000 \ (\text{rad/s})^2$$

For critical damping, $\mu = \omega$

$$\therefore \text{actual value of } \mu = \frac{\omega}{5} \text{ rad/s}$$

i.e.
$$\frac{c}{2 \times 10^3} = \frac{\sqrt{14\,000}}{5}$$

$$\therefore c = 47\,300 \text{ N s/m}$$

Maximum velocity of bedplate $= pa = \left(\frac{2\pi}{60} \times 2000\right) \times 0 \cdot 06 \times 10^{-3}$

$$= 0 \cdot 012\,58 \text{ m/s}$$

$\therefore$ maximum damping force $= 47\,300 \times 0 \cdot 012\,58 = 595$ N

Maximum spring force $= 4 \times 3 \cdot 5 \times 10^6 \times 0 \cdot 06 \times 10^{-3} = 840$ N

The maximum value of the dynamic force transmitted to the ground is the vector sum of the spring and damping forces. These forces are 90° out of phase (see equation 13.20), so that

maximum dynamic force $= \sqrt{(840^2 + 595^2)} = 1029$ N

$\therefore$ maximum total force $= 1029 + 1000 \times 9 \cdot 81 = 10\,839$ N

$\therefore$ force transmitted through each mounting $= \frac{1}{4} \times 10\,839 = \underline{2710 \text{ N}}$

7. *The system shown in Fig. 19.12 consists of a lever AB, hinged at A, and a mass of 8 kg at B. A spring CD of stiffness 20 kN/m is connected to the lever at C, and a dash-pot giving a viscous resistance of 35 N s/m is connected at B. The end D of the spring is given a vertical sinusoidal motion of amplitude 1·25 mm at a frequency of 3 Hz. Find the amplitude of the resulting steady motion of the mass B and the dynamic force required at D.*

(U. Glas.)

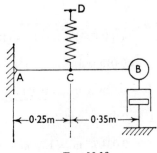

Fig. 19.12

$$I_A = 8 \times 0 \cdot 6^2 = 2 \cdot 88 \text{ kg m}^2$$

Let the instantaneous angular displacement of AB be θ.

Then   displacement of C = $0 \cdot 25\theta$ m

$\therefore$ extension of spring = $0 \cdot 25\theta - 0 \cdot 001\,25 \cos 6\pi t$ m

$\therefore$ restoring moment about A = $20 \times 10^3 (0 \cdot 25\theta - 0 \cdot 001\,25 \cos 6\pi t)$
$$\times 0 \cdot 25 \text{ N m}$$

$$= 1250\theta - 6 \cdot 25 \cos 6\pi t \text{ N m}$$

Velocity of B = $0 \cdot 6\dfrac{d\theta}{dt}$ m/s

$\therefore$ damping force = $0 \cdot 6\dfrac{d\theta}{dt} \times 35$ N

$\therefore$ moment of damping force about A = $0 \cdot 6\dfrac{d\theta}{dt} \times 35 \times 0 \cdot 6$

$$= 12 \cdot 6\dfrac{d\theta}{dt} \text{ N m}$$

Therefore the equation of motion of AB is

$$2 \cdot 88 \frac{d^2\theta}{dt^2} + 12 \cdot 6 \frac{d\theta}{dt} + 1250\theta - 6 \cdot 25 \cos 6\pi t = 0$$

or

$$\frac{d^2\theta}{dt^2} + 4 \cdot 375 \frac{d\theta}{dt} + 434\theta = 2 \cdot 17 \cos 18 \cdot 85t$$

Therefore from equation (19.8),

$$\text{amplitude of } \theta = \frac{2 \cdot 17}{\sqrt{[4 \cdot 375^2 \times 18 \cdot 85^2 + (434 - 18 \cdot 85^2)^2]}}$$

$$= 0 \cdot 019 \text{ rad}$$

$\therefore$ amplitude of B $= 0 \cdot 019 \times 0 \cdot 6 = \underline{0 \cdot 0114 \text{ m}}$

Amplitude of C $= 0 \cdot 019 \times 0 \cdot 25$

$$= 0 \cdot 004 \ 75 \text{ m} = 4 \cdot 75 \text{ mm}$$

$$\alpha = \tan^{-1} \frac{4 \cdot 375 \times 18 \cdot 85}{434 - 18 \cdot 85^2} \quad \text{from equation (19.2)}$$

$$= 46° \ 21'$$

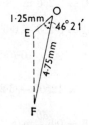

Fig. 19.13

Referring to Fig. 19.13, maximum extension of spring,

$$EF = \sqrt{(1 \cdot 25^2 + 4 \cdot 75^2 - 2 \times 1 \cdot 25 \times 4 \cdot 75 \cos 46° \ 21')}$$

$$= 3 \cdot 99 \text{ mm}$$

$\therefore$ maximum dynamic force at D $= 0 \cdot 003 \ 99 \times 20 \times 10^3 = \underline{79 \cdot 8 \text{ N}}$

8. *A generator, of moment of inertia 300 kg m², is driven by a three-cylinder oil engine through a shaft which may be taken as equivalent to a length of 2·2 m of 140 mm uniform diameter. The engine, with its flywheel, has a total moment of inertia of 400 kg m², and its torque fluctuation is given by 3500 cos 3Ωt N m where Ω is the speed in rad/s. At a speed of 280 rev/min torsiograph records show that the vibrational angle of twist in the equivalent shaft is 1·15°.*

Determine : (a) *the third order critical speed of the system,*

(b) *the damping torque per unit of vibrational velocity of twist,*

(c) *the cyclic stress in the equivalent shaft at the full speed of 300 rev/min assuming the same damping factor. G = 80 GN/m².*   (U. Lond.)

The damping is proportional to the relative angular velocities of the two rotors.

Let the instantaneous angular displacements of A and B be θ_A and θ_B respectively, Fig. 19.14.

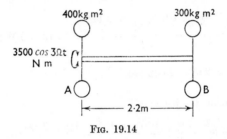

FIG. 19.14

Then the equation of motion of A is

$$I_A \frac{d^2\theta_A}{dt^2} + c\left(\frac{d\theta_A}{dt} - \frac{d\theta_B}{dt}\right) + q(\theta_A - \theta_B) = 3500 \cos 3\Omega t$$

or $$\frac{d^2\theta_A}{dt^2} + \frac{c}{I_A}\left(\frac{d\theta_A}{dt} - \frac{d\theta_B}{dt}\right) + \frac{q}{I_A}(\theta_A - \theta_B) = \frac{3500}{I_A} \cos 3\Omega t \qquad (1)$$

and the equation of motion of B is

$$I_B \frac{d^2\theta_B}{dt^2} - c\left(\frac{d\theta_A}{dt} - \frac{d\theta_B}{dt}\right) - q(\theta_A - \theta_B) = 0$$

or $$\frac{d^2\theta_B}{dt^2} - \frac{c}{I_B}\left(\frac{d\theta_A}{dt} - \frac{d\theta_B}{dt}\right) - \frac{q}{I_B}(\theta_A - \theta_B) = 0 \qquad . \qquad (2)$$

If the relative displacement between the rotors is ψ then

$$\psi = \theta_A - \theta_B$$

$$\therefore \frac{d\psi}{dt} = \frac{d\theta_A}{dt} - \frac{d\theta_B}{dt}$$

and

$$\frac{d^2\psi}{dt^2} = \frac{d^2\theta_A}{dt^2} - \frac{d^2\theta_B}{dt^2}$$

Therefore, subtracting equation (2) from equation (1),

$$\frac{d^2\psi}{dt^2} + c\left(\frac{1}{I_A} + \frac{1}{I_B}\right)\frac{d\psi}{dt} + q\left(\frac{1}{I_A} + \frac{1}{I_B}\right)\psi = \frac{3500}{I_A}\cos 3\Omega t \quad (3)$$

but

$$I_A = 400 \text{ kg m}^2 \quad \text{and} \quad I_B = 300 \text{ kg m}^2$$

$$\therefore \frac{1}{I_A} + \frac{1}{I_B} = \frac{7}{1200}$$

$$q = \frac{GJ}{l} = \frac{80 \times 10^9 \times \frac{\pi}{32} \times 0{\cdot}14^4}{2{\cdot}2} = 1{\cdot}372 \times 10^6 \text{ N m/rad}$$

and $\quad \Omega = \frac{2\pi}{60} \times 280 = 29{\cdot}32 \text{ rad/s}$

Therefore equation (3) becomes

$$\frac{d^2\psi}{dt^2} + \frac{7c}{1200}\frac{d\psi}{dt} + \frac{1{\cdot}372 \times 10^6 \times 7\psi}{1200} = \frac{3500}{400}\cos 3\times 29{\cdot}32t$$

i.e.

$$\frac{d^2\psi}{dt^2} + 0{\cdot}005\,83c\frac{d\psi}{dt} + 8000\psi = 8{\cdot}75\cos 87{\cdot}96t \quad . \quad (4)$$

(a) The third order critical speed is the engine speed, Ω, corresponding to its natural frequency of vibration, ω, resonance being produced by the third harmonic, 3Ω.

Thus the third order speed $= \omega = \sqrt{8000}$

$$= 89{\cdot}45 \text{ rad/s} = \underline{854 \text{ rev/min}}$$

(b) The amplitude of $\psi = \dfrac{1{\cdot}15 \times \pi}{180}$

$$= \frac{8{\cdot}75}{\sqrt{[(0{\cdot}005\,83c)^2 \times 87{\cdot}96^2 + (8000 - 87{\cdot}96^2)^2]}} \quad \text{from equation (19.8)}$$

from which $\qquad\qquad c = \underline{677{\cdot}5 \text{ N m s/rad}}$

(c) At 300 rev/min, $\Omega = \dfrac{2\pi}{60} \times 300 = 10\pi$ rad/s

$$\therefore \; 3\Omega = 30\pi \text{ rad/s}$$

$\therefore$ amplitude of ψ

$$= \frac{8 \cdot 75}{\sqrt{[(0 \cdot 005\ 83 \times 677 \cdot 5)^2 \times (30\pi)^2 + (8000 - (30\pi)^2)^2]}}$$

i.e. $\Psi = 0 \cdot 009\ 14$ rad

The vibration stress τ is given by $\dfrac{\tau}{r} = \dfrac{G\Psi}{l}$

i.e. $\tau = \dfrac{80 \times 10^9 \times 0 \cdot 009\ 14 \times 0 \cdot 07}{2 \cdot 2}$ N/m²

$$= 23 \cdot 3 \text{ MN/m}^2$$

9. *A specimen for a torsional fatigue testing machine consists of a hollow shaft 225 mm outside diameter, 150 mm bore and 0·6 m long, with a heavy disc attached at each end. The two discs are identical, and a fluctuating torque of 10 kN m maximum value is applied at one of them, with a frequency of 2500 per min, the latter coinciding with the frequency of free undamped vibration of the system.*

(a) Neglecting the motion of the shaft, find the moment of inertia of each disc.

(b) Derive an expression for the amplitude of the forced vibration, assuming viscous damping. If, under that, the dynamic magnifier is found to be 30, find the value of the damping coefficient.

(c) Calculate the power absorbed in damping.

(d) If the speed of the exciting motion were allowed to increase by 1 per cent, what would be the percentage change in maximum stress in the shaft?
G = 80 GN/m².   (U. Lond.)

(a) $$J = \frac{\pi}{32}(0 \cdot 225^4 - 0 \cdot 15^4)$$

$$= 0 \cdot 000\ 202 \text{ m}^4$$

When the system is vibrating freely, the node is at the centre of the shaft, Fig. 19.15.

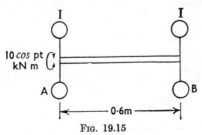

Fɪɢ. 19.15

Hence $\qquad n = \dfrac{1}{2\pi}\sqrt{\dfrac{GJ}{Ia}}$.   .   from equation (16.3)

i.e. $\qquad \dfrac{2500}{60} = \dfrac{1}{2\pi}\sqrt{\left[\dfrac{80 \times 10^9 \times 0 \cdot 000\ 202}{I \times 0 \cdot 3}\right]}$

$$\therefore\ I = \underline{786\ \text{kg m}^2}$$

(*b*) As in **Example 8** the equations of motion of A and B are respectively

$$I\frac{d^2\theta_A}{dt^2} + c\left(\frac{d\theta_A}{dt} - \frac{d\theta_B}{dt}\right) + q(\theta_A - \theta_B) = 10\ 000 \cos pt \quad . \qquad (1)$$

and $\quad I\dfrac{d^2\theta_B}{dt^2} - c\left(\dfrac{d\theta_A}{dt} - \dfrac{d\theta_B}{dt}\right) - q(\theta_A - \theta_B) = 0$.   .   .   (2)

Subtracting equation (2) from equation (1) and letting $\psi = \theta_A - \theta_B$,

$$I\frac{d^2\psi}{dt^2} + 2c\frac{d\psi}{dt} + 2q\psi = 10\ 000 \cos pt$$

or $\qquad \dfrac{d^2\psi}{dt^2} + \dfrac{2c}{I}\dfrac{d\psi}{dt} + \dfrac{2q}{I}\psi = \dfrac{10\ 000}{I} \cos pt$.   (3)

but $\qquad p = \dfrac{2\pi}{60} \times 2500 = 261 \cdot 7$ rad/s

and $\qquad q = \dfrac{GJ}{l} = \dfrac{80 \times 10^9 \times 0 \cdot 000\ 202}{0 \cdot 6} = 26 \cdot 95 \times 10^6$ N m/rad

Therefore equation (3) becomes

$$\frac{d^2\psi}{dt^2} + \frac{2c}{786}\frac{d\psi}{dt} + \frac{2 \times 26 \cdot 95 \times 10^6}{786}\psi = \frac{10\ 000}{786} \cos 261 \cdot 7t$$

or $\qquad \dfrac{d^2\psi}{dt^2} + 0 \cdot 002\ 545c\dfrac{d\psi}{dt} + 68\ 530\psi = 12 \cdot 74 \cos 261 \cdot 7t$.   (4)

Therefore from equation (19.8),

$$\text{amplitude of } \psi = \frac{12 \cdot 74}{\sqrt{[(0 \cdot 002\ 545c)^2 \times 261 \cdot 7^2 + (68\ 530 - 261 \cdot 7^2)^2]}}$$

i.e. $\qquad \Psi = \underline{\dfrac{19 \cdot 15}{c}\ \text{rad}}$.   .   .   .   .   .   (5)

Static angular displacement under a torque of 10 000 N m

$$= \frac{10\,000}{q} = \frac{10\,000}{26 \cdot 95 \times 10^6} = 0 \cdot 000\,377$$

$$\therefore \frac{19 \cdot 15}{c} = 30 \times 0 \cdot 000\,377$$

$$\therefore \underline{c = 1720 \text{ N m/rad}}$$

(c) $\psi = \Psi' \cos{(pt - \alpha)}$, where $\alpha = \tan^{-1} \dfrac{2\mu p}{\omega^2 - p^2}$ from equation (19.2)

$$\therefore \frac{d\psi}{dt} = -\Psi' p \sin{(pt - \alpha)}$$

$$\therefore \text{damping torque} = c\frac{d\psi}{dt} = -c\Psi' p \sin{(pt - \alpha)}$$

$$\therefore \text{energy absorbed per cycle} = \int \left(c\frac{d\psi}{dt} \right) d\psi$$

$$= \int_0^{\frac{2\pi}{p}} \{-c\Psi' p \sin{(pt - \alpha)}\}\{-\Psi' p \sin{(pt - \alpha)}\}\, dt$$

$$= c\Psi'^2 p^2 \int_0^{\frac{2\pi}{p}} \sin^2{(pt - \alpha)}\, dt = c\Psi'^2 p^2 \times \frac{\pi}{p} = \pi c\Psi'^2 p$$

$$\therefore \text{power absorbed} = \frac{\pi c\Psi'^2 p}{\left(\dfrac{2\pi}{p}\right)} = \frac{c\Psi'^2 p^2}{2}$$

but $$\Psi' = \frac{19 \cdot 15}{1720} = 0 \cdot 011\,14 \text{ rad} \qquad \text{from equation (5)}$$

$$\therefore \text{power absorbed} = \frac{1720 \times 0 \cdot 011\,14^2 \times 261 \cdot 7^2}{2} = \underline{7310 \text{ W}}$$

(d) When the speed is increased by 1 per cent,

$$\Psi' = \frac{12 \cdot 74}{\sqrt{[(0 \cdot 002\,545 \times 1720)^2 \times (1 \cdot 01 \times 261 \cdot 7)^2 + (68\,530 - (1 \cdot 01 \times 261 \cdot 7)^2)^2]}}$$

$$= 0 \cdot 007\,23 \text{ rad}$$

Percentage change in stress = percentage change in Ψ'

$$= \frac{0 \cdot 011\,14 - 0 \cdot 007\,23}{0 \cdot 011\,14} \times 100$$

$$= \underline{35 \cdot 1 \text{ per cent}}$$

10. A body of mass 18 kg is suspended from a spring which deflects 16 mm under this load. Calculate the frequency of free vibration and verify that a viscous damping force amounting to approximately 900 N at a speed of 1 m/s is just sufficient to make the motion aperiodic.

If when damped to this extent the body is subjected to a disturbing force with a maximum value of 100 N making 8 Hz, find the amplitude of the ultimate motion. (*I. Mech. E.*)   (*Ans.:* 3·94 Hz ; 1·774 mm)

11. A mass of 3 kg is supported by an elastic structure and causes a static extension of 60 mm. The mass is acted upon by a simple harmonic disturbing force having a maximum value of 10 N and a frequency of 1·5 Hz. The system experiences a damping force proportional to the velocity of motion and equal to 36 N at a velocity of 1 m/s. Determine the amplitude of steady motion of the forced vibration and show that this is approximately the maximum value to which the system is subject whatever the value of the frequency of the disturbing force. (*U. Lond.*)   (*Ans.:* 24·55 mm)

12. A simple vibrating system with one degree of freedom is under spring control and the motion is subjected to a damping force which is directly proportional to the velocity. The equivalent mass is 2 kg, the stiffness of the spring is 15 kN/m, and the damping force is 7 N at 1 m/s. Find the amplitude of the forced vibrations produced when a periodic force of $25 \cos 100t$ N acts on the mass. (*I. Mech. E.*)   (*Ans.:* 4·95 mm)

13. A piece of rotating machinery of mass $\frac{1}{2}$ t is mounted on flexible supports, causing these supports to deflect by 6 mm. The supports exert a total damping force equal to 15% of the critical, the damping being proportional to the velocity of motion. With the machine running at 500 rev/min an amplitude of vibration of 5 mm is measured. Calculate the maximum value of the disturbing force operating within the machinery. (*U. Lond.*)   (*Ans.:* 3·19 kN)

14. A mass of 50 kg suspended from a spring produces a statical deflection of 18·5 mm and when in motion it experiences a viscous damping force with a value of 720 N at a velocity of 1 m/s. Calculate the periodic time of damped vibration.

If the mass is then subjected to a periodic disturbing force, having a maximum value of 180 N and making 2 Hz, find the amplitude of the ultimate motion. (*I. Mech. E.*)   (*Ans.:* 0·2875 s ; 8·7 mm)

15. A mass hanging from a spring is observed to make one complete oscillation in $\frac{1}{2}$ s, and the amplitude of the fifth oscillation is half that of the first. If the top of the spring be compelled to make vertical oscillations of period 2 s and amplitude 25 mm, find the amplitude of the motion of the mass. The damping may be assumed proportional to the velocity. (*U. Lond.*)
                    (*Ans.:* 26·68 mm)

16. An instrument is mounted on a table, the table being suspended from the roof of a building by a set of springs. The effective mass of the table with instrument is 100 kg and the effective stiffness of the set of springs is 10 kN/m. Motion of the table is damped by a frictional resistance proportional to the velocity, the resistance being equal to 220 N s/m.

If the roof vibrates vertically $\pm 2\cdot5$ mm about a mean position with a frequency of 20 Hz, find from first principles the amplitude of the forced vibrations of the table : (*a*) taking account of the damping force ; (*b*) neglecting the damping force.

Comment on the difference between these values. (*U. Lond.*)
       (*Ans.:* 0·015 93 mm ; 0·015 93 mm [damping is negligible])

17. An automobile whose riding characteristics are to be investigated is placed on a platform which can be moved up and down with simple harmonic motion. For the purpose of making preliminary calculations it is assumed that the body of the automobile has one degree of freedom in the vertical direction and that it is supported by suspension springs having a total stiffness of 50 kN/m.

The total damping force offered by the shock absorbers is linearly viscous and 6 kN s/m. The body has a mass of 1 t and the tyres are assumed to be rigid.

If the platform vibrates with an amplitude of 25 mm and with a frequency equal to that of the undamped free frequency of the body on the suspension springs, determine the amplitude of motion of the body. (*U. Lond.*)

(*Ans. :* 29·5 mm (relative) ; 38·65 mm (absolute))

18. An engine of mass 240 kg is supported symmetrically on springs. When the engine is not running, it is observed that any free vertical vibrations are subject to viscous damping forces which cause a reduction of 20% in amplitude during each complete oscillation. The frequency of these oscillations is 9 Hz.

When the engine is running there is a harmonically varying vertical out-of-balance force whose amplitude is proportional to the square of the engine speed and is 140 N at 9 Hz. Find the amplitude of the steady-state forced vibrations when the engine is running at (*a*) 9 rev/s, (*b*) 18 rev/s. (*U. Lond.*)

(*Ans. :* 2·57 mm ; 0·243 mm)

19. A vibrating system consists of a 500-kg mass supported on a beam arrangement. The static deflection under the mass is 8·17 mm. Damping effects are small and are estimated from observing the amplitude of vibration of the mass decreasing from 5 mm to 1·25 mm in approximately 4 s.

If the supports of the beam are now given a vibratory movement of amplitude 2·5 mm and frequency equal to the natural frequency of the system, estimate the amplitude of the forced vibration of the mass. (*U. Glas.*) (*Ans :* 125 mm)

20. An electric motor of mass 120 kg is attached to the centre of a supported beam which deflects 0·75 mm under this mass. The armature of the motor has a mass of 40 kg. As the motor is run up gradually to a speed of 1500 rev/min it is observed that the maximum amplitude of vertical oscillation is 2·5 mm, diminishing to 0·5 mm at the running speed of 1500 rev/min.

Calculate the value of the damping factor *c* and the distance of the centre of gravity of the armature from the axis of rotation. The mass of the beam may be neglected. (*U. Lond.*)   (*Ans. :* 1·305 kN s/m ; 0·713 mm)

21. An unbalanced engine, mounted on an elastic support, imposes a periodic vertical force on the support. An increase of engine speed from its normal value of 600 rev/min to 900 rev/min was observed to treble the amplitude of the resulting forced vibration. A damping device is required to reduce the amplitude at normal speed by 10%. Find the necessary damping coefficient expressed in terms of the mass of the engine. (*U. Lond.*)   (*Ans. :* 152*m*)

22. In a viscous damped vibrating system the suspended body has a mass of 70 kg and the stiffness of the suspension is 9 kN/m of deflection. The damping force is proportional to velocity and is equal to 300 N at a velocity of 1 m/s. An external force of magnitude 180 cos ωt N is applied directly to the suspended body, ω being variable.

Determine, justifying any equation used, the maximum possible value of the vibration amplitude, the frequency at which it occurs, and the phase angle difference between the applied force and the motion. (*U. Lond.*)

(*Ans. :* 53·9 mm ; 1·74 Hz ; 78° 54′)

23. A single-cylinder engine has an out-of-balance force of 500 N at the engine speed of 300 rev/min. The complete mass of the engine and base is 140 kg and this is carried on a set of springs of total stiffness 25 kN/m.

Find (a) the maximum oscillating force transmitted to the ground, (b) the maximum oscillating force transmitted through the springs if a viscous damper is fitted between the base and the ground, the damping force being 840 N at 1 m/s of velocity. (*U. Lond.*)　　　　　(*Ans.:* 110·5 N; 107·6 N)

24. Write down the equation of S.H.M. for a body subject to damping proportional to the velocity and acted upon by a periodic simple harmonic force. Find the solution.

If the maximum value of the impressed force would produce a deflection of 48 mm, find the value of the damping coefficient if the deflection is not to exceed 6 mm when the impressed periodicity is the same as the natural frequency, 5 Hz. Mass of body 4 kg. Express the damping coefficient as a force on the body at 1 m/s. (*U. Glas.*)　　　　　(*Ans.:* 1005 N s/m)

25. A machine mounted on elastic supports is free to vibrate vertically. The machine has a mass of 40 kg and rotor out-of-balance effects are equivalent to 2 kg at 150 mm radius. Resonance occurs when the machine is run at 621 rev/min, the amplitude of vibration at this speed being 45 mm. Determine the amplitude when running at 500 rev/min and show, in a sketch of the rotor, the angular position of the out-of-balance mass when the machine is at its highest position during vibration. (*U. Glas.*)

(*Ans:* 12·95 mm; machine lags mass by 20° 53′)

26. A mass of 20 kg is attached to a frame by a flexible connection which is equivalent to a spring of stiffness 12 kN/m and a damper with a viscous damping coefficient c. The frame has simple harmonic motion with amplitude 50 mm.

An accelerometer is mounted on the mass. It consists of a spring of stiffness 500 N/m and a mass of 0·1 kg, with means for recording the displacement of the accelerometer mass relative to the main mass. Its operation may be assumed not to affect the motion of the main mass.

When the frame is vibrated at a frequency of 3 Hz the amplitude of the record from the accelerometer is 6 mm. Find (a) the amplitude of the main mass, (b) the value of the damping coefficient c. (*U. Lond.*)

(*Ans.:* 84·4 mm; 198·4 N s/m)

27. In an experimental vibration system, shown in Fig. 19.16, the point A is given a sinusoidal motion of amplitude 6 mm at a frequency of 2 Hz. The mass of 1 kg at B is connected through the spring of stiffness 85 N/m, and its

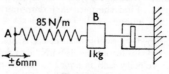

Fɪɢ. 19.16

motion is opposed by a dash-pot giving a viscous resistance of 5 N s/m. Find the amplitude of the mass B, the maximum force in the spring and the work done per cycle by the driving force at A. (*U. Glas.*)

(*Ans.:* 5·3 mm; 0·9 N; 5·55 mJ)

28. A body of mass m attached to a light spring of stiffness k is acted upon by a periodic force $P_0 \sin pt$, where P_0 is constant. Viscous damping, of magnitude c per unit velocity, is present.

Show, from first principles, that the mean power dissipation due to damping during the steady-state forced vibration is given by

$$\tfrac{1}{2}P_0{}^2 cp^2 / [(k - mp^2)^2 + c^2 p^2] \qquad (U.\ Lond.)$$

29. The link shown in Fig. 19.17 has a mass of 0·25 kg and a radius of gyration of 25 mm about its centre of mass which is at A. The spring attached at B has a stiffness of 2·5 kN/m. The link BD has a mass of 0·1 kg and the viscous damper provides 80% critical damping.

If the piston C is oscillated with S.H.M. at 30 Hz and amplitude 1·25 mm, find the amplitude at D and the phase relationship between the two motions. (*U. Lond.*) (*Ans.:* 0·42 mm; 121° lagging)

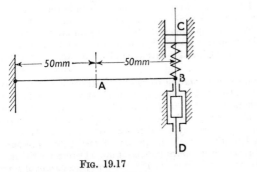

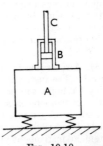

Fig. 19.17   Fig. 19.18

30. Fig. 19.18 shows a block of metal A mounted on springs on a rigid floor. The cylinder of an oil-filled damper B is attached rigidly to the block. A rod C attached to the piston of the damper projects upwards. The total mass of the block and the damper cylinder is 72 kg. The total stiffness of the springs is 160 kN/m. The force exerted between the piston and cylinder of the damper is proportional to the relative velocity of these two parts and is 60 kN when the relative velocity is 1 m/s.

By using an equation of motion, calculate the maximum and minimum values of the force in the springs when the rod C is made to move vertically with a simple harmonic motion of frequency 11·5 Hz and amplitude ±4 mm. (*U. Lond.*) (*Ans. :* 79 to 1347 N)

31. A mass of 55 kg is supported by a spring. When the system vibrates freely it is found that the frequency of vibration is 3 Hz and that in 10 complete oscillations the amplitude diminishes in the ratio 100 : 35. Assuming that the damping force is proportional to the velocity, calculate the amount of the damping force corresponding to a velocity of 1 m/s.

Find also the power which must be supplied to the system in order to force the mass to vibrate with the same frequency but with a constant amplitude, the amplitude being 50 mm. (*U. Lond.*) (*Ans.* 34·68 N s/m; 15·4 W)

32. A light steel reed of 25 mm by 0·8 mm section is fixed at one end and carries a mass of 0·1 kg at the other, which is free. The effective length of the reed is 80 mm. Vibrations are caused by a harmonic movement of 2·5 mm at the fixed end. Damping at the mass is 3·5 N s/m. Find the maximum stress in the reed at a frequency 10% above the natural undamped frequency. $E = 200$ GN/m². (*U. Glas.*) (*Ans.*: 262 MN/m²)

33. Two rotors, with moments of inertia, I_1 and I_2, are mounted co-axially on a shaft of torsional stiffness k. A torsional vibration damper, with viscous damping coefficient c, is fitted between the masses, and a fluctuating torque $T \cos pt$ is applied at rotor 1. Find the value of p at which power absorption is a maximum and find the power absorbed at this frequency. (*U. Lond.*)

$$\left(Ans.: \quad \sqrt{\frac{k(I_1 + I_2)}{I_1 I_2}} \; ; \quad \frac{T^2 I_2^2}{2c(I_1 + I_2)^2} \right)$$

34. The shaft of a light high-speed air compressor is directly coupled to a second shaft whose relatively heavy masses justify the assumption of infinite moment of inertia. The moment of inertia of the compressor is 0·4 kg m². The existence of the fourth harmonic of maximum value 7 N m in the compressor torque curve is responsible for a critical speed at 480 rev/min, the maximum deflection being limited by damping influences' to 2°. Assuming the damping torque proportional to the vibration speed, determine the vibration amplitude at 300 rev/min produced by a sixth order harmonic of maximum value 2 N m. (*U. Glas.*) (*Ans.*: 0·059°)

35. A recording instrument has rotating parts of effective moment of inertia 1 g m². The instrument is driven from an engine shaft through a light spiral spring having a torque characteristic of 0·02 N m/degree of twist. The drum of the instrument is slightly out of balance, and this gives rise to a torque fluctuation of ±1 mN m at a frequency equal to that of rotation. When the speed of rotation is equal to the natural frequency of the instrument-spring system, an instrument vibration of 10° amplitude is set up. Determine the range of speed within which the amplitude exceeds 1°. Assume the damping forces proportional to the vibration velocity. (*U. Glas.*)

(*Ans.*: 315·5 to 331·6 rev/min)

36. In order to measure the amplitude of torsional oscillations at a particular place on a shaft, a small flywheel of 0·04 kg m² angular inertia is mounted with a running fit co-axially on the shaft and connected to it by a spiral spring of stiffness 0·06 N m/degree. The viscous damping between the flywheel and shaft is estimated by observing an amplitude reduction of 20% per cycle of a transient natural oscillation of the flywheel, produced when the shaft is stationary. It may be taken as proportional to the relative velocity between shaft and flywheel. When the shaft is running, the relative motion between shaft and flywheel has an amplitude of 1·5° at a frequency of 3·5 Hz. Determine the amplitude of the shaft itself. (*U. Glas.*) (*Ans.*: 1·303°)